MODULAR REPRESENTATION THEORY OF FINITE AND p-ADIC GROUPS

LECTURE NOTES SERIES
Institute for Mathematical Sciences, National University of Singapore

Series Editors: Chitat Chong and Wing Keung To
Institute for Mathematical Sciences
National University of Singapore

ISSN: 1793-0758

Published

Vol. 20 Mathematical Horizons for Quantum Physics
edited by H Araki, B-G Englert, L-C Kwek & J Suzuki

Vol. 21 Environmental Hazards:
The Fluid Dynamics and Geophysics of Extreme Events
edited by H K Moffatt & E Shuckburgh

Vol. 22 Multiscale Modeling and Analysis for Materials Simulation
edited by Weizhu Bao & Qiang Du

Vol. 23 Geometry, Topology and Dynamics of Character Varieties
edited by W Goldman, C Series & S P Tan

Vol. 24 Complex Quantum Systems: Analysis of Large Coulomb Systems
edited by Heinz Siedentop

Vol. 25 Infinity and Truth
edited by Chitat Chong, Qi Feng, Theodore A Slaman & W Hugh Woodin

Vol. 26 Notes on Forcing Axioms
by Stevo Todorcevic, edited by Chitat Chong, Qi Feng, Theodore A Slaman, W Hugh Woodin & Yue Yang

Vol. 27 *E*-Recursion, Forcing and C*-Algebras
edited by Chitat Chong, Qi Feng, Theodore A Slaman, W Hugh Woodin & Yue Yang

Vol. 28 Slicing the Turth: On the Computable and Reverse Mathematics of Combinatorial Principles
by Denis R. Hirschfeldt, edited by Chitat Chong, Qi Feng, Theodore A Slaman, W Hugh Woodin & Yue Yang

Vol. 30 Modular Representation Theory of Finite and *p*-Adic Groups
edited by Wee Teck Gan & Kai Meng Tan

*For the complete list of titles in this series, please go to
http://www.worldscientific.com/series/LNIMSNUS

Lecture Notes Series, Institute for Mathematical Sciences,
National University of Singapore
Vol. 30

MODULAR REPRESENTATION THEORY OF FINITE AND p-ADIC GROUPS

Editors

Wee Teck Gan
Kai Meng Tan
National University of Singapore, Singapore

World Scientific

NEW JERSEY • LONDON • SINGAPORE • BEIJING • SHANGHAI • HONG KONG • TAIPEI • CHENNAI

Published by

World Scientific Publishing Co. Pte. Ltd.
5 Toh Tuck Link, Singapore 596224
USA office: 27 Warren Street, Suite 401-402, Hackensack, NJ 07601
UK office: 57 Shelton Street, Covent Garden, London WC2H 9HE

Library of Congress Cataloging-in-Publication Data
Modular representation theory of finite and p-adic groups / edited by Wee Teck Gan (NUS, Singapore), Kai Meng Tan (NUS, Singapore).
pages cm. -- (Lecture notes series, institute for mathematical sciences, national university of singapore ; volume 30)
NUS represents National University of Singapore.
Includes bibliographical references and index.
ISBN 978-9814651806 (hardcover : alk. paper)
1. Group theory. 2. p-adic groups. 3. Representations of groups. I. Gan, Wee Teck, editor. II. Tan, Kai Meng, editor.
QA174.2.M63 2015
512'.23--dc23

2015000042

British Library Cataloguing-in-Publication Data
A catalogue record for this book is available from the British Library.

Front cover: Image courtesy of Andrew Mathas.
Back cover: Image of the Bruhat–Tits tree courtesy of Thomas Nyberg.

CONTENTS

FOREWORD

The Institute for Mathematical Sciences (IMS) at the National University of Singapore was established on 1 July 2000. Its mission is to foster mathematical research, both fundamental and multidisciplinary, particularly research that links mathematics to other efforts of human endeavor, and to nurture the growth of mathematical talent and expertise in research scientists, as well as to serve as a platform for research interaction between scientists in Singapore and the international scientific community.

The Institute organizes thematic programs of longer duration and mathematical activities including workshops and public lectures. The program or workshop themes are selected from among areas at the forefront of current research in the mathematical sciences and their applications.

Each volume of the *IMS Lecture Notes Series* is a compendium of papers based on lectures or tutorials delivered at a program/workshop. It brings to the international research community original results or expository articles on a subject of current interest. These volumes also serve as a record of activities that took place at the IMS.

We hope that through the regular publication of these *Lecture Notes* the Institute will achieve, in part, its objective of reaching out to the community of scholars in the promotion of research in the mathematical sciences.

November 2014

Chitat Chong
Wing Keung To
Series Editors

PREFACE

A month-long program on "Modular Representations of Finite and p-Adic Groups" was held during April 2013 at the Institute for Mathematical Sciences (IMS) in Singapore. The program was organised by Joseph Chuang (City University, London, UK), Karin Erdmann (University of Oxford, UK), Wee Teck Gan (NUS), Florian Herzig (Toronto, Canada), Kay Jin Lim (NUS), Alberto Mínguez (Jussieu, France) and Kai Meng Tan (NUS). The goal of the program is to bring together leading researchers in the areas of modular representation theory of finite groups and p-adic groups to discuss the latest developments in the field, chart out new directions for research and explore possible collaboration. It was well attended by about 50 participants worldwide.

By "modular representations", one means the representations of a group G on a vector space over a field of nonzero characteristic ℓ, when ℓ divides the (pro-)order of G. It is with such ℓ-modular representations that the program is concerned. The groups G which feature in the program include the finite groups of Lie type, their Weyl groups and related Lie or Hecke algebras, as well as the reductive p-adic groups which are infinite topological groups that play an important role in the Langlands program and number theory. While the modular representation theory of finite groups have been pursued for about a hundred years, the case of p-adic groups is a relatively young field, where it began its life about 20 years ago. There is so far relatively little interaction between the two areas and it is the hope of the program to stimulate such interactions.

The program began with six short lecture series (aka tutorials) given by leading experts, followed by a two-week conference. The notes of five of these tutorials are collected in this volume. The chapter by M. Cabanes introduces the finite groups of Lie type and discusses recent results and outstanding conjectures about their modular representation theory. The understanding of the representation theory of these finite groups of Lie type provides a crucial starting point for the study of p-adic groups. The chapters by V. Sécherre and F. Herzig (with notes prepared by K. Kozioł) give an exposition of the state-of-the-art in the ℓ-modular representation theory of p-adic groups when $\ell \neq p$ and $\ell = p$, respectively. They convey the exciting developments in the ℓ-modular and p-modular Langlands program. Finally, the chapters by A. Kleshchev and A. Mathas touch upon the representation theory of the increasingly popular and important Khovanov–Lauda–Rouquier (KLR) algebras. Specifically, Kleshchev discusses the basic representation theory of these KLR algebras and their homological properties, whereas the chapter by Mathas is devoted to the emerging graded representation theory in the special case of type A. These KLR algebras arise from the introduction of the idea of "categorification" in representation theory. The last tutorial, given by R. Rouquier, is an introduction to this revolutionary idea of "categorification". Unfortunately, due to circumstances beyond our control, the content of Rouquier's talks is not included in this volume.

We take this opportunity to thank all the participants of the program for contributing to its success. We especially want to thank the six tutorial lecturers for their excellent presentations and for taking the time and effort to write up their notes for publication. We hope that the chapters contained in this volume will serve as a useful reference for researchers and students in these and related areas.

Finally, we would like to convey our deepest appreciation to IMS for generously and graciously providing both financial and administrative support for our program, without which the program would not have been successful.

November 2014

Wee Teck Gan and Kai Meng Tan
National University of Singapore
Singapore

Volume Editors

MODULAR REPRESENTATIONS OF FINITE REDUCTIVE GROUPS

Marc Cabanes
Institut de Mathématiques de Jussieu
Université Paris Diderot
Bâtiment Sophie Germain
75205 Paris Cedex 13, France
marc.cabanes@imj-prg.fr

We report on the main results about linear representations of finite reductive groups or finite groups of Lie type. Following the historical order, we comment on representations in the defining characteristic, ordinary characters and representations in non-defining characteristic.

Contents

INTRODUCTION

This survey deals with linear representations of a class of finite groups strongly related to linear algebra itself, finite reductive groups. Finite reductive groups also provide almost all finite simple groups and their essential central extensions. They are therefore omnipresent, explicitly or not, in finite group theory.

Finite reductive groups are finite analogues of reductive algebraic groups whose structure is in turn close to that of Lie groups. This also explains the terminology "groups of Lie type" which is often used. Let us mention also the term "Chevalley groups", which pays tribute to Chevalley's construction of those finite groups from integral forms of semi-simple complex Lie algebras [Ch55]. The later approach is also described in [Cart1], though the prevailing approach is now to see them as fixed point subgroups $\mathbf{G}^F$ under a Frobenius endomorphism $F\colon \mathbf{G} \to \mathbf{G}$ of some algebraic group $\mathbf{G}$, an approach due to Steinberg and allowing to take advantage of the geometry of algebraic groups.

The pace of this survey is intended as quite slow, giving details necessary to understand most definitions. This should suit beginners or more experienced readers from other branches of representation or group theory. We tried to comment on some examples, mainly in type A (general linear groups, finite or not), and we strongly recommend Bonnafé's book [Bn] where much of our matter, and more, is thoroughly explored for SL_2.

In Part I, we describe roughly the main features of representation theory, that is mainly the study of module categories for rings that are mostly finite groups algebras or close analogues. This also leads to the study of associated categories, most famously the derived and the homotopic categories.

We did not comment on the already spectacularly successful methods of categorification, referring instead to Rouquier's series of talks.

Part II comments on the constructions and structure of those groups both from the elementary point of view of split BN-pairs and of algebraic groups.

Part III deals with linear representations most naturally associated with finite reductive groups since performed over the field $\overline{\mathbb{F}}_p$ defining the ambient algebraic group. The classical theorems of Chevalley and Steinberg are supplemented by more recent contributions by Lusztig, though it seems that some questions remain as mysterious as they were 25 years ago.

Part IV, by far the longest, comments on both the theory by Deligne-Lusztig leading to a very precise description of ordinary characters of finite groups of Lie type and more recent investigations on representations in characteristics different from p. The first is presented in many references (see [Sri], [DiMi], [Cart2]) and we quickly recall the main results. The second part was largely initiated by the papers of Fong-Srinivasan ([FoSr82], [FoSr89]) relating ordinary characters of finite classical groups $\mathbf{G}^F$ with blocks of $\overline{\mathbb{F}}_\ell$ ($\ell \neq p$). Many contributions followed by Dipper (decomposition numbers [Dip85a], [Dip85b]), and Geck-Hiss-Malle (ℓ-modular Harish Chandra theory [GeHiMa94], [GeHiMa96]). Broué in [Br90] gave several conjectures, the one on Jordan decomposition of characters being later proved by Bonnafé-Rouquier (see [BnRo03]). We report on all those subjects, which leads us to many theorems on ℓ-blocks, decomposition numbers and modular Harish Chandra series.

I. MODULAR REPRESENTATIONS

We refer to the first chapter of [Be] or the short book [Sch] for most of the information we need.

1. Representations

The framework is the one of representations of non commutative rings A that are mainly finite dimensional algebras over a commutative field K. The finite dimensional representations can be seen as objects of the category A-**mod** of finitely generated A-modules. It may also be that A is an $\mathcal{O}$-free algebra of finite rank over a local ring $\mathcal{O}$, in which case we require that the objects of A-**mod** to be $\mathcal{O}$-free of finite rank.

We do not recall here the terminology of *simple, indecomposable,* or *projective* A-modules. The *Jacobson radical* of A is denoted by $J(A)$, the group of invertible elements of A is denoted by $A^\times$. Simple modules are equivalently called *irreducible representations*, and $\mathrm{Irr}(A)$ denotes their set of isomorphism types. When A is over a field K, we denote by $\mathrm{K}_0(A)$ the *Grothendieck ring* of A, which may be seen as the commutative group $\mathbb{Z}\mathrm{Irr}(A)$ endowed with the multiplication induced by tensor product of modules over K. The multiplicity of a simple A-module S as a composition factor in the Jordan-Hölder series of an A-module M is denoted by $[M : S]$.

We also use the notion of *blocks* as minimal indecomposable two-sided ideal direct summands of A. Note that this also makes sense when A is an $\mathcal{O}$-free algebra of finite rank over a local ring $\mathcal{O}$.

Our modules are left modules but we need also the notion of *bimodule.* It is defined as follows. We have two rings A, B, and T is an A-module on the left, and a B-module on the right, with both actions commuting, that is $a(tb) = (at)b$ for any $a \in A$, $t \in T$, $b \in B$. This of course coincides with the notion of an $A \times B^{\rm opp}$-module where $B^{\rm opp}$ is the opposite ring of B.

An object T of $A \times B^{\rm opp}$-**mod** defines a functor B-**mod** $\to$ A-**mod** by $M \mapsto T \otimes_B M$. The notion of a *Morita equivalence* is a typical example of such a functor (see [Be] Sect. 2.2).

In connection with representations of finite groups, one may be led to consider the category $\mathrm{C^b}(A)$ of bounded complexes $\ldots M^i \xrightarrow{\partial^i} M^{i+1} \ldots$ of A-modules ($\partial^{i+1}\partial^i = 0$ for all $i \in \mathbb{Z}$, and only finitely many i's are such that $M^i \neq \{0\}$) and the related *homotopic category* $\mathrm{Ho^b}(A)$.

2. Group Algebras

We are mostly interested in the cases where A is a *group algebra* RG of a (multiplicative) finite group G over a commutative ring R. Recall that this is the R-free R-module of all sums $\sum_{g\in G} r_g g$ (for $(r_g)_{g\in G}$ any family of elements of R indexed by G) with R-bilinear multiplication extending the one of G. For us, an R-linear representation of G is an R-free RG-module M of finite rank or equivalently a group morphism $G \to \mathrm{GL}_n(R)$.

When $H \leq G$ is a subgroup, restriction Res^G_H has an adjoint on both sides Ind^G_H which is the functor RH-**mod** $\to$ RG-**mod** associated with RG seen as a bimodule with translation actions of G on the left and H on the right, i.e. $\mathrm{Ind}^G_H(N) = RG \otimes_{RH} N$ whenever N is an RH-module.

When $R = K$ is a field where $|G|$ inverts and with primitive $|G|$-th roots of 1, then KG is a split semi-simple algebra $\cong \prod_i \mathrm{Mat}_{d_i}(K)$. If moreover the characteristic of K is 0, then the isomorphism type of a KG-module M is given by its trace character $\chi_M\colon G \to K$ sending g to the trace of its action on M. *Irreducible characters* are the ones of simple KG-modules, and one denotes by $\mathrm{Irr}(G)$ the corresponding set of functions on G. This has values in $\mathbb{Z}[\omega_{|G|}]$ where $\omega_{|G|}$ is a primitive $|G|$-th root of 1, it is therefore independent of K. Note that $\mathrm{Irr}(KG) \xrightarrow{\sim} \mathrm{Irr}(G)$. For most of character theory, see [Isa].

When R is a field k of characteristic a prime divisor of $|G|$ with a primitive $|G|_{p'}$-th root of 1, the algebra $kG/J(kG)$ is split semi-simple. Then $\mathrm{Irr}(kG)$ is often denoted as $\mathrm{IBr}(G)$ and identifies with the central functions on $G_{p'}$ (p-regular elements of G) obtained by Brauer's method of lifting p'-roots of 1 in k into an extension of $\mathbb{Z}_p$ (see [Sch] Sect. 3.1).

When p is a prime number and G is a finite group, a p-modular system $(\mathcal{O}, K, k)$ is a triple where $\mathcal{O}$ is a complete discrete valuation ring containing the $|G|$-th roots of 1, free of finite rank over $\mathbb{Z}_p$, with K denoting the field of fractions of $\mathcal{O}$ (a finite extension of $\mathbb{Q}_p$) and $k = \mathcal{O}/J(\mathcal{O})$ is its (finite) residue field.

The decomposition of $\mathcal{O}G$ into blocks $\mathcal{O}G = \oplus_i B_i$ (sometimes called the *p-blocks* of G) gives the decomposition of kG into blocks $kG = \oplus_i B_i \otimes k$ and induces a partition of both $\mathrm{Irr}(G)$, by $\mathrm{Irr}(G) = \mathrm{Irr}(KG) = \sqcup_i \mathrm{Irr}(B_i \otimes K)$, and $\mathrm{IBr}(G)$ by $\mathrm{IBr}(G) = \mathrm{Irr}(kG) = \sqcup_i \mathrm{Irr}(B_i \otimes k)$.

The *principal block* of G in characteristic p is defined as the one not in the kernel of the one-dimensional trivial representation of G (where each element of G acts by 1).

II. THE GROUPS

3. Coxeter Groups

3.A. Definitions

References for what follows are the corresponding chapters of [Bou], [GePf].

Definition 3.1. A **Coxeter graph** on a set S is a non-oriented graph without loops, with nodes the elements of S, and valued edges carrying a number $m_{st} \in \{3, \ldots, \infty\}$ (of course $m_{st} = m_{ts}$) for any $s, t \in S$. One omits m_{st} when $m_{st} = 3$ and one completes the matrix $(m_{st})_{s,t \in S}$ by putting $m_{st} = 2$ whenever $s \neq t$ and there is no edge between s and t.

The associated **Coxeter group** is the group generated by the elements of S subject to the relations:

(quadratic) $s^2 = 1$ for all $s \in S$.

(braid) $sts \cdots = tst \ldots$ (with m_{st} terms on each side) for all $s \neq t \in S$ with $m_{st} \neq \infty$.

One denotes by $l_S \colon W \to \mathbb{N}$ the length function with regard to the generating set S.

Note that the relation for $s \neq t$ and $m_{st} = 2$ specifies that s and t commute. So the connected components of the Coxeter graph gives a partition of S into pairwise commuting sets.

Example 3.2. (a) The graphs

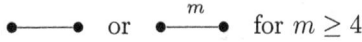

 or 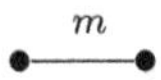 for $m \geq 4$

give rise to the dihedral groups of order $2m$ (6 in the first case, infinite in the case when $m = \infty$).

(b) Another important example is the one of the graph of so-called type A_n as follows ($n \geq 1$ nodes)

whose associated group is isomorphic with the symmetric group $\mathfrak{S}_{n+1}$ by the map sending the i-th generator s_i in the above list (from left to right) to the transposition $(i, i+1)$ (Moore's presentation of the symmetric group).

Several other basic properties are as follows.

Fact 3.3. *S injects in W and m_{st} is the order of the product st in W.*

Fact 3.4. *If W is finite, it has a single element $w_S \in W$ of maximal l_S.*

3.B. Parabolic subgroups, finite Coxeter groups

One calls **parabolic subgroups** of W the subgroups generated by a subset of S. If $I \subseteq S$, one denotes $W_I := <I>$. The following is a consequence of the so-called *exchange condition* governing the way to obtain minimal decompositions (see [Bou] IV.1.5).

Fact 3.5. *The map $I \mapsto W_I$ is a bijection between subsets of S and parabolic subgroups of W. It satisfies $W_I \cap W_J = W_{I \cap J}$ for all $I, J \subseteq S$. Moreover $S \cap W_I = I$ and the restriction of l_S to W_I is l_I.*

An abstract group W with a subset S of involutions is called a **Coxeter group** if W is isomorphic through the canonical map with the group presented by S subject to the relations of Definition 3.1 for m_{st} being defined as the order of the product st (which requires that S generates W). A Coxeter group is said to be **irreducible** if, and only if, it comes from a connected Coxeter graph.

Theorem 3.6. *An irreducible Coxeter graph gives rise to a finite group W if and only if it is among the following (the index n recalls the number of nodes):*

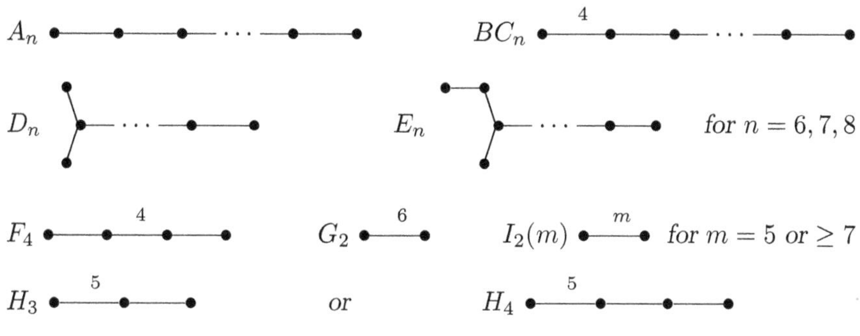

On the other hand a typical graph producing an infinite group is as follows

$\widetilde{A}_n$ 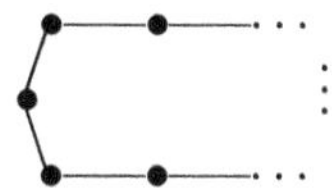 ($n + 1$ nodes).

3.C. Reflection representations

Coxeter groups have a **reflection representation**. It is defined as follows. Let us start from the set S and integers m_{st} for $s \neq t \in S$. One then defines a real vector space $V = \oplus_{s \in S} \mathbb{R} e_s$ whose basis is in bijection with S by $s \mapsto e_s$. One defines on it a symmetric bilinear form sending (e_s, e_s) to 1 and (e_s, e_t) to $-\cos(\pi/m_{st})$ when $s \neq t$. Denote by $\mathrm{Isom}(V) \leq \mathrm{GL}_{\mathbb{R}}(V)$ the corresponding orthogonal group. Then it is easy to see that the map sending s to the reflection through the vector e_s is a group morphism

$$W \to \mathrm{Isom}(V).$$

A more remarkable fact is that it is *injective*. Another fact is that W is finite if and only if the above bilinear symmetric form is definite positive (see [Bou] Sect. V.4.8). The latter is a key fact in the proof of the above Theorem 3.6.

An important generalization of the finite case above is when V is a finite dimensional complex vector space and $W \leq \mathrm{GL}_{\mathbb{C}}(V)$ is a finite subgroup generated by elements $r \in W$ such that the image of $r - \mathrm{Id}_V$ is a line ("pseudo-reflections", remembering that r has finite order hence is semi-simple). Such a W is called a *finite reflection group*, a good survey is provided by [GeMa]. A defining property is that the action of W on $S(V)$, the ring of polynomials on V is such that the invariant subring $S(V)^W$ is isomorphic to a polynomial ring (Chevalley, see [Bou] V.5.3).

4. Iwahori-Hecke Algebras

A deformation of the group algebra of a Coxeter groups, Iwahori-Hecke algebras were defined in connection with representations of finite reductive groups.

4.A. Generalities

All that follows can be taken from [GePf].

Definition 4.1. If R is a commutative ring, $q \in R$ and W is a Coxeter group (with subset S and integers m_{st}), one denotes by $\mathcal{H}_{R,q}(W)$ the R-algebra with generators $(A_s)_{s\in S}$ subject to the relations:
(quadratic) $(A_s + 1)(A_s - q) = 0$ for any $s \in S$,
(braid) $A_sA_t \cdots = A_tA_s \ldots$ (m_{st} terms on each side) for any distinct $s, t \in S$.

Note that the choice $q = 1$ gives us the group algebra RW.
More generally, one has freeness over R for an R-basis indexed by W.

Theorem 4.2. *If $w = s_1 \ldots s_{l_S(w)}$ is a minimal expression of w as a product of elements of S, the product $A_{s_1} \ldots A_{s_k}$ is $\neq 0$ and depends only on w. One denotes it by $A_w := A_{s_1} \ldots A_{s_k}$ (with $A_1 := 1$). Then*

$$\mathcal{H}_{R,q}(W) = \oplus_{w\in W} RA_w.$$

Many properties ensue, among them the relation with parabolic subgroups of W, or with scalar extension.

4.B. Semi-simplicity in characteristic zero

The first assertion below is due to Gyoja-Uno ([GyUn89], see also a generalization with several parameters in [*GePf*]), the second is due to Tits.

Theorem 4.3. *Assume $R = \mathbb{C}$ and W is finite. Let $q \in \mathbb{C}$.*
(a) $\mathcal{H}_{\mathbb{C},q}(W)$ is semi-simple if, and only if, $q \cdot \sum_{w\in W} q^{l_S(w)} \neq 0$.
(b) (Deformation theorem). If $\mathcal{H}_{\mathbb{C},q}(W)$ is semi-simple, then $\mathcal{H}_{\mathbb{C},q}(W) \cong \mathbb{C}W$.

Note that the so-called *Poincaré polynomial* $P_W(X) = \sum_{w\in W} X^{l_S(w)}$ has a factorization $P_W(X) = \Pi_{i=1}^{|S|} \frac{X^{d_i+1}-1}{X-1}$ where the d_i are the *exponents* of W, that is the degrees of the generators of the invariants of the action of W on the symmetric powers of the reflection representation. So we see that complex Iwahori-Hecke algebras over W are isomorphic to the group algebra $\mathbb{C}W$ except for finitely many values of q that are either 0 or a root of unity.

Variants of Iwahori-Hecke algebras have been defined in the case of complex (finite) reflection groups (see [BrMaM93]).

4.C. The Kazhdan-Lusztig polynomials

Let (W, S) be a Coxeter group, possibly with W finite. If $w = s_1 \dots s_{l_S(w)}$ is a minimal expression of w as a product of elements of S, then the set $[1, w] \subseteq W$ defined by all sub-products $s_{i_1} \dots s_{i_t}$ for $1 \leq i_1 < \cdots < i_t \leq l_S(w)$, is independent of the minimal expression chosen for w. One then defines the **Bruhat order** $\leq$ on W by $x \leq y$ if and only if $x \in [1, y]$. See [GePf] Ex. 1.7 for this. See also Sec. 6.B below for an interpretation in terms of Zariski closure in reductive groups.

Let $q^{1/2}$ be an indeterminate whose powers are denoted by $(q^{1/2})^m = q^{m/2}$. Let $R = \mathbb{Z}[q^{-1/2}, q^{1/2}]$ the ring of Laurent polynomials in those indeterminates. Let $\mathcal{H}$ be the Iwahori-Hecke algebra $\mathcal{H}_{R,q}(W)$ (see Definition 4.1 above). Note that q being invertible, each generator A_s ($s \in S$) is invertible with $(A_s)^{-1} = q^{-1}A_s + (q^{-1} - 1)A_1$, so that any A_w ($w \in W$) is also invertible.

Definition 4.4. Let $\mathcal{H} \to \mathcal{H}$, $x \mapsto \overline{x}$ be the $\mathbb{Z}$-linear map sending $q^{m/2}$ to $q^{-m/2}$ and A_w to $(A_{w^{-1}})^{-1}$. This is a ring endomorphism.

Here is the existence theorem for Kazhdan-Lusztig's polynomials (see [KL79]).

Theorem 4.5. *There exists a unique set of polynomials $P_{v,w} \in \mathbb{Z}[q]$ for $v \leq w$ with $P_{w,w} = 1$, $P_{v,w}$ has degree $\leq (l_S(w) - l_S(v) - 1)/2$ for $v < w$ and such that $C'_w := q^{-l_S(w)/2} \sum_{v \leq w} P_{v,w} A_w$ satisfies*

$$\overline{C'_w} = C'_w$$

for any $w \in W$.

It has long been conjectured that the above polynomials have coefficients in $\mathbb{N}$. This was proved by Kazhdan-Lusztig for finite W – and also affine type like $\widetilde{A}_n$ above – by showing the relation with cohomology of Schubert varieties $\overline{\mathbf{B}w\mathbf{B}}/\mathbf{B}$ (see Sec. 6B below). The positivity in the general case was proved recently by Elias-Williamson [EW12].

The polynomials allow Kazhdan-Lusztig to define certain subsets of the group W, called *cells*, and associated *cell representations* of the Iwahori-Hecke algebra. See more generally the notion of *cellular algebras* [GraL].

The new basis $\{C'_w\}$ of Theorem 4.5 also allows a more explicit version of the isomorphism in Theorem 4.3 above.

More general Iwahori-Hecke algebras can be given a family of parameters $(q_s)_s$ indexed by elements of S modulo W-conjugacy. See [Lus03] Ch. 14 for several conjectures about those algebras – some of them having been checked since then.

5. BN-Pairs and Simple Groups

BN-pairs, first axiomatized by Tits, are a basic trait common to all groups of Lie type, be them algebraic, p-adic or finite.

5.A. The axioms and basic properties

Definition 5.1. A group G is said to have a BN-pair if it possesses two subgroups B and N, and a subset $S \subseteq N/B \cap N$ such that

- $T := B \cap N$ is normal in N and one has $s^2 = 1$ in the group $W := N/T$ for any $s \in S$,
- $G = \langle B, N \rangle$ and $W = \langle S \rangle$,
- for all $s \in S$, $sBs \neq B$,
- for all $s \in S$, $w \in N/T$, one has $BsBwB = BswB \cup BwB$.

The groups B, T, W are often called the **Borel subgroup**, a **maximal torus** and the **Weyl group** of G, respectively.

Fact 5.2. *The elements of S are the only non-trivial elements of W such that $B \cup BsB$ is a subgroup of G (this is why they can be considered implicit in the definition of the "BN"-pair).*

Fact 5.3. *(W, S) is a Coxeter group.*

Fact 5.4. Bruhat decomposition. *The double cosets BwB for $w \in W$ are pairwise disjoint and are all the elements of $B \backslash G/B$:*

$$G = \sqcup_{w \in W} BwB.$$

Fact 5.5. *The set of subgroups of G containing B is in bijection with the set of subsets of S by $I \mapsto P_I := BW_IB = B\langle I \rangle B$. We get in particular $P_I \cap P_J = P_{I \cap J}$ for any $I, J \subseteq S$.*

5.B. Split BN-pairs and Levi decompositions

The group G with a BN-pair is said to be **split** whenever the group B is a semi-direct product $B = U \rtimes T$. Some authors sometimes add the condition that T is abelian. An important property of the same nature is the following.

Assume G has a split BN-pair $B = UT$ with *finite* W.

Definition 5.6. For $I \subseteq S$, denote by w_I the element of maximal length of (W_I, I) and $U_I := U \cap w_I U w_I$. Also $B^- = w_S B w_S$ and $L_I := BW_IB \cap B^-W_IB^- \subseteq P_I = BW_IB$. The latter is called the **standard Levi subgroup** associated to I. One says that P_I satisfies a *Levi decomposition* if $U_I \lhd P_I$ and $P_I = U_I \rtimes L_I$.

5.C. Examples, finite and p-adic

Example 5.7. The case of $\mathrm{GL}_n(\mathbb{F})$. Let $\mathbb{F}$ be a field, $n \geq 1$ an integer, and $G := \mathrm{GL}_n(\mathbb{F})$. Let B (resp. U, resp. T) be the subgroup of upper triangular (resp. unipotent upper triangular, resp. diagonal) matrices in G. Let N be the subgroup of monomial matrices in G (invertible matrices with only one non-zero element on each row and on each column). Then $N/T \cong \mathfrak{S}_n$ (permutation matrices) and the permutations $(1,2), \ldots, (n-1,n)$ used in Example 3.2.(b) above give a subset $S \subseteq N/T$ with properties of a BN-pairs.

Whenever $I \subseteq S$ generates $W_I \cong \mathfrak{S}_{n_1} \times \cdots \times \mathfrak{S}_{n_k}$, one gets for $P_I = BW_IB$ the subgroup of matrices in the form $\begin{pmatrix} A_1 & * & * & * \\ 0 & A_2 & * & * \\ & & \ddots & \\ 0 & 0 & & * \\ 0 & 0 & 0 & A_k \end{pmatrix}$ with $A_i \in \mathrm{GL}_{n_i}(\mathbb{F})$. We also have the Levi decomposition $P_I = U_I \rtimes L_I$ where U_I (resp. $L_I \cong \mathrm{GL}_{n_1}(\mathbb{F}) \times \cdots \times \mathrm{GL}_{n_k}(\mathbb{F})$) is the subgroup of matrices in the form $\begin{pmatrix} I_{n_1} & * & * & * \\ 0 & I_{n_2} & * & * \\ & & \ddots & \\ 0 & 0 & & * \\ 0 & 0 & 0 & I_{n_k} \end{pmatrix}$ (resp. $\begin{pmatrix} A_1 & 0 & 0 & 0 \\ 0 & A_2 & 0 & 0 \\ & & \ddots & \\ 0 & 0 & & 0 \\ 0 & 0 & 0 & A_k \end{pmatrix}$ with $A_i \in \mathrm{GL}_{n_i}(\mathbb{F})$).

Example 5.8. A p-adic example. Let $\mathcal{O}$ be a finite extension of some $\mathbb{Z}_p$, with fraction field K and residual field $k = \mathcal{O}/J(\mathcal{O})$. The group $G = \mathrm{SL}_n(K)$ has the following BN-pair (Iwahori-Matsumoto).

Reduction of matrix entries modulo $J(\mathcal{O})$ gives a surjective group morphism $\mathrm{SL}_n(\mathcal{O}) \to \mathrm{SL}_n(k)$. Let $B \leq \mathrm{SL}_n(\mathcal{O})$ be the inverse image of the upper triangular subgroup, a Borel subgroup, of $\mathrm{SL}_n(k)$. Let N be the subgroup of monomial matrices in G. Then $T := B \cap N$ is the subgroup of diagonal matrices with coefficients in $\mathcal{O}$ and determinant 1. The quotient $W = N/T$ is an affine Coxeter group of type $\widetilde{A}_{n-1}$ when $n \geq 3$ (see Sec. 3.B above), $\widetilde{A}_1 = I_2(\infty)$ if $n = 2$. Note that the proper parabolic subgroups for this BN-pair (sometimes called "parahoric" subgroups of $G = \mathrm{SL}_n(K)$) are finite unions of double cosets BwB since they are associated to finite subgroups of W.

5.D. Simple groups

Recall that a simple group is any group having no other normal subgroup than itself and the trivial subgroup. A perfect group is any group G generated by its commutators, i.e. $G = [G, G]$. We have the following important criterion of simplicity.

Theorem 5.9. *Let G a group with a BN-pair B, N, S and associated Weyl group (W, S). Assume the hypotheses:*

- *(W, S) is irreducible,*
- *B is solvable and G is perfect.*

Then $G/\bigcap_{g\in G} gBg^{-1}$ is simple non-abelian.

The proof is remarkably easy. It suffices to show that any normal subgroup of G is either G or a subgroup of B. So let $H \trianglelefteq G$. Then BH is a subgroup of G containing B, therefore $BH = BW_IB$ for some $I \subseteq S$ (see Fact 5.5 above).

Let us show that if $s \in S \setminus I$ and $w \in I$, then they commute. We have

$$\begin{aligned} sws^{-1} = sws \subseteq sBHs \subseteq sBsH \subseteq BH \cup BsBH \quad & \text{since } H \trianglelefteq G, \\ \subseteq BW_IB \cup BsBW_IB \subseteq B(W_I \cup sW_I)B & \end{aligned}$$

by the axioms of BN-pairs. The uniqueness of the Bruhat decomposition (see Fact 5.4) then implies that sws^{-1} or $s \in W_I$, and therefore $sws^{-1} \in W_I$ in both cases. Then $sws^{-1} \in P_I \cap P_{\{s,w\}} = P_{\{w\}} = B \cup BwB$ by Fact 5.5. Again by Bruhat decomposition, we get now $sws^{-1} = w$.

In view of the irreducibility hypothesis we must have $I = S$ or $\emptyset$, that is $HB = G$ or $H \leq B$. If $HB = G$, then $G/H \cong B/B \cap H$, so the last assumptions of the theorem imply that this factor group is at the same time solvable and equal to its derived subgroup. So it is indeed trivial and $H = G$. ∎

So BN-pairs can be seen as a way to construct simple groups, finite or not. Understandably, finite BN-pairs have been classified (see [FoSe74], [HeKaSe72]) as a (small) part of the classification of finite simple groups.

This in turn can be seen as a (very difficult) converse of the above, see below Sec. 7.C.

6. Reductive Groups

We fix $\mathbb{F}$ an algebraically closed field.

One considers $\mathbb{F}$-varieties essentially as locally closed subvarieties of affine varieties. Algebraic groups over $\mathbb{F}$ are affine $\mathbb{F}$-varieties such that multiplication $(x, y) \mapsto xy$ and inversion $x \mapsto x^{-1}$ are $\mathbb{F}$-algebraic morphisms. Much of their abstract theory was done by Borel (see the reference [Bo] and the textbooks [MT], [Spr]).

The group $\mathrm{GL}_n(\mathbb{F})$ is clearly such an algebraic group ($n \geq 0$ an integer), and in fact all algebraic groups over $\mathbb{F}$ are closed subgroups of some $\mathrm{GL}_n(\mathbb{F})$. The latter is related with the existence of linear representations of algebraic groups, that is algebraic morphisms $\mathbf{G} \to \mathrm{GL}_n(\mathbb{F})$ (see [MT] 5.5).

An important property of algebraic groups is the notion of unipotent and semi-simple elements, along with the Jordan decomposition of elements of $\mathbf{G}$, $x = us = su$ where u, resp. s, is sent to a unipotent (resp. semi-simple) element by any linear representation of $\mathbf{G}$.

6.A. Reductive groups

Definition 6.1. The unipotent radical $\mathrm{R}_u(\mathbf{G})$ of an algebraic group $\mathbf{G}$ is its largest normal connected subgroup whose elements are all unipotent. The group $\mathbf{G}$ is said to be **reductive** if and only if it is connected and $\mathrm{R}_u(\mathbf{G}) = 1$. It is said to be semi-simple if in addition its center $\mathrm{Z}(\mathbf{G})$ is finite (this is also equivalent to being perfect, i.e. $[\mathbf{G}, \mathbf{G}] = \mathbf{G}$, and connected).

When $\mathbb{F}$ is the field of complex numbers, the above notion coincides with the one of semi-simple Lie groups. Those are classified essentially by use of Lie algebras and the classification of root systems (see [Hum1]). Some similar results can be obtained for reductive groups over algebraically closed fields $\mathbb{F}$ (Chevalley, 1956), leading to presentations by generators and relations (Steinberg, 1962) and identification for non exceptional Dynkin types to classical groups. We recall below some main notions and steps for this classification, mainly because they are also crucial to the description of linear representations of both reductive groups and their finite analogues.

6.B. Borel subgroups, tori and root data

In what follows $\mathbf{G}$ is a reductive groups over $\mathbb{F}$.

A Borel subgroup $\mathbf{B} \leq \mathbf{G}$ is any connected solvable closed subgroup maximal for inclusion. They form a single $\mathbf{G}$-conjugacy class.

Definition 6.2. A torus of $\mathbf{G}$ is any closed subgroup isomorphic (as an algebraic group) to a direct product of copies of $\mathrm{GL}_1(\mathbb{F}) = \mathbb{F}^\times$.

Maximal tori of $\mathbf{G}$ are a single $\mathbf{G}$-conjugacy class.

It is now customary to fix a maximal torus $\mathbf{T}$ in $\mathbf{G}$ and introduce several notions attached to this. Each Borel subgroup can be written $\mathbf{B} = \mathrm{R}_u(\mathbf{B}) \rtimes \mathbf{T}$ for a (non-unique) maximal torus of $\mathbf{G}$.

Let $X(\mathbf{T}) = \mathrm{Hom}(\mathbf{T}, \mathbb{F}^\times)$ (morphisms of algebraic groups), which is a lattice isomorphic to $\mathbb{Z}^r$ whenever $\mathbf{T} \cong (\mathbb{F}^\times)^r$. Similarly one defines $Y(\mathbf{T}) = \mathrm{Hom}(\mathbb{F}^\times, \mathbf{T})$ and gets a pairing

$$X(\mathbf{T}) \times Y(\mathbf{T}) \to \mathbb{Z} \cong \mathrm{Hom}(\mathbb{F}^\times, \mathbb{F}^\times), \quad (x, y) \mapsto \langle x, y\rangle = x \circ y.$$

Definition 6.3. A root subgroup is any minimal unipotent subgroup normalized by $\mathbf{T}$.

Each root subgroup associated to $\mathbf{T}$ is isomorphic to $\mathbb{F}$ by some map $\mathbb{F} \ni a \mapsto u(a) \in \mathbf{G}$ and (consequently) there exists $\alpha \in X(\mathbf{T})$ such that $t.u(a).t^{-1} = u(\alpha(t)a)$ for all $a \in \mathbb{F}$, $t \in \mathbf{T}$. The set of $\alpha \in X(\mathbf{T})$ giving rise to such root subgroups is denoted by $\Phi(\mathbf{G}, \mathbf{T})$ and called the **set of roots of G with respect to T**.

One also defines a set $\Phi(\mathbf{G}, \mathbf{T})^\vee \subseteq Y(\mathbf{T})$ and a bijection $\Phi(\mathbf{G}, \mathbf{T}) \to \Phi(\mathbf{G}, \mathbf{T})^\vee$ with $\alpha \mapsto \alpha^\vee$ actually obtained as follows. For $\alpha \in \Phi(\mathbf{G}, \mathbf{T})$ and associated root subgroup $\mathbf{U}_\alpha \leq \mathbf{G}$, there is some morphism $\varphi_\alpha \colon \mathrm{SL}_2(\mathbb{F}) \to \mathbf{G}$ sending the unipotent upper triangular matrices into $\mathbf{U}_\alpha$, the diagonal matrices into $\mathbf{T}$. Then $\alpha^\vee \in Y(\mathbf{T}) = \mathrm{Hom}(\mathbb{F}^\times, \mathbf{T})$ is defined by $\alpha^\vee(\lambda) = \varphi_\alpha \begin{pmatrix} \lambda & 0 \\ 0 & \lambda^{-1} \end{pmatrix}$ for any $\lambda \in \mathbb{F}^\times$. This $\alpha^\vee$ is called the *coroot* associated with the root α. It is well-defined thanks to the fact that φ_α above is unique up to $\mathbf{T}$-conjugacy. The relation with the usual notion of (crystallographic) root system is that $\Phi(\mathbf{G}, \mathbf{T}) \subseteq X(\mathbf{T}) \subseteq X(\mathbf{T}) \otimes_{\mathbb{Z}} \mathbb{R}$ endowed with the above pairing $X(\mathbf{T}) \times Y(\mathbf{T}) \to \mathbb{Z}$ and $\alpha \mapsto \alpha^\vee$ allow to define reflections

$$s_\alpha \colon X(\mathbf{T}) \otimes_{\mathbb{Z}} \mathbb{R} \to X(\mathbf{T}) \otimes_{\mathbb{Z}} \mathbb{R}$$

associated with any α and to prove that they preserve the set $\Phi(\mathbf{G}, \mathbf{T})$. A positive subsystem $\Phi(\mathbf{G}, \mathbf{T})^+$ is obtained by selecting the α such that the corresponding root subgroups $\mathbf{U}_\alpha$ are subgroups of a given Borel subgroup $\mathbf{B}$. Recall that a positive subsystem of a root system always contains a unique basis Δ of the root system (this allows to define many notions related with this positivity phenomenon, such as the notion of *highest positive root*). Such a basis Δ determines the Dynkin type of the root system and the corresponding reflections are a generating set $S := \{s_\delta \mid \delta \in \Delta\}$ for the group $W \leq \mathrm{GL}(X(\mathbf{G},\mathbf{T}) \otimes_{\mathbb{Z}} \mathbb{R})$ generated by all reflections above. The pair (W, S) is a Coxeter system (see Sec. 3 above).

The (infinite) group $\mathbf{G}$ has a BN-pair for the Borel subgroup $\mathbf{B}$ and $N = \mathrm{N}_{\mathbf{G}}(\mathbf{T})$. The quotient group $N/\mathbf{T}$ identifies with the group W described above (associate with $\alpha \in \Delta$ the class of $\varphi_\alpha\left(\begin{pmatrix} 0 & 1 \\ -1 & 0 \end{pmatrix}\right)$.

Example 6.4. (a) The case of $\mathbf{G} = \mathrm{GL}_n(\mathbb{F})$ is as follows. Let $\mathbf{T} \cong (\mathbb{F}^\times)^n$ be the subgroup of diagonal matrices. Then $X(\mathbf{T}) = \oplus_{i=1}^n \mathbb{Z} e_i$ where e_i is the morphism sending that n-tuple $(t_1, \dots, t_n)$ to t_i. With that notation, for any $i \neq j$ in $\{1, \dots, n\}$, $\alpha := e_i - e_j$ sends $(t_1, \dots, t_n)$ to $t_i(t_j)^{-1}$ and this is an element of $\Phi(\mathbf{G},\mathbf{T})$ corresponding to the subgroup $\mathbf{U}_\alpha$ of matrices in u such that $u - I_n$ has only zeros except possibly the element at row i and column j.

The positive roots corresponding to the Borel subgroup $\mathbf{B} \leq \mathbf{G}$ of upper triangular matrices correspond to pairs (i, j) with $j > i$. The associated basis is $\{e_1 - e_2, \dots, e_{n-1} - e_n\}$. The type of Coxeter system is A_{n-1} (see Theorem 3.6 above) corresponding with the above ordering of roots.

(b) The case of $\mathbf{G} = \mathrm{Sp}_{2n}(\mathbb{F})$. Let us denote by $M \mapsto M^T$ the transposition of matrices. Let $J_n \in \mathrm{GL}_n(\mathbb{F})$ be the matrix whose only non-zero elements are 1's on the second diagonal, that is the (i, j) element is $\delta_{n+1-i,j}$. Let $\mathrm{Sp}_{2n}(\mathbb{F}) \leq \mathrm{GL}_{2n}(\mathbb{F})$ be defined by the equation $M\begin{pmatrix} 0 & -J_n \\ J_n & 0 \end{pmatrix} M^T = \begin{pmatrix} 0 & -J_n \\ J_n & 0 \end{pmatrix}$. One can choose as maximal torus the subgroup $\mathbf{T}$ of diagonal matrices with diagonals of type

$$(t_1, \dots, t_n, t_n^{-1}, \dots, t_1^{-1}),$$

and one obtains a Borel subgroup as $\mathbf{UT}$ where $\mathbf{U}$ is the subgroup of matrices of type $\begin{pmatrix} X & XJ_nS \\ 0 & J_n(X^{-1})^T J_n \end{pmatrix}$ where $X \in \mathrm{U}_n(\mathbb{F})$ (unipotent upper triangular matrices of $\mathrm{GL}_n(\mathbb{F})$) and $S \in \mathrm{Sym}_n(\mathbb{F})$ (symmetric $n \times n$ matrices).

It is not difficult to single out "one-parameter" unipotent subgroups normalized by $\mathbf{T}$. They come in two types corresponding with the roots α sending the above diagonal matrix to $t_i t_j$ for some $1 \leq i, j \leq n$ ($\alpha = e_i + e_j$, first type) and to $t_i t_j^{-1}$ for some $i \neq j$ ($\alpha = e_i - e_j$, second type). Then $e_1 - e_2, \dots, e_{n-1} - e_n, 2e_n$ gives a basis of the root system and the corresponding Coxeter diagram is of type BC_n

Relation with the Bruhat order. It is easy to prove the following interpretation of the Bruhat order $\leq$ in $W(\mathbf{G},\mathbf{T})$ in terms of Zariski closure of double cosets, namely

$$\overline{\mathbf{B}y\mathbf{B}} = \cup_{1\leq x\leq y}\mathbf{B}x\mathbf{B}.$$

Following the model of Lie algebras, one is often led to consider the Borel subgroups "opposite" the one already used. In the above case, noting that $\mathbf{U} = <\mathbf{U}_\alpha \mid \alpha \in \Phi(\mathbf{G},\mathbf{T})^+>$, one defines

$$\mathbf{U}^- := <\mathbf{U}_\alpha \mid \alpha \in \Phi(\mathbf{G},\mathbf{T})^- = -\Phi(\mathbf{G},\mathbf{T})^+>.$$

Note that $\mathbf{U}^- = w_S\mathbf{U}w_S$ where w_S is the element of W of maximal length with regard to the basis above. The group $\mathbf{B}^- := \mathbf{T}\mathbf{U}^-$ is a Borel subgroup. The product $\mathbf{B}w_S\mathbf{B}w_S = \mathbf{B}\mathbf{B}^- = \mathbf{U}\mathbf{T}\mathbf{U}^-$ is called the *big cell* and is open, hence dense in $\mathbf{G}$ (use the closure property recalled above).

6.C. Classification in terms of root data

One defines a notion of **root datum** which stands for quadruples $(X,\Phi,Y,\Phi^\vee)$ where X, Y are lattices endowed with a perfect pairing over $\mathbb{Z}$, and $\Phi\subseteq X$, $\Phi^\vee\subseteq Y$ are subsets endowed with a bijection $\alpha\mapsto\alpha^\vee$ satisfying axioms producing root systems in the same fashion as seen above in the case of reductive groups where a maximal torus has been chosen.

Then the classification theorem of Chevalley ensures that the choice of an algebraically closed field $\mathbb{F}$ and a root datum $(X,\Phi,Y,\Phi^\vee)$ selects a reductive group $\mathbf{G}$ over $\mathbb{F}$ such that one of its maximal torus $\mathbf{T}$ is such that $(X,\Phi,Y,\Phi^\vee)\cong(X(\mathbf{T}),\Phi(\mathbf{G},\mathbf{T}),Y(\mathbf{T}),\Phi^\vee(\mathbf{G},\mathbf{T}))$. Such a $\mathbf{G}$ is unique up to an isomorphism which itself is unique up to $\mathbf{T}$-conjugacy.

The group is said to be of simply-connected type whenever it is semi-simple and $Y=\mathbb{Z}\Phi^\vee$. From the isomorphism theorem it is clear enough that, once Φ a root system and $\mathbb{F}$ are chosen, then there is only one group $\mathbf{G}_{\rm sc}$ which is of simply connected type. Any semi-simple $\mathbf{G}$ with same Φ and $\mathbb{F}$ is a central quotient $\mathbf{G}_{\rm sc}\to\mathbf{G}$ (see [Cart2] p 25).

The group is said to be of adjoint type whenever it is semi-simple and $X=\mathbb{Z}\Phi$. This is also equivalent to having trivial center. Any semi-simple $\mathbf{G}$ with same Φ and $\mathbb{F}$ satisfies $\mathbf{G}/\mathrm{Z}(\mathbf{G})\cong\mathbf{G}_{\rm ad}$.

Example 6.5. For the root system of type A_{n-1}, one has $\mathbf{G}_{\text{sc}} = \text{SL}_n(\mathbb{F})$ and $\mathbf{G}_{\text{ad}} = \text{PGL}_n(\mathbb{F})$. In type C_n, $\mathbf{G}_{\text{sc}}$ is the symplectic group $\text{Sp}_{2n}(\mathbb{F})$ (see Example 6.4.b above) and $\mathbf{G}_{\text{ad}}$ is its quotient by $\pm\text{Id}_{2n}$.

In certain types, the simply connected covering is less naturally found. In type B_n, $\mathbf{G}_{\text{ad}}$ is $\text{SO}_{2n+1}(\mathbb{F})$ (special orthogonal group for the quadratic form defined by the sum of squares of coordinates in $\mathbb{F}^{2n+1}$) but $\mathbf{G}_{\text{sc}}$ is a spin group $\text{Spin}_{2n+1}(\mathbb{F})$ defined by means of the Clifford algebra.

7. Finite Reductive Groups

Let p be a prime and $\mathbb{F}$ the algebraic closure of the field with p elements.

7.A. Definition and Lang's theorem

Let $f \geq 1$ and $q := p^f$. Let

$$\text{Frob}\colon \text{GL}_n(\mathbb{F}) \to \text{GL}_n(\mathbb{F})$$

be the raising of matrix entries to the q-th power. This turns the Frobenius automorphism of the field into a so-called **Frobenius endomorphism** of the algebraic group $\text{GL}_n(\mathbb{F})$.

Definition 7.1. A finite reductive group is any group of fixed points $\mathbf{G}^F := \{g \in \mathbf{G} \mid F(g) = g\}$ where $F\colon \mathbf{G} \to \mathbf{G}$ is an algebraic endomorphism such that, for at least one $k \geq 1$, F^k is the restriction to $\mathbf{G}$ of a Frobenius endomorphism $\text{Frob}\colon \text{GL}_n(\mathbb{F}) \to \text{GL}_n(\mathbb{F})$ with $\text{Frob}(\mathbf{G}) = \mathbf{G}$.

The study of finite reductive groups is made easier by the following important theorem due to Lang and generalized by Steinberg (see [CaE] 7.1).

Theorem 7.2. *In the above setting, if* $\mathbf{S}$ *is a closed connected* F*-stable subgroup of* $\mathbf{G}$, *one has* $\mathbf{S} = \{g^{-1}F(g) \mid g \in \mathbf{S}\}$.

In particular this implies the existence of F-stable Borel subgroups and F-stable maximal tori in $\mathbf{G}$.

Fact 7.3. *Finite reductive groups* $\mathbf{G}^F$ *have BN-pairs of type* $\mathbf{B}^F, N^F$ *where* $\mathbf{B}$, $N = \text{N}_{\mathbf{G}}(\mathbf{T})$ *make the BN-pair of the reductive group* $\mathbf{G}$ *and* $\mathbf{T} \leq \mathbf{B}$ *are* F*-stable.*

7.B. Examples

Taking the examples of 6.4 above, and with $k = 1$ in Definition 7.1 above, one finds the finite groups $\text{GL}_n(q)$ and $\text{Sp}_{2n}(q)$.

The case $k = 2$ can account for finite unitary groups $\mathrm{GU}_n(q)$, the latter a subgroup of $\mathrm{GL}_n(q^2)$. For this we take $F = \mathrm{Frob} \circ \sigma$ where Frob is the raising of matrix entries to the q-th power and $\sigma\colon \mathrm{GL}_n(\mathbb{F}) \to \mathrm{GL}_n(\mathbb{F})$ is the automorphism of order two sending a matrix to its transpose-inverse.

Cases where $k \neq 1$ is needed in Definition 7.1 are often called *twisted.* The above case where Frob is composed with a so-called *graph automorphism* also occurs in type D_n and E_6.

In types B_2 and F_4 with $p = 2$, and in type G_2 with $p = 3$, one may build an endomorphism $F\colon \mathbf{G} \to \mathbf{G}$ which behaves like a Frobenius composed with a graph automorphism on certain root subgroups (changing the root and applying the Frobenius to the parameter) and like a graph automorphism on others (changing the root but not the parameter), see [Cart1] Sect. 12.3, 12.4, [Cart2] 1.19. Then F^2 is a Frobenius endomorphism. The corresponding groups $\mathbf{G}_{\mathrm{sc}}^F$ are denoted by ${}^2B_2(q)$ or ${}^2F_4(q)$ for q a power of 2, and ${}^2G_2(q)$ for q a power of 3.

7.C. Classification: Finite simple groups, quasi-simple groups

The classification of finite simple groups is the collective work of dozens of mathematicians and was completed at the start of the 80s.

Theorem 7.4. *The finite simple groups are either*

- *any cyclic group of prime order*
- *any alternating group $\mathfrak{A}_n$ with $n \geq 5$*
- *a finite group with a BN-pair of type $G/\mathrm{Z}(G)$ where G is a finite reductive group, or the Tits group $[{}^2F_4(2), {}^2F_4(2)]$*
- *one of the 26 sporadic groups*

The members of the third item above are called *simple groups of Lie type*, see [GLS] for an in-depth analysis of their properties.

As is well-known, perfect groups G (i.e. such that $[G, G] = G$) have a unique maximal central extension

$$\hat{G} \to G$$

(a surjective morphism with central kernel and perfect $\hat{G}$), maximality being here defined by the property that any other would be covered by that one. The kernel of the above universal map is called the *Schur multiplier*, it is finite when G is. Finite **quasi-simple groups** are defined as perfect groups G such that $G/\mathrm{Z}(G)$ is simple. They are central quotients of the maximal central extensions of non-abelian finite simple groups. Apart from a finite number of exceptions, the situation for simple groups of Lie type parallels the one of semi-simple groups $\mathbf{G}$. Namely, the maximal quasi-simple groups are the finite reductive groups $E = (\mathbf{G}_{\mathrm{sc}})^F$ for $\mathbf{G}$ of irreducible Weyl group (or root system). Then the simple quotient is of course $E/\mathrm{Z}(E)$. Note that the latter is not necessarily a finite reductive group, see the example below.

Example 7.5. $\mathrm{PSL}_n(q) := \mathrm{SL}_n(q)/\mathrm{Z}(\mathrm{SL}_n(q))$ while $\mathrm{SL}_n(q) = (\mathbf{G}_{\mathrm{sc}})^F$ for $\mathbf{G}$ of type A_{n-1} over $\mathbb{F} = \bar{\mathbb{F}}_q$.

III. REPRESENTATIONS IN DEFINING CHARACTERISTIC

8. Rational G-Modules and Weights

In this section, we consider a reductive group $\mathbf{G}$ over an algebraically closed field $\mathbb{F}$ and we assume chosen $\mathbf{T} \leq \mathbf{B}$ a maximal torus and Borel subgroup of $\mathbf{G}$. We recall the set of roots $\Phi(\mathbf{G}, \mathbf{T})$ (see Sec. 6.C above). We denote $\mathbf{U} = \mathrm{R}_u(\mathbf{B})$ and we recall $\mathbf{U}^-$.

We consider the rational representations of a reductive group $\mathbf{G}$.

Definition 8.1. A rational representation of $\mathbf{G}$ is any rational group morphism $\mathbf{G} \to \mathrm{GL}_{\mathbb{F}}(M)$ where M is a finite dimensional $\mathbb{F}$-vector space.

Note that M is then a $\mathbb{F}\mathbf{G}$-module for the (infinite dimensional) group algebra $\mathbb{F}\mathbf{G}$, and we denote simply by $g.m$ or gm the outcome of the action of $g \in \mathbf{G}$ on $m \in M$.

Fact 8.2. *One has* $\mathrm{Res}_{\mathbf{T}} M = \oplus_{\lambda \in X(\mathbf{T})} M_\lambda$ *where* $M_\lambda = \{m \in M \mid t.m = \lambda(t)m \text{ for all } t \in \mathbf{T},\ m \in M\}$*. The latter are called the* ***weight subspaces*** *of* M.

Recall the exponential notation for fixed points: for instance $M^{\mathbf{T}} = M_0 = \{m \in M \mid t.m = m \text{ for any } t \in \mathbf{T}\}$.

Fact 8.3. *If* $M \neq \{0\}$*, then* $M^{\mathbf{U}} \neq \{0\}$

Fact 8.4. $\mathbb{F}\mathbf{G}.M^{\mathbf{U}} = \mathbb{F}\mathbf{U}^-.M^{\mathbf{U}}$.

The character of M, $\mathrm{ch}(M)$ keeps track of the dimensions of the weight subspaces.

Definition 8.5. One defines formal symbols e^λ indexed by elements of $X(\mathbf{T})$ and satisfying $e^\lambda e^\mu = e^{\lambda+\mu}$. The formal character of M is

$$\mathrm{ch}(M) = \sum_{\lambda \in X(\mathbf{T})} \dim_{\mathbb{F}}(M_\lambda) e^\lambda.$$

Let us recall the pairing $X(\mathbf{T}) \times Y(\mathbf{T}) \to \mathbb{Z}$ that we denote below as $\langle -, - \rangle$.

Definition 8.6. The dominant weights are defined by $X^+(\mathbf{T}) := \{\lambda \in X(\mathbf{T}) \mid \langle \lambda, \alpha^\vee \rangle \geq 0$ for all $\alpha \in \Phi(\mathbf{G}, \mathbf{T})^+\}$.

9. Defining Characteristic: The Simple Modules

We keep the same notations as in the section above. The following is due to Chevalley.

Theorem 9.1. *The (isomorphism types of) simple rational representations of* $\mathbf{G}$ *are in bijection with dominant weights.*

One denotes by $L(\lambda)$ *the simple rational* $\mathbb{F}\mathbf{G}$*-module associated with* $\lambda \in X^+(\mathbf{T})$*. It is characterized by the property that* $L(\lambda)^{\mathbf{U}} = L(\lambda)_\lambda$ *and is a line. Moreover* $L(\lambda)_\mu \neq \{0\}$ *implies that* $\lambda - \mu \in \mathbb{N}\Phi(\mathbf{G}, \mathbf{T})^+$.

Example 9.2. The case of $\mathbf{G} = \mathrm{SL}_2(\mathbb{F})$. In this case we take for $\mathbf{T}$ the diagonal torus of matrices $\begin{pmatrix} t & 0 \\ 0 & t^{-1} \end{pmatrix}$ for $t \in \mathbb{F}^\times$ and Borel subgroup $\mathbf{TU}$ where $\mathbf{U} = \{\begin{pmatrix} 1 & a \\ 0 & 1 \end{pmatrix} \mid a \in \mathbb{F}\}$. Then $X(\mathbf{T}) \cong \mathbb{Z}$ by the map associating to $m \in \mathbb{Z}$ the map $\begin{pmatrix} t & 0 \\ 0 & t^{-1} \end{pmatrix} \mapsto t^m$. The dominant weights correspond to $\mathbb{N}$.

Consider the natural action of $\mathbf{G}$ on $\mathbb{F}[X, Y] = S(\mathbb{F}^2)$ where $P(X, Y)$ is sent to $P(aX + bY, cX + dY)$ by $\begin{pmatrix} a & c \\ b & d \end{pmatrix} \in \mathrm{SL}_2(\mathbb{F})$. For $p - 1 \geq m \geq 0$, let $L'(m) \subseteq \mathbb{F}[X, Y]$ be the subspace of homogeneous polynomials of degree m. Then $L'(m)^{\mathbf{U}}$ is clearly the line $\mathbb{F}X^m = L'(m)_m$ while $L'(m) = \mathbb{F}\mathbf{U}^-.X^m$ (this is where the hypothesis that $m \leq p - 1$ is used). So we get part of the parametrization of Theorem 9.1. Note that $L'(m)$ has dimension $m + 1$ in that case. For the whole parametrization of simple rational representations, see Theorem 10.3 below. Note that when m is any integer ≥ 0, then $\mathbb{F}\mathbf{G}.X^m = \mathbb{F}\mathbf{U}^-.X^m$ provides indeed a simple rational representation of $\mathbf{G}$, that we should rename $L(m)$. This is the parametrization of Theorem 9.1 by dominant weights. But we no longer have $L(m) = L'(m)$ in general and the dimension can be smaller than $m + 1$, (though computable, see Theorem 10.2 below).

Example 9.3. $L(0) = \mathbb{F}$ the trivial module. In the notations of Example 6.4.a above, $L(e_1)$ for $\mathrm{GL}_n(\mathbb{F})$ is the natural representation, of dimension n.

10. Steinberg's Tensor Product and Restriction Theorems

We now turn to relations of simple rational modules with simple modules in the same characteristic for finite reductive groups. The theorems are by Steinberg (see [St63]) and are the main source of information on representations of finite reductive group in the defining characteristic.

We keep $\mathbf{G}$ a reductive group over $\mathbb{F}$ which is assumed to be of characteristic the prime p. We keep $\mathbf{T} \leq \mathbf{B} \leq \mathbf{G}$ a maximal torus and Borel with corresponding definition of root system $\Phi(\mathbf{G}, \mathbf{T})$ with basis Δ.

Definition 10.1. If q is a power of p, we denote $X_q^+(\mathbf{T}) := \{\lambda \in X(\mathbf{T}) \mid 0 \leq \langle\lambda, \delta^\vee\rangle \leq q-1$ for any $\delta \in \Delta\}$.

We now assume that $\mathbf{G} \leq \mathrm{GL}_n(\mathbb{F})$ is a subgroup invariant under Frob the raising of matrix entries to the p-th power. If M is the underlying vector space of a rational representation of $\mathbf{G}$, and $i \geq 0$, one denotes by $M^{[i]}$ the same vector space but where the action of $\mathbf{G}$ is twisted by Frob^i. With the new action being denoted by $*$, we have $g * m = \mathrm{Frob}^i(g).m$ for any $g \in \mathbf{G}$, $m \in M$.

The following is Steinberg's tensor product theorem.

Theorem 10.2. *Let $\lambda \in X^+(\mathbf{T})$ written as $\lambda = \sum_{i\geq 0} p^i\lambda_i$ where $\lambda_i \in X_p^+(\mathbf{T})$. Then*

$$L(\lambda) \cong \otimes_{i\geq 0} L(\lambda_i)^{[i]}.$$

Assume $F = (\mathrm{Frob})^f$ where $q = p^f$. The following gives the simple $\mathbb{F}\mathbf{G}^F$-modules.

Theorem 10.3. *For any $\lambda \in X_q(\mathbf{T})$ the module $L(\lambda)$ restricts into an irreducible representation of $\mathbf{G}^F$.*

Moreover, if $Y(\mathbf{T}) = \mathbb{Z}\Phi^\vee$, then all simple $k\mathbf{G}^F$-modules are of that type.

Definition 10.4. When $Y(\mathbf{T}) = \mathbb{Z}\Phi^\vee$, one defines the **Steinberg module** of $\mathbf{G}^F$ as $\mathrm{Res}^{\mathbf{G}}_{\mathbf{G}^F} L(\lambda)$ where $\lambda \in X(\mathbf{T})$ is defined by $\langle\lambda, \alpha^\vee\rangle = q-1$ for all $\alpha \in \Delta$.

The Steinberg $\mathbb{F}\mathbf{G}^F$-module is projective and has dimension $|\mathbf{U}^F|$, the order of the Sylow p-subgroup of $\mathbf{G}^F$.

11. Weyl Modules and Lusztig's Conjecture

We return to the study of $\mathbb{F}\mathbf{G}$-modules $L(\lambda)$ for $\lambda \in X^+(\mathbf{T}_0)$. We assume in this section that $\mathbf{G} = \mathbf{G}_{\mathrm{sc}}$, that is $Y(\mathbf{T}_0) = \mathbb{Z}\Phi^\vee$. *Weyl* $\mathbf{G}$*-modules* $V(\lambda)$ are a construction borrowed from the characteristic zero case. In the case of characteristic zero the formal characters of those modules are known thanks to Weyl's so-called character formula (see [Hum1] Sect. 24). So it is enough to determine the multiplicities $[V(\lambda) : L(\mu)]$ for any $\lambda, \mu \in X^+(\mathbf{T}_0)$ to have the characters of the simple modules $L(\lambda)$.

11.A. Weyl modules

Let $\mathfrak{g}$ be the semi-simple Lie algebra over $\mathbb{C}$ with root system Φ. Any $\lambda \in X^+(\mathbf{T}_0)$ can be considered as a dominant weight of $\mathfrak{g}$, so it gives rise to a simple $\mathfrak{g}$-module with highest weight λ, $V(\lambda)_{\mathbb{C}}$, which corresponds to $L(\lambda)$ for the simply-connected algebraic group $\mathbf{G}_{\mathbb{C}}$ of root system Φ over $\mathbb{C}$. Then, by a construction of Chevalley (see [Hum1] Ch. VII), it is possible to choose a lattice $V(\lambda)_{\mathbb{Z}}$ stable under the action of a Lie subalgebra $\mathfrak{g}_{\mathbb{Z}}$ such that $V(\lambda) := \mathbb{F} \otimes_{\mathbb{Z}} V(\lambda)_{\mathbb{Z}}$ has a compatible algebraic action of $\mathbf{G} = \mathbf{G}_{\mathrm{sc}}$.

The formal character of $V(\lambda)$ is known from the one of $V(\lambda)_{\mathbb{C}}$ which in turn is given by Weyl's character formula (see [Hum2] Sect. 3.2).

11.B. Affine Weyl group action on $X(\mathbf{T})$ and Lusztig's conjecture

Let α_0 be the highest root of Φ with regard to the basis Δ. Let $\rho \in X(\mathbf{T}_0)$ be the half sum of all positive roots, so that $\langle \rho, \alpha^\vee \rangle = 1$ for all $\alpha \in \Delta$.

Definition 11.1. Let W_{a} be the subgroup of $\mathrm{GL}(X(\mathbf{T}))$ generated by $W(\mathbf{T})$ and the translations by elements of $p\Phi = \{p\alpha \mid \alpha \in \Phi\}$. This is a Coxeter (affine) group for the generators $\{s_\delta \mid \delta \in \Delta\} \cup \{r_0\}$ where r_0 is the reflection in the hyperplane $\{x \in X(\mathbf{T}) \mid \langle x, \alpha_0^\vee \rangle = p\}$. Let $W_{\mathrm{dom}} := \{w \in W_{\mathrm{a}} \mid -w(\rho) - \rho \in X^+(\mathbf{T})\}$.

Since W_{a} is a Coxeter group for the indicated set of generating reflections, one has an associated Bruhat order $\leq$, length map $l\colon W_{\mathrm{a}} \to \mathbb{N}$, and Kazhdan-Lusztig polynomials $P_{v,w} \in \mathbb{Z}[x]$ ($v \leq w$ in W_{a}). Lusztig's conjecture, formulated as a "Problem" in [Lus80], is as follows

Problem 11.2. *Assume* $p > \langle \alpha_0^\vee, \rho \rangle$ *and let* $w \in W_{\mathrm{dom}}$ *with* $\langle \alpha_0^\vee, -w(\rho) \rangle \leq p(p - h + 2)$ *where* h *is the Coxeter number of* Φ *(see for instance [GePf] p. 29), then*

$$\mathrm{ch}(L(-w(\rho)-\rho)) = \sum_{\substack{v\in W_{\mathrm{dom}}\\ v\leq w}} (-1)^{l(w)-l(v)} P_{v,w}(1)\,\mathrm{ch}(V(-v(\rho)-\rho)).$$

The above, combined with Andersen-Humphreys' "linkage principle" and Jantzen's "translation principle " (see [Jn] II.6 and II.7), allows to express any $\mathrm{ch}(L(\lambda))$ ($\lambda \in X^+(\mathbf{T}_0)$) in terms of $\mathrm{ch}(V(\mu))$'s as long as $p > h$. The latter $\mathrm{ch}(V(\mu))$ is known in turn by Weyl's character formula.

It has been proved by Andersen-Jantzen-Soergel (see [AJS94]) with additional restriction on p.

Theorem 11.3. *Lusztig's problem has a positive answer for large p.*

In this work the bound on p is not explicit. An explicit bound exponential in the rank was given by Fiebig [F12]. Using Soergel's bimodules and Juteau-Mautner-Williamson's theory of parity sheaves, G. Williamson has shown that part of Lusztig's conjecture for $p > h$ is equivalent to absence of p-torsion in certain cohomology of Schubert varieties. In [W13], Williamson gives a method to find such torsion numbers. In particular Fibonacci numbers F_n and F_{n+1} are shown to be such numbers for $\mathbf{G} = \mathrm{SL}_{4n+7}(\mathbb{F})$, which readily excludes that Lusztig's conjecture could be proved for $p \geq f(h)$ with f a linear function (for SL_n). It is expected that any polynomial bound can be excluded.

12. More Modules

We show here some relations between simple modules for finite reductive groups in defining characteristic and representations of modular Iwahori-Hecke algebras (Green-Tinberg-Sawada, see [CaE] Ch. 6). This leads to define certain indecomposable modules with interesting properties (see [Gre78]).

We are back to a general finite reductive group $\mathbf{G}^F$ where $\mathbf{G}$ is a reductive group over $\mathbb{F}$ of characteristic p. Then we have seen in Fact 7.3 that $G = \mathbf{G}^F$ has a split BN-pair with subgroups $B = \mathbf{B}^F$, $T = \mathbf{T}^F$, $N = \mathrm{N}_{\mathbf{G}}(\mathbf{T})^F$. Recall that $\mathbf{U} = \mathrm{R}_u(\mathbf{B})$ is such that $U = \mathbf{U}^F$ is a Sylow p-subgroup of G. Note that an immediate corollary of Bruhat decomposition is that $G = \sqcup_{n\in N} UnU$ (disjoint union).

12.A. Modular Iwahori-Hecke algebra

The group G acts by translation on the set G/U of left cosets with regard to U, so the vector space $Y := \mathbb{F}G/U$ is a $\mathbb{F}G$-module. It is isomorphic with the induced module $\mathrm{Ind}_U^G\mathbb{F}$ where $\mathbb{F}$ denotes here the trivial $\mathbb{F}U$-module.

Definition 12.1. Let $\mathcal{H}_{\mathbb{F}}(G,U) = \mathrm{End}_{\mathbb{F}G}(Y)$. This is a finite dimensional algebra with basis $(a_n)_{n\in N}$ defined by $a_n(U) = \sum_{x\in U/U\cap nUn^{-1}} xnU$ (where we have given the image of the class U in $G/U \subseteq \mathbb{F}G/U$, the image of the other classes being then easily determined).

Proposition 12.2. *(i) The simple $\mathcal{H}_{\mathbb{F}}(G,U)$-modules are 1-dimensional.*
(ii) $\mathcal{H}_{\mathbb{F}}(G,U)$ is a self-injective algebra (i.e. the regular representation is injective).

12.B. Fixed point functor and simple modules

We now define the following classical functor H_Y from (left) $\mathbb{F}G$-modules to right $\mathcal{H}_{\mathbb{F}}(G,U)$-modules. If M is an $\mathbb{F}G$-module,

$$H_Y(M) = \mathrm{Hom}_{\mathbb{F}G}(Y,M) \cong M^U.$$

The second equality results from $Y = \mathrm{Ind}_U^G\mathbb{F}$ and Frobenius reciprocity. Note that this second equality implies that $H_Y(M) \neq \{0\}$ whenever $M \neq \{0\}$ since U is a p-group for p the characteristic of $\mathbb{F}$.

Using self-injectivity from Proposition 12.2 in a crucial way, one proves the following.

Theorem 12.3. *(Green) One has a decomposition*

$$Y = \oplus_{i=1}^{a} Y_i$$

with each Y_i an indecomposable module. Each Y_i had a simple head and simple socle (though not isomorphic in general). Moreover if $1 \leq i,j \leq a$, then $Y_i \cong Y_j \Leftrightarrow i = j \Leftrightarrow \mathrm{hd}(Y_i) \cong \mathrm{hd}(Y_j) \Leftrightarrow \mathrm{soc}(Y_i) \cong \mathrm{soc}(Y_j) \Leftrightarrow \mathrm{hd}(H_Y(Y_i)) \cong \mathrm{hd}(H_Y(Y_j)) \Leftrightarrow \mathrm{soc}(H_Y(Y_i)) \cong \mathrm{soc}(H_Y(Y_j))$.

Corollary 12.4. *([CE] 1.25.(i)) Let $\mathbb{F}G$-$\mathbf{mod}_Y$ be the full subcategory of the category of $\mathbb{F}G$-modules whose objects are the $\mathbb{F}G$-modules in the form $e(Y^m)$ for $m \geq 1$ and $e \in \mathrm{End}_{\mathbb{F}G}(Y^m)$. Then H_Y induces an equivalence between $\mathbb{F}G$-$\mathbf{mod}_Y$ and the category of right $\mathcal{H}_{\mathbb{F}}(G,U)$-modules.*

12.C. The p-blocks

The following is due to Humphreys (see [CaE] 6.18).

Theorem 12.5. *If* $S = \mathbf{G}^F/\mathrm{Z}(\mathbf{G}^F)$ *is simple non-abelian, then* $\mathbb{F}S = \mathbb{F}_{St} \oplus B_0(\mathbb{F}S)$ *a sum of the one-dimensional block corresponding to the Steinberg module and the principal block (see Sec. 2 above) of* S.

Remark 12.6. So we see that simple groups of Lie type have only two blocks in the defining characteristic. Certain groups have even less, that is just one (principal) block. This is the case for the simple Mathieu groups M_{22} and M_{24} with regard to the prime $p = 2$. On the other hand, groups with just one p-block are never a simple group of Lie type. It can even be proved that for p an *odd* prime and G a finite group, the group algebra $\bar{\mathbb{F}}_p G$ is just one block if and only if G has a normal p-subgroup $P \lhd G$ such that $\mathrm{C}_G(P) \leq P$ (see [Ha85] Th. 1).

IV. OTHER CHARACTERISTICS

We take $G = \mathbf{G}^F$ a finite reductive group with $\mathbf{G}$ over $\mathbb{F}$ of characteristic p. We let k be an algebraically closed field of characteristic $\ell \neq p$. The goal is to study the category kG-**mod**. The case $\ell = 0$ is very developed thanks to the work of Deligne-Lusztig and subsequent work of Lusztig, Shoji and others. Much less is known in the case of a positive ℓ but Harish Chandra philosophy can be used and provides a partition of simple modules.

13. Harish Chandra Philosophy

We assume $\mathbf{G}^F$ is equipped with its BN-pair $\mathbf{B}^F$, $\mathbf{T}^F$, N, etc... as in Fact 7.3. A *Levi decomposition* is any $\mathbf{G}^F$-conjugate of some standard Levi decomposition of a standard parabolic subgroup $\mathbf{P}_I = \mathbf{B}W_I\mathbf{B} = \mathrm{R}_u(\mathbf{P}_I) \rtimes \mathbf{L}_I$ (see Sec. 5.B above) where I is an F-stable subset of Δ, the basis of $\Phi(\mathbf{G}, \mathbf{T})$ associated with $\mathbf{B}$. We then write $\mathbf{P}^F = \mathrm{R}_u(\mathbf{P})^F \rtimes \mathbf{L}^F$ where $\mathbf{P} = g\mathbf{P}_I g^{-1}$, $\mathbf{L} = g\mathbf{L}_I g^{-1}$, for some $g \in \mathbf{G}^F$.

13.A. Harish Chandra induction and restriction

For $\mathbf{P}^F = \mathrm{R}_u(\mathbf{P})^F \rtimes \mathbf{L}^F$ a Levi decomposition, we define

$$e = |\mathrm{R}_u(\mathbf{P})^F|^{-1} \sum_{x \in \mathrm{R}_u(\mathbf{P})^F} x \quad \in k\mathbf{G}^F,$$

an idempotent commuting with the elements of $\mathbf{L}^F$. One defines two functors.

Definition 13.1. Let

$$\mathrm{R}_{\mathbf{L}}^{\mathbf{G}}\colon k\mathbf{L}^F\text{-}\mathbf{mod} \to k\mathbf{G}^F\text{-}\mathbf{mod} \quad \text{defined by}$$
$$N \mapsto k\mathbf{G}^F e \otimes_{k\mathbf{L}^F} N$$

and

$$^*\mathrm{R}_{\mathbf{L}}^{\mathbf{G}}\colon k\mathbf{G}^F\text{-}\mathbf{mod} \to k\mathbf{L}^F\text{-}\mathbf{mod} \quad \text{defined by}$$
$$M \mapsto M^{\mathrm{R}_u(\mathbf{P})^F} = eM = ek\mathbf{G}^F \otimes_{k\mathbf{G}^F} M.$$

The first is clearly the inflation of $\mathbf{L}^F$-modules to $\mathbf{P}^F$-modules followed by induction from $\mathbf{P}^F$ to $\mathbf{G}^F$. The two functors $\mathrm{R}_{\mathbf{L}}^{\mathbf{G}}$ and $^*\mathrm{R}_{\mathbf{L}}^{\mathbf{G}}$ are exact and are clearly adjoint to each other on both sides. Note that $\mathrm{R}_{\mathbf{G}}^{\mathbf{G}} = {}^*\mathrm{R}_{\mathbf{G}}^{\mathbf{G}}$ is the identical functor.

We have a transitivity formula $\mathrm{R}_{\mathbf{H}}^{\mathbf{G}} = \mathrm{R}_{\mathbf{L}}^{\mathbf{G}} \circ \mathrm{R}_{\mathbf{H}}^{\mathbf{L}}$ whenever $\mathbf{Q} = \mathrm{R}_u(\mathbf{Q})\mathbf{H}$ is another F-stable Levi decomposition in $\mathbf{G}$ with $\mathbf{Q} \leq \mathbf{P}$ and $\mathbf{H} \leq \mathbf{L}$ (so that $\mathbf{Q} \cap \mathbf{L} = (\mathrm{R}_u(\mathbf{Q}) \cap \mathbf{L})$. $\mathbf{H}$ is a Levi decomposition in the reductive group $\mathbf{L}$).

We have a Mackey formula for the compound $^*\mathrm{R}_{\mathbf{L}}^{\mathbf{G}} \circ \mathrm{R}_{\mathbf{L}'}^{\mathbf{G}}$:

Theorem 13.2. *Whenever* $\mathbf{P}' = \mathrm{R}_u(\mathbf{P}')\mathbf{L}'$ *is another* F*-stable Levi decomposition, one has*

$$^*\mathrm{R}_{\mathbf{L}}^{\mathbf{G}} \circ \mathrm{R}_{\mathbf{L}'}^{\mathbf{G}} = \sum_g \mathrm{R}_{\mathbf{L}\cap g\mathbf{L}'g^{-1}}^{\mathbf{L}} \circ \mathrm{ad}_g \circ {}^*\mathrm{R}_{\mathbf{L}'\cap g^{-1}\mathbf{L}g}^{\mathbf{L}'}$$

where g *ranges over a representative system of the double cosets* $\mathbf{P}^F\backslash\mathbf{G}^F/\mathbf{P}'^F$.

In order to simplify our notation we have omitted the parabolic subgroup used to define $\mathrm{R}_{\mathbf{L}}^{\mathbf{G}}$. In fact we have invariance with regard to the choice of $\mathbf{P}$, for a given $\mathbf{L}$. The following is due simultaneously to Dipper-Du and Howlett-Lehrer ([DipDu93], [HowL94], see also [CaE] 3.10):

Theorem 13.3. *Whenever* $\mathbf{P}' = \mathrm{R}_u(\mathbf{P}')\mathbf{L}$ *is another* F*-stable Levi decomposition with same Levi subgroup* $\mathbf{L}$*, and one denotes*

$$e' := |\mathrm{R}_u(\mathbf{P}')^F|^{-1} \sum_{x\in \mathrm{R}_u(\mathbf{P}')^F} x \qquad \in k\mathbf{G}^F$$

one has an isomorphism

$$kGe \to kGe'$$

by the map sending x *to* xe'.

Corollary 13.4. *Assume* $\mathbf{P}' = \mathrm{R}_u(\mathbf{P}')\mathbf{L}$ *and* $\mathbf{P} = \mathrm{R}_u(\mathbf{P})\mathbf{L}$ *as above are two* F*-stable Levi decompositions with same Levi subgroup* $\mathbf{L}$*, then* $\mathbf{P}$ *and* $\mathbf{P}'$ *give the same* $\mathrm{R}_{\mathbf{L}}^{\mathbf{G}}$.

Remark 13.5. While at least ${}^*\mathrm{R}_{\mathbf{L}}^{\mathbf{G}}$ as fixed point functor can be defined easily in characteristic p, it would no longer be independent of the parabolic used to define it. See [CaE] Ex. 5 p. 99.

13.B. Cuspidality and partition of simple modules

Definition 13.6. A non-zero $k\mathbf{G}^F$-module M is said to be **cuspidal** if and only if ${}^*\mathrm{R}_{\mathbf{L}}^{\mathbf{G}}(M) = 0$ for any $\mathbf{L} \neq \mathbf{G}$. A simple $k\mathbf{G}^F$-module is said to be **supercuspidal** if and only if its projective cover is cuspidal.

Note that by exactness of the ${}^*\mathrm{R}$ functors, a module is cuspidal if and only if all its simple composition factors are cuspidal. This shows that supercuspidal simple modules are cuspidal. It is also easily seen that a simple $k\mathbf{G}^F$-module M is cuspidal (resp. supercuspidal) if and only if it is not a submodule (resp. a composition factor) of any $\mathrm{R}_{\mathbf{L}}^{\mathbf{G}}N$ for $\mathbf{L} \neq \mathbf{G}$ and N a $k\mathbf{L}^F$-module.

Given a simple $k\mathbf{G}^F$-module S one may define the minimal pair $(\mathbf{L}, N)$ such that there is a non-zero map $\mathrm{R}_{\mathbf{L}}^{\mathbf{G}}(N) \to S$ (for instance with minimal $|\mathbf{L}^F| + \dim_k(N)$). Mackey formula, see Theorem 13.2 allows to show that N is simple cuspidal and that the pair $(\mathbf{L}, N)$ is unique up to $\mathbf{G}^F$-conjugacy. This gives a partition

$$\mathrm{Irr}(k\mathbf{G}^F) = \coprod_{(\mathbf{L},N)} \mathrm{Irr}(k\mathbf{G}^F, \mathbf{L}, N) \tag{13.7}$$

indexed by conjugacy classes of the above "cuspidal pairs" $(\mathbf{L}, N)$ where $\mathrm{Irr}(k\mathbf{G}^F, \mathbf{L}, N)$ is the set of simple $k\mathbf{G}^F$-module with a non-zero map $\mathrm{R}_{\mathbf{L}}^{\mathbf{G}}(N) \to S$. When $(\mathbf{L}, N)$ is such a pair, $\mathrm{Irr}(k\mathbf{G}^F, \mathbf{L}, N)$ is parametrized by the irreducible representations of the endomorphism algebra $\mathrm{End}_{k\mathbf{G}^F}(\mathrm{R}_{\mathbf{L}}^{\mathbf{G}}(N))$. For a field k of characteristic zero the latter is a Iwahori-Hecke algebra of the type studied by Howlett-Lehrer and Lusztig (see [HowL80], [Lus76] Sect. 5). In characteristic $\ell > 0$, the knowledge is less complete. The following is due to Geck-Hiss-Malle (see [GeHiMa96] 2.4.(a)):

Theorem 13.8. *If* $\mathbf{P} = \mathrm{R}_u(\mathbf{P})\mathbf{L}$ *is an* F*-stable Levi decomposition of an* F*-stable parabolic subgroup of* $\mathbf{G}$*, and* N *is a simple cuspidal* $k\mathbf{L}^F$*-modules, then as a* $k\mathbf{G}^F$*-module*

$$\mathrm{R}_{\mathbf{L}}^{\mathbf{G}}(N) = Y_1 \oplus \cdots \oplus Y_a$$

where the Y_i's are indecomposable and

- *for each i, $\mathrm{soc}(Y_i) \cong \mathrm{hd}(Y_i)$ and they are simple modules*
- *for any $1 \leq i < j \leq a$, one has $\mathrm{soc}(Y_i) \ncong \mathrm{soc}(Y_j)$ (and therefore also $Y_i \ncong Y_j$).*

The case of $\mathbf{G}^F = \mathrm{GL}_n(q)$ (q a power of p) has been worked out early on by Dipper (see [Dip85b]). One has the following (see also [CaE] 19.11). We assume that k is the residual field of a valuation ring $\mathcal{O}$, whose fraction field (of characteristic 0) is denoted by K.

Theorem 13.9. *Take $G = \mathrm{GL}_n(q)$. Any simple cuspidal kG-module is the reduction modulo $J(\mathcal{O})$ of a G-stable $\mathcal{O}$-lattice inside a simple cuspidal KG-module.*

Remark 13.10. Let us go back to an arbitrary finite reductive group G. Let us explain briefly how each simple kG-module is a composition factor of at least one $\mathrm{R}_L^G\pi$ where π is a supercuspidal simple kL-module. If a simple kG-module S is not supercuspidal, this means that ${}^*\mathrm{R}_L^G\mathcal{P}_S \neq \{0\}$ for some proper standard Levi $L \leq G$ and where $\mathcal{P}_S$ denotes the projective cover of S. Then there is a simple kL-module S' such that $\mathrm{Hom}_L({}^*\mathrm{R}_L^G\mathcal{P}_S, S') \neq \{0\}$, or equivalently by adjunction $\mathrm{Hom}_G(\mathcal{P}_S, \mathrm{R}_L^G S') \neq \{0\}$, which means that S is a composition factor of the module $\mathrm{R}_L^G S'$. By induction hypothesis, S' is a composition factor of some $\mathrm{R}_M^L S''$ for a supercuspidal simple kM-module and therefore by exactness S is a composition factor of $\mathrm{R}_M^G S''$. So indeed

$$\mathrm{Irr}(kG) = \bigcup_{(L,\pi)} \mathrm{Irr}(kG, L, \pi)$$

where (L, π) ranges over the (G-conjugacy classes of) pairs of a standard Levi L of G and π is a simple supercuspidal kL-module, and where $\mathrm{Irr}(kG, L, \pi)$ denotes the set of simple composition factors of $\mathrm{R}_L^G\pi$.

However, in contrast to the union (13.7) above, it is not clear whether the above union is always disjoint. See however Theorem 19.3 below.

14. Deligne-Lusztig Theory

In the next two sections we give the basic definitions introduced by Deligne-Lusztig's seminal paper ([DeLu76]) and some of the main consequences on the parametrization of irreducible characters of finite reductive groups (see also [Lus84]).

We keep $\mathbf{G}^F$ a finite reductive group associated with an F-stable reductive group $\mathbf{G}$ (see Definition 7.1 above).

14.A. Deligne-Lusztig varieties and functors

A key innovation brought by Deligne-Lusztig consists in generalizing Harish Chandra functors $\mathrm{R}_{\mathbf{L}}^{\mathbf{G}}$ to Levi subgroups that are F-stable but not necessarily present in the Levi decomposition of an F-stable parabolic subgroup. Let us recall that Levi subgroups of $\mathbf{G}$ are simply $\mathbf{G}$-conjugates of standard Levi subgroups $\mathbf{L}_I$, or more intrinsically centralizers $\mathrm{C}_{\mathbf{G}}(\mathbf{S})$ of tori (non maximal in general) of $\mathbf{G}$ (see for instance [DiMi] 1.22).

Example 14.1. A subclass of the class of F-stable Levi subgroups is the one of F-stable maximal tori. Assume as before that a pair $\mathbf{T}_0 \leq \mathbf{B}$ of a maximal tori and Borel subgroup, both F-stable, have been chosen (see Sec. 7.A above). Any other maximal torus has to be in the form $g\mathbf{T}_0g^{-1}$. It is F-stable if and only if $g^{-1}F(g) \in \mathrm{N}_{\mathbf{G}}(\mathbf{T}_0)$. Since any element of $\mathbf{G}$ can be written in the form $g^{-1}F(g)$ (see Theorem 7.2 above), we may have maximal tori $g\mathbf{T}_0g^{-1}$ such that $g^{-1}F(g)$ is in any class of $\mathrm{N}_{\mathbf{G}}(\mathbf{T}_0)/\mathbf{T}_0$. Arguing this way one finds easily that

$$g\mathbf{T}_0g^{-1} \mapsto g^{-1}F(g)\mathbf{T}_0$$

induces a bijection between $\mathbf{G}^F$-conjugacy classes of F-stable maximal tori and the so-called *F-classes* of the group $W = \mathrm{N}_{\mathbf{G}}(\mathbf{T}_0)/\mathbf{T}_0$ (the equivalence relation is defined by $w \sim_F w'$ in W if and only if there exists $v \in W$ such that $w = v^{-1}w'F(v)$). The class of the maximally split torus $\mathbf{T}_0$ just corresponds to $1 \in W$.

On étale topology and ℓ-adic cohomology groups, see [CaE] A.3, [Cart2], [Sri]. One chooses a prime $\ell \neq p$.

Definition 14.2. If $\mathbf{P} = \mathrm{R}_u(\mathbf{P}) \rtimes \mathbf{L}$ is a Levi decoposition with $F(\mathbf{L}) = \mathbf{L}$, one defines the **Deligne-Lusztig variety**

$$Y_{\mathbf{L},\mathbf{G}} := \{g\mathrm{R}_u(\mathbf{P}) \mid g^{-1}F(g) \in \mathrm{R}_u(\mathbf{P})\} \leq \mathbf{G}/\mathrm{R}_u(\mathbf{P})$$

which is stable by left translation under $\mathbf{G}^F$ and right translation under $\mathbf{L}^F$.

The étale topology on $Y_{\mathbf{L},\mathbf{G}}$ allows to define the so-called groups of ℓ-adic cohomology with compact support $R^i\Gamma_c(Y_{\mathbf{L},\mathbf{G}})$ (for $i \in \mathbb{Z}$), some K-vector spaces with inherited actions of $\mathbf{G}^F$ on the left and $\mathbf{L}^F$ on the right. Their alternating sum gives an element of $\mathrm{K}_0(K\mathbf{G}^F \times \mathbf{L}^{F\mathrm{opp}})$. By tensor product one then gets two adjoint functors

$$\mathrm{R}_{\mathbf{L}}^{\mathbf{G}} \colon \mathrm{K}_0(K\mathbf{L}^F) \to \mathrm{K}_0(K\mathbf{G}^F) \text{ and } {}^*\mathrm{R}_{\mathbf{L}}^{\mathbf{G}} \colon \mathrm{K}_0(K\mathbf{G}^F) \to \mathrm{K}_0(K\mathbf{L}^F).$$

It can be shown that those functors are independent of the prime $\ell \neq p$ chosen to define ℓ-adic cohomology (identifying simple $K\mathbf{G}^F$-modules with complex characters of $\mathbf{G}^F$).

When the parabolic subgroup $\mathbf{P}$ is F-stable, then $Y_{\mathbf{L},\mathbf{G}}$ is just the finite set $\mathbf{G}^F/\mathrm{R}_u(\mathbf{P})^F$ and the above functors are the images in the Grothendieck groups of the Harish Chandra functors denoted the same in the preceding section.

As in the case of the Harish Chandra functors, one has a transitivity formula associated with inclusions $\mathbf{Q} \leq \mathbf{P}$ of parabolic subgroups, see [DiMi] 11.5. One may also ask if $\mathrm{R}_{\mathbf{L}}^{\mathbf{G}}$ and ${}^*\mathrm{R}_{\mathbf{L}}^{\mathbf{G}}$ are independent of the parabolic $\mathbf{P}$ used to define $Y_{\mathbf{L},\mathbf{G}}$. This is related to the Mackey formula that can be expected for that functor, see Theorem 13.2. It is known from [DeLu76] when $\mathbf{L}$ is a torus (see also [DiMi] 11.15.(i)). For a general proof leaving out a small number of cases ($q = 2$ in exceptional types), see [BnMi11].

15. Unipotent Characters and Lusztig Series

15.A. Unipotent characters

Definition 15.1. The irreducible components of various $\mathrm{R}_{\mathbf{T}}^{\mathbf{G}}(1)$ for $\mathbf{T}$ an F-stable maximal torus of $\mathbf{G}$ and 1 the trivial representation of $\mathbf{T}^F$ make a subset $\mathcal{E}(\mathbf{G}^F, 1)$ of $\mathrm{Irr}(\mathbf{G}^F)$ called the set of **unipotent characters**.

It is relatively easy to show that

$$\mathcal{E}(\mathbf{G}^F, 1) \xleftarrow{\sim} \mathcal{E}(\mathbf{G}_{\mathrm{ad}}^F, 1)$$

through natural maps. More strongly, one can show that the set of unipotent characters "does not depend on q" but only on the type of the Dynkin diagram of $\mathbf{G}$ along with the action of F on it. Their degrees are polynomials in q.

Example 15.2. Let us discuss that case of type A_{n-1}. In this case $\mathcal{E}(\mathbf{G}^F, 1)$ is in bijection with the set $\mathfrak{P}_n$ of partitions of n. The $\mathbf{G}^F$-classes of F-stable maximal tori of $\mathbf{G}$ are in bijection $w \mapsto \mathbf{T}_w$ with elements w of the symmetric group $\mathfrak{S}_n$ (see Example 14.1 above). If $\rho \in \mathrm{Irr}(\mathfrak{S}_n)$ denotes a character of the symmetric group, then $\frac{1}{n!}\sum_{w\in\mathfrak{S}_n} \rho(w)\mathrm{R}_{\mathbf{T}_w}^{\mathbf{G}}(1)$ is $\pm$ an element of $\mathcal{E}(\mathbf{G}^F, 1)$ and this induces a bijection $\mathrm{Irr}(\mathfrak{S}_n) \to \mathcal{E}(\mathbf{G}^F, 1)$ in that case (see for instance [DiMi] 15.8). It is in turn well-known that the irreducible characters of $\mathfrak{S}_n$, like its conjugacy classes, are parametrized by $\mathfrak{P}_n$.

15.B. Lusztig series

An important feature of character theory of finite reductive groups is the notion of dual groups. If $\mathbf{G}$ is a group over $\mathbb{F}$, its dual is also one, so this notion is quite intrinsic and slightly different from Langland's dual (a group over $\mathbb{C}$). Here $\mathbf{G}^*$ is the reductive group over $\mathbb{F}$ whose root datum is $(Y, \Phi^\vee, X, \Phi)$ if the one of $\mathbf{G}$ was $(X, \Phi, Y, \Phi^\vee)$. One has a Frobenius endomorphism $F^* \colon \mathbf{G}^* \to \mathbf{G}^*$ inducing the same action as F on the groups X and Y.

Then it is elementary to check that $\mathbf{G}^F$-classes of pairs $(\mathbf{T}, \theta)$ where $\mathbf{T}$ is an F-stable maximal torus of $\mathbf{G}$ and $\theta \in \mathrm{Irr}(\mathbf{T}^F)$, are in bijection with ${\mathbf{G}^*}^F$-classes of pairs $(\mathbf{T}^*, s)$ where $\mathbf{T}^*$ is an F-stable maximal torus of $\mathbf{G}^*$ and $s \in (\mathbf{T}^*)^F$. This, along with a deep disjunction theorem on the generalized characters $\mathrm{R}_{\mathbf{T}}^{\mathbf{G}}(\theta)$ due to Deligne-Lusztig (see [DiMi] 13.3) allows to prove the following

Theorem 15.3. *There exists a partition*

$$\mathrm{Irr}(\mathbf{G}^F) = \sqcup_s \mathcal{E}(\mathbf{G}^F, s)$$

where s ranges over the ${\mathbf{G}^*}^F$-classes of semi-simple elements of ${\mathbf{G}^*}^F$. Moreover, one has a bijection (sometimes called a ***Jordan decomposition*** *of characters)*

$$\mathcal{E}(\mathbf{G}^F, s) \xleftarrow{\sim} \mathcal{E}(\mathrm{C}_{\mathbf{G}^*}(s)^F, 1)$$

where $\mathcal{E}(\mathrm{C}_{\mathbf{G}^*}(s)^F, 1)$ denotes the set of irreducible components of characters of $\mathrm{C}_{\mathbf{G}^*}(s)^F$ induced from a unipotent character of the finite reductive group $\mathrm{C}_{\mathbf{G}^*}(s)^{\circ F}$.

The "Jordan decomposition" of characters has many additional properties. In particular the character degrees are multiplied by $|{\mathbf{G}^*}^F : \mathrm{C}_{\mathbf{G}^*}(s)^F|_{p'}$ when going from right to left. This is a consequence of a more general compatibility with generalized characters of type $\mathrm{R}_{\mathbf{T}}^{\mathbf{G}}\theta$ for $\mathbf{T}$ a torus. This compatibility also ensures uniqueness of the Jordan decomposition when $\mathrm{Z}(\mathbf{G})$ is connected. Moreover the relation with $\mathrm{R}_{\mathbf{L}}^{\mathbf{G}}$ functors is such that when $\mathbf{L}^* := \mathrm{C}_{\mathbf{G}^*}(s)$ is a Levi subgroup of $\mathbf{G}^*$ (hence connected) dual to some F-stable Levi subgroup $\mathbf{L} \leq \mathbf{G}$, the above bijection is induced, up to a sign, by $\zeta \mapsto \mathrm{R}_{\mathbf{L}}^{\mathbf{G}}(\hat{s}\zeta)$ where $\hat{s}$ is the linear character of $\mathbf{L}^F$ induced by s (an element in the center of $\mathbf{L}^* = \mathrm{C}_{\mathbf{G}^*}(s)$) and duality.

16. ℓ-Blocks

Let us assume now that $\mathbf{G}^F$ is a finite reductive group of characteristic p, that ℓ is a prime $\neq p$ and that $(\mathcal{O}, K, k)$ is an ℓ-modular system for $\mathbf{G}^F$ and its subgroups in the sense of section 2 above.

Along with $(\mathbf{G}, F)$, we consider its dual group $(\mathbf{G}^*, F)$, see Sec. 15.B.

16.A. ℓ-blocks and Lusztig series

Definition 16.1. When s is a semi-simple ℓ-regular element of $\mathbf{G}^{*F}$, one denotes

$$\mathcal{E}_\ell(\mathbf{G}^F, s) := \cup_{t \in s\mathrm{C}_{\mathbf{G}^*}(s)^F_\ell} \mathcal{E}(\mathbf{G}^F, t)$$

where t actually ranges over the semi-simple elements of $\mathbf{G}^{*F}$ with ℓ-regular part s.

The following theorem is due to Broué-Michel (see [BrMi89] and [CaE] 9.12).

Theorem 16.2. *There is a sum of blocks of the group ring $\mathcal{O}\mathbf{G}^F$, denoted by $B_\ell(\mathbf{G}^F, s)$ such that*

$$\mathrm{Irr}(B_\ell(\mathbf{G}^F, s) \otimes_{\mathcal{O}} K) = \mathcal{E}_\ell(\mathbf{G}^F, s).$$

Remark 16.3. Relations between the functor $\mathrm{R}_{\mathbf{L}}^{\mathbf{G}}$ and ℓ-modular questions can be expected from the definition of $\mathrm{R}_{\mathbf{L}}^{\mathbf{G}}$, and actually a decisive argument for the above is given by the character formula of Deligne-Lusztig (see [DiMi] 12.2) whose proof indeed relies on projectivity arguments for $\mathcal{O}\mathbf{G}^F$-modules (see for instance [Sri] 6.4, [DeLu76] 3.5).

A related more elementary statement is as follows (Hiss, see [CaE] 9.12):

Proposition 16.4. *Any ℓ-block B' in $B_\ell(\mathbf{G}^F, s)$ satisfies* $\mathrm{Irr}(B' \otimes_{\mathcal{O}} K) \cap \mathcal{E}(\mathbf{G}^F, s) \neq \emptyset$.

From general theory, we know that irreducible Brauer characters in characteristic ℓ generate the subspace of the space $K\mathrm{Irr}(\mathbf{G}^F)$ of central functions on $\mathbf{G}^F$ that vanish outside ℓ-regular elements. Here $\mathcal{E}(\mathbf{G}^F, s)$ seems to play a similar role within $\mathcal{E}_\ell(\mathbf{G}^F, s)$ and relates strongly with irreducible Brauer characters (see Theorem 18.3 below).

16.B. Jordan decomposition of characters as a Morita equivalence

Bonnafé-Rouquier's theorem on Jordan decomposition (2003, see [BnRo03]) is the deep proof of a result conjectured by Broué.

Theorem 16.5. *Assume $s \in \mathbf{G}^{*F}$ is a semi-simple ℓ-regular element. Let $\mathbf{L}^*$ be an F-stable Levi subgroup of $\mathbf{G}^*$ such that $\mathrm{C}_{\mathbf{G}^*}(s) \leq \mathbf{L}^*$. Then considering an F-stable Levi subgroup $\mathbf{L} \leq \mathbf{G}$ dual to $\mathbf{L}^*$, one has a Morita equivalence between the module categories*

$$B_\ell(\mathbf{L}^F, s)\text{-}\mathbf{mod} \xrightarrow{\sim} B_\ell(\mathbf{G}^F, s)\text{-}\mathbf{mod}$$

such that the induced bijection between ordinary characters over K is induced by Lusztig's functor $\mathcal{E}_\ell(\mathbf{L}^F, s) \ni \zeta \mapsto \pm\mathrm{R}_{\mathbf{L}}^{\mathbf{G}}(\zeta) \in \mathcal{E}_\ell(\mathbf{G}^F, s)$.

The proof in [BnRo03] involves a thorough analysis of Deligne-Lusztig varieties and their finite coverings, along with deep considerations on the derived category of $k\mathbf{G}^F$-**mod**, even though the theorem is a Morita equivalence (see also [CaE] 10–12). More derived equivalences will be given in the next section.

16.C. Unipotent blocks and generalized Harish Chandra theory

Taking the minimal case in Theorem 16.5 where $\mathbf{L}^*$ can be taken to be $\mathrm{C}_{\mathbf{G}^*}(s)$, Bonnafé-Rouquier's theorem suggests strongly that the study of ℓ-blocks of $\mathbf{G}^F$ reduces to the study of blocks of some $B_\ell(\mathbf{G}^F, s)$ where s is central in $\mathbf{G}^*$, and in turn easily to blocks of some $B_\ell(\mathbf{G}^F, 1)$. The latter are called the **unipotent blocks** of the group $\mathbf{G}^F$ (see Proposition 16.4).

The cardinality of any finite reductive group $\mathbf{G}^F$ is a polynomial expression of q, with coefficients in $\mathbb{Z}$ (see tables in [Car2] Sect. 2.9), often called the *polynomial order* of $(\mathbf{G}, F)$. In this polynomial $P_{(\mathbf{G},F)}(x)$, the prime divisors $\neq x$ are cyclotomic polynomials.

If $\mathbf{S}$ is an F-stable subgroup of $\mathbf{G}$ whose polynomial order is a power of the d-th cyclotomic polynomial ϕ_d, then $\mathbf{S}$ is a torus. It is natural to study those "ϕ_d-tori" like ℓ-elements of a finite group (ℓ a prime). This leads to an analogue of Sylow's theorem, due to Broué–Malle, see ([BrM92], [CaE] 13.18). If $\mathbf{S}$ is a ϕ_d-torus, then $\mathrm{C}_{\mathbf{G}}(\mathbf{S})$ is an F-stable Levi subgroup, called a "d-split" Levi subgroup of $\mathbf{G}$. More important, Broué-Malle-Michel ([BrMaM93], see also [FoSr86]) have shown the existence of a generalized d-Harish Chandra theory where d-split Levi subgroups replace the standard Levi subgroups used in the Harish Chandra theory described in Sec. 13 above.

The following relation with unipotent ℓ-blocks has been shown by Cabanes-Enguehard (see [CaE] 22.9).

Let us denote by $e(q,\ell)$ the order of q in $(\mathbb{Z}/\ell\mathbb{Z})^\times$.

Theorem 16.6. *Assume $\ell \geq 7$. One has*

$$B_\ell(\mathbf{G}^F, 1) = \oplus_{\mathbf{L},\zeta} B_{\mathbf{L},\zeta}$$

where the $B_{\mathbf{L},\zeta}$'s are ℓ-blocks of $\mathbf{G}^F$ and $(\mathbf{L},\zeta)$ ranges over the pairs where $\mathbf{L}$ is an $e(q,\ell)$-split Levi subgroup of $\mathbf{G}$ and $\zeta \in \mathcal{E}(\mathbf{L}^F, 1)$ is such that ${}^*\mathrm{R}_{\mathbf{L}'}^{\mathbf{L}}(\zeta) = 0$ for any proper $e(q,\ell)$-split Levi subgroup of $\mathbf{L}$. Moreover $\mathrm{Irr}(K \otimes B_{\mathbf{L},\zeta}) \cap \mathcal{E}(\mathbf{G}^F, 1)$ is the set of irreducible components of the generalized character $\mathrm{R}_{\mathbf{L}}^{\mathbf{G}}(\zeta)$.

Remark 16.7. When ℓ divides $q-1$, the partition of unipotent characters into ℓ-blocks correspond with the one induced by Harish Chandra theory.

Example 16.8. Let us illustrate the above theorem in the case of $\mathbf{G}^F = \mathrm{GL}_n(q)$ (the theorem is then due to Fong-Srinivasan, see [FoSr82]). We assume $\mathrm{GL}_n(\mathbb{F})$ endowed with its usual Frobenius endomorphism F raising matrix entries to the q-th power.

Let us abbreviate $e = e(q,\ell)$. A typical e-split Levi subgroup of $\mathrm{GL}_e(\mathbb{F})$ is the maximal torus $\mathbf{C}_e$ of type the cycle of order e in the Weyl group $\mathfrak{S}_e$ with respect to the diagonal torus (see Example 14.1 above). Note that $(\mathbf{C}_e)^F$ is a cyclic group of order $q^e - 1$. Then we can define the e-split Levi subgroups $\mathbf{L}_m \leq \mathrm{GL}_{n-me}(\mathbb{F}) \times \mathrm{GL}_e(\mathbb{F})^m \leq \mathrm{GL}_n(\mathbb{F})$ for $0 \leq me \leq n$ with $\mathbf{L}_m \cong \mathrm{GL}_{n-me}(\mathbb{F}) \times (\mathbf{C}_e)^m$.

It can be shown that the pairs $(\mathbf{L},\zeta)$ as in Theorem 16.6 are the pairs $(\mathbf{L}_m,\zeta)$ where $0 \leq me \leq n$ and $\zeta \in \mathcal{E}(\mathbf{L}_m, 1)$ correspond to an element of $\mathcal{E}(\mathrm{GL}_{n-me}(q), 1)$ whose associated partition of $n - me$ (see Example 15.2 above) is a so-called *e-core*. (Recall from the combinatorics of partitions that a partition $\lambda_1 \geq \lambda_2 \geq \cdots \geq \lambda_k > 0$ is an e-core if and only if, for all i, $\lambda_i - i - e \in \{\lambda_j - j \mid j = 1,\ldots,k\}$ or $\lambda_i - i - e \leq -k$, see for instance [JaKe] 2.7 or [CaE] 5.11.)

The above theorem describes the splitting of $B_\ell(\mathbf{G}^F, 1)$ into ℓ-blocks from the point of view of ordinary (unipotent) characters. In fact much more is known about the blocks of $\mathbf{G}^F$. From a similar point of view, a splitting of an arbitrary $B_\ell(\mathbf{G}^F, s)$ into blocks and the ordinary characters of those blocks are known (see [CaE99] for $\ell \geq 7$, and [KM13] for the remaining cases). Those references give also information on the so-called defect ℓ-subgroup of $\mathbf{G}^F$ associated to each block.

The notion of defect group of an ℓ-block is due to R. Brauer and has been reformulated by J.A. Green (see [N]). In the following definition, let B be a block of the ring $\mathcal{O}G$ of a finite group G, with $(\mathcal{O}, K, k)$ an ℓ-modular system for G.

Definition 16.9. The **defect subgroup** of B is any minimal subgroup $D \leq G$ such that the $\mathcal{O}G$-bimodule map $B \otimes_{\mathcal{O}D} \mathcal{O}G \to B$, $x \otimes y \mapsto xy$ splits. It is a single G-conjugacy class of ℓ-subgroups of G.

Defect subgroup can be seen as measuring the non-simplicity of the algebra $B \otimes_{\mathcal{O}} k$. It is $\{1\}$ when B has projective module such that its reduction modulo $J(\mathcal{O})$ is simple. On the other extreme, the principal block (see Sec. 2 above) has defect the Sylow ℓ-subgroup of G. The Steinberg module gives a p-block with defect $\{1\}$ in defining characteristic ("$\ell = p$") for any group $\mathbf{G}^F$.

The ℓ-block of $\mathrm{GL}_n(q)$ defined by a pair $(\mathbf{L}_m, \zeta)$ as in Remark 16.8 above has defect any Sylow ℓ-subgroup of $\mathrm{GL}_{me}(q)$ (Fong-Srinivasan 1982, see for instance [CaE] 22.10, [FoSr82]).

16.D. Cyclic defect

When the defect group D of a block B is cyclic, $B \otimes_{\mathcal{O}} k$ has only a finite number of indecomposable modules and the whole category $B \otimes_{\mathcal{O}} k$-**mod** can be described by a combinatorial object called the *Brauer tree* of the block. This is a tree whose *edges* (not nodes !) correspond with simple $B \otimes_{\mathcal{O}} k$-modules. The projective cover of a given simple module corresponding with the edge ϵ can be described in terms of the edges adjacent to ϵ.

The determination of Brauer trees occurring for blocks of finite reductive groups was started by Fong-Srinivasan and is now close to completion thanks to recent work of Craven, Dudas and Rouquier (see [Cr12] and its references). The methods make use of modules in characteristic ℓ defined by ℓ-adic cohomology of Deligne-Lusztig varieties. Craven has proposed in [CR12] 1.3 a new conjecture on the cohomology of those varieties.

17. Some Derived Equivalences

In [B90], M. Broué issued several conjectures on derived equivalences between module categories of blocks of finite groups. In many cases that have been verified, not only a derived equivalence but a homotopic equivalence has been checked. This is a stronger result and since the homotopic category is easier to describe we concentrate on this case. The homotopic category $\mathrm{Ho}^{\mathrm{b}}(A)$ of some A-**mod** (or any abelian category) is obtained from the category $C^b(A\text{-}\mathbf{mod})$ of bounded complexes of A-modules by keeping the same objects and replacing the additive groups of morphisms $\mathrm{Hom}_A(C, C')$ between complexes of A-**mod** with their quotients $\mathrm{Hom}_A(C, C')/I(C, C')$ where $I(C, C')$ is the subgroup of morphisms homotopic to the null morphism.

When A is a semi-simple ring, for instance a division ring, every bounded complex $C = (C^i, \partial^i)_i$ is isomorphic with its homology

$$H(C) = (\mathrm{Ker}(\partial^i)/\partial^{i-1}(C^{i-1}))_i,$$

seen as a complex with morphisms all equal to zero.

If B and B' are two blocks of finite group algebras $\mathcal{O}G$ and $\mathcal{O}G'$ and one has an equivalence

$$\mathrm{Ho}^b(B) \cong \mathrm{Ho}^b(B')$$

this clearly implies an equivalence

$$\mathrm{Ho}^b(B \otimes_{\mathcal{O}} K) \cong \mathrm{Ho}^b(B' \otimes_{\mathcal{O}} K)$$

and therefore a bijection

$$\mathrm{Irr}(B) \to \mathrm{Irr}(B')$$

between irreducible characters with signs attached to each element of $\mathrm{Irr}(B)$. Broué's conjectures often consist in asking if some given correspondence between characters is coming from such an equivalence of homotopic categories.

17.A. Chuang-Rouquier's theorems

We have seen in Example 16.8 that the unipotent blocks of $\mathrm{GL}_n(q)$ are indexed by pairs (m, ζ) where $0 \leq me \leq n$ and $\zeta \in \mathcal{E}(\mathrm{GL}_{n-me}(q), 1)$ is associated with a partition of $n - me$ with no e-hook.

In [ChR08] 7.18, Chuang-Rouquier proved the following.

Theorem 17.1. *For $i = 1, 2$, let B_i be the unipotent ℓ-block of $\mathrm{GL}_{n_i}(q)$ associated with (m_i, ζ_i). Assume $m_1 = m_2$, then*

$$\mathrm{Ho}^b(B_1) \cong \mathrm{Ho}^b(B_2).$$

This theorem is in fact an application of the study of the series of Grothendieck groups of modular representations of symmetric groups, through actions of Lie algebras $\mathfrak{sl}_2$ on those categories - or *"categorifications"*. This approach proves a similar theorem as the above for blocks of symmetric groups (also conjectured by Broué in the form of a derived equivalence) and for cyclotomic Iwahori-Hecke algebras (see [ChR08] 7.2 and 7.12).

17.B. Alvis-Curtis duality as a self-equivalence

Let us consider the BN-pair of $G = \mathbf{G}^F$ (see Sec. 7.3 above). In particular we have the Weyl group (W, S) and the parabolic subgroups $P_I = U_I \rtimes L_I$ associated with subsets $I \subseteq S$.

Keep $(\mathcal{O}, K, k)$ an ℓ-modular system for G. Alvis-Curtis duality is the map

$$D_G := \sum_{I \subseteq S} (-1)^{|I|} \mathrm{R}_{L_I}^G \circ {}^*\mathrm{R}_{L_I}^G : \mathrm{K}_0(KG) \to \mathrm{K}_0(KG).$$

Alvis-Curtis showed that there is a bijection $\chi \mapsto \chi^*$ from $\mathrm{Irr}(G)$ to itself, such that $D_G(\chi) = \pm\chi^*$ for any $\chi \in \mathrm{Irr}(G)$ (see [DiMi] 8).

Theorem 17.2. *There is an equivalence*

$$\mathrm{Ho}^{\mathrm{b}}(\mathcal{O}G) \xrightarrow{\sim} \mathrm{Ho}^{\mathrm{b}}(\mathcal{O}G)$$

such that the induced equivalence $\mathrm{K}_0(KG) \xrightarrow{\sim} \mathrm{K}_0(KG)$ *is Alvis-Curtis duality* D_G.

A version of the above with derived categories instead of homotopic categories was first conjectured by Broué and proved by Cabanes-Rickard (2001, see [CaE] 4.19), the version with the homotopic categories is due to Okuyama (2006, see [Ca09]). The methods used there are fairly elementary, through a complex related with the coefficient system on subsets of S such that $X_I = \mathcal{O}Ge_I \otimes_{P_I} e_I\mathcal{O}G$ and $X_I \to X_J$ for $I \subseteq J$ is the map sending $x \otimes_{P_I} y$ to $x \otimes_{P_J} y$. The independence Theorem 13.3 above is used in a crucial way.

18. Decomposition Numbers

Given the better knowledge we have of ordinary characters of our groups, it is reasonable to expect information on modular characters from them. The coefficients expressing this are the so-called *decomposition numbers.*

18.A. Decomposition matrices

Let us recall the notion of a decomposition matrix, relating Brauer's modular characters with the ordinary characters.

Let G be a finite group, let ℓ be a prime and $(\mathcal{O}, K, k)$ be an ℓ-modular system for G (see Sec. 2 above). The decomposition matrix $\mathrm{Dec}(\mathcal{O}G) =$

$(d_{\chi\phi})_{\chi,\phi}$ of G in characteristic ℓ is defined as a matrix whose rows are indexed by $\mathrm{Irr}(G)$ and columns by $\mathrm{IBr}(G)$ and the element $d_{\chi,\phi} \in \mathcal{O}$ is defined by

$$\chi(g) = \sum_{\phi \in \mathrm{IBr}(G)} d_{\chi,\phi}\phi(g)$$

for any ℓ-regular element g of G.

$$\begin{array}{c} \\ \chi \end{array}\begin{pmatrix} & S & \\ & \vdots & \\ \cdots & d_{\chi,S} & \cdots \\ & \vdots & \\ & \vdots & \end{pmatrix} = \mathrm{Dec}(\mathcal{O}G).$$

Each row may be seen as representing the multiplicities of simple kG-modules in the reduction modulo $J(\mathcal{O})$ of an $\mathcal{O}G$-lattice inside a fixed simple KG-module of character $\chi \in \mathrm{Irr}(G)$. Each column can be seen as $\mathcal{P}_S \otimes_{\mathcal{O}} K$ for $\mathcal{P}_S$ the projective cover of a simple kG-module S. This could of course be defined in the same fashion for any $\mathcal{O}$-algebra, $\mathcal{O}$-free of finite rank, but a group algebra $\mathcal{O}G$ has the property that its decomposition matrix with regard to ℓ has always more rows than column when ℓ divides the order of G.

From those definitions it is not difficult to see that the partitions of $\mathrm{Irr}(G)$ and $\mathrm{IBr}(G)$ induced by blocks of $\mathcal{O}G$ allow to write the decomposition matrix "diagonally" (note that the matrices are not square) as

$$\mathrm{Dec}(\mathcal{O}G) = \begin{pmatrix} \mathrm{Dec}(B_1) & & \\ & \mathrm{Dec}(B_2) & \\ & & \ddots \end{pmatrix}.$$

In other words $d_{\chi,\phi} = 0$ when $\chi \in \mathrm{Irr}(G)$ and $\phi \in \mathrm{IBr}(G)$ select distinct blocks of $\mathcal{O}G$.

One also sees easily that if two blocks B of $\mathcal{O}G$, resp. B' of $\mathcal{O}G'$, are Morita equivalent $B \sim_{\mathrm{M}} B'$, then we have induced Morita equivalences $B \otimes_{\mathcal{O}} K \sim_{\mathrm{M}} B' \otimes_{\mathcal{O}} K$ and $B \otimes_{\mathcal{O}} k \sim_{\mathrm{M}} B' \otimes_{\mathcal{O}} k$, hence bijections between simple modules $\mathrm{Irr}(B) \to \mathrm{Irr}(B')$ and $\mathrm{IBr}(B) \to \mathrm{IBr}(B')$, yielding equality $\mathrm{Dec}(B) = \mathrm{Dec}(B')$ between decomposition matrices.

18.B. Triangular decomposition matrices: Symmetric groups

The work on decomposition matrices of finite reductive groups was initiated by Dipper for general linear groups (see [Dip85a] and [Dip85b]). The model is the corresponding older theorem on symmetric groups.

We have (see [JaKe] 6.3.60):

Theorem 18.1. *There are orderings of the rows and columns of the decomposition matrix of* $\mathfrak{S}_n$ *such that* $\mathrm{Dec}(\mathcal{O}\mathfrak{S}_n) = \begin{pmatrix} D_1 \\ D_2 \end{pmatrix}$ *where* D_1 *is square lower unitriangular.*

18.C. Triangular decomposition matrices: Finite reductive groups

Concerning now finite reductive groups in non-defining characteristics, Theorem 16.5 and the above generalities imply that we can somehow reduce the study of blocks to the study of the sum of unipotent blocks $B_\ell(\mathbf{G}^F, 1)$ (see Theorem 16.2). Let us denote by $\mathrm{Dec}_{\mathrm{uni}}$ the corresponding part of the ℓ-decomposition matrix.

Building on many examples we have the following conjecture (Geck, see [GeHi97] 3.4).

Conjecture 18.2. *Under mild conditions on the prime* $\ell \neq p$*, one has* $\mathrm{Dec}_{\mathrm{uni}} = \begin{pmatrix} D_1 \\ D_2 \end{pmatrix}$ *where* D_1 *is square lower unitriangular.*

Note that a consequence of the unitriangularity of $\mathrm{Dec}(G)$, once the subset of $\mathrm{Irr}(G)$ corresponding to D_1 has been chosen (a so-called *basic set*), then we have a unique injection $\mathrm{IBr}(G) \hookrightarrow \mathrm{Irr}(G)$.

The following is due to Geck-Hiss [GeHi91] (see also [CaE] 14.4) and somehow relates with Proposition 16.4 above.

Theorem 18.3. *If* $\mathrm{Z}(\mathbf{G})$ *is connected and* $\ell \geq 7$*, then putting first the lines of* $\mathrm{Dec}_{\mathrm{uni}}$ *corresponding to unipotent characters, one gets a square submatrix* D_1 *with determinant* ± 1*. In other words* $\mathcal{E}(\mathbf{G}^F, 1)$ *has same cardinality as the set of Brauer characters of* $B_\ell(\mathbf{G}^F, 1)$ *and their restrictions to* ℓ*-regular classes generate the same space of class functions.*

Apart form Dipper's work ([Dip85a] and [Dip85b]), Geck's conjecture was proved for any prime $\ell \neq p$ by Geck for unitary groups, by Geck-Hiss-Miyachi for type E_6, by Köhler for type F_4, by Hiss-Shamash for type G_2. When $\mathbf{G}^F$ is a classical group and the prime ℓ is linear (i.e. a Sylow

ℓ-subgroup of $\mathbf{G}^F$ is included in a Levi subgroup of type A), then Gruber-Hiss have proved the unitriangularity property of Geck's conjecture (see [GrHi97]).

Note that, combined with the fact that ℓ-solvable groups also satisfy unitriangularity of the whole decomposition matrix (with D_1 an identity matrix, by a theorem of Fong-Swan, 1961), this implies that actually many finite groups satisfy this statement for almost all primes.

18.D. The q-Schur algebra

The strategy followed by Dipper and formalized in subsequent work by Dipper-James consists in choosing certain projective $\mathcal{O}\mathbf{G}^F$-modules of type $\mathrm{R}_{\mathbf{L}}^{\mathbf{G}}(\Gamma_{\mathbf{L}})$ where $\mathbf{L}$ is a split Levi subgroup whose root system has only types A and $\Gamma_{\mathbf{L}}$ is a so-called Gelfand-Graev representation of $\mathbf{L}^F$. Gelfand-Greav representations of $\mathbf{G}^F$ are induced from one-dimensional representations of the maximal unipotent subgroup $\mathbf{U}^F$ of $\mathbf{G}^F$ (hence their projectivity), see [DeLu76] 10 or [DiMi] 14.29. This leads in turn to study the Iwahori-Hecke algebra $\mathcal{H}_{\mathcal{O}}(\mathfrak{S}_n, q)$ of type A and parameter q over $\mathcal{O}$, and certain ideals $x_\lambda \mathcal{H}_{\mathcal{O}}(\mathfrak{S}_n, q)$ defined as follows.

Definition 18.4. For $\lambda = (\lambda_1 \geq \lambda_2 \geq \cdots \geq \lambda_k > 0) \vdash n$, let $\mathfrak{S}_\lambda \cong \mathfrak{S}_{\lambda_1} \times \cdots \times \mathfrak{S}_{\lambda_k}$ the corresponding parabolic subgroup of $\mathfrak{S}_n$ and $x_\lambda = \sum_{w \in \mathfrak{S}_\lambda} (-q)^{l(w)} A_w \in \mathcal{H} := \mathcal{H}_{\mathcal{O}}(\mathfrak{S}_n, q)$ in the notations of Theorem 4.2 above. The q-Schur algebra of $\mathfrak{S}_n$ and parameter $q \in \mathcal{O}^\times$ is

$$\mathcal{S}_{\mathcal{O}}(n, q) := \mathrm{End}_{\mathcal{H}}(\coprod_{\lambda \vdash n} x_\lambda \mathcal{H}).$$

Dipper-James have shown that $\mathcal{S}_{\mathcal{O}}(n, q) \otimes_{\mathcal{O}} K$ and $\mathcal{S}_{\mathcal{O}}(n, q) \otimes_{\mathcal{O}} k$ have the same number of simple modules, hence a square decomposition matrix $D_{\mathcal{S}}$, which is also proved to be unitriangular (see [DipJam89]). Their proof of Geck's conjecture shows namely that the part of $\mathrm{Dec}_{\mathrm{uni}}$ corresponding to the unipotent characters of $\mathrm{GL}_n(q)$ is precisely the decomposition matrix of $\mathcal{S}_{\mathcal{O}}(n, q)$.

18.E. Genericity

Several theorems show that for a given type of group $\mathbf{G}^F$, the decomposition matrix of the unipotent blocks $\mathrm{Dec}_{\mathrm{uni}}$ is as generic as one can expect. For instance it is shown in [DipJam89] and [GrHi97] that the entries of $\mathrm{Dec}_{\mathrm{uni}}$ are bounded independently of q and ℓ for $\mathbf{G}^F$ a group as considered there.

A conjecture by G. James is about the case of $\mathbf{G}^F = \mathrm{GL}_n(q)$. Recall that $e(q, \ell)$ denotes the multiplicative order of q modulo ℓ. James' conjecture asserts that if $e(q,\ell)\ell < n$, then $\mathrm{Dec}_{\mathrm{uni}}$ only depends on $e(q,\ell)$. This was proved for $n \leq 10$ by James [Ja90] but Williamson's recent work on Lusztig's conjecture implies that counter-examples to James' statement exist (for quite large but explicit n, see [W13] Sect. 6).

19. Some Harish Chandra Series

19.A. The case of $\mathrm{GL}_n(q)$

We describe here the partition of $\mathrm{IBr}(B_\ell(\mathbf{G}^F,1))$ in terms of Harish Chandra series for $\mathbf{G}^F = \mathrm{GL}_n(q)$. In what follows, denote $G := \mathrm{GL}_n(q)$.

Recall the set $\mathcal{E}(\mathbf{G}^F,1)$ in bijection with partitions of n by the bijection

$$\mathrm{Irr}(\mathfrak{S}_n) \to \mathcal{E}(\mathbf{G}^F,1), \rho \mapsto \chi_\rho := \pm\frac{1}{n!}\sum_{w\in\mathfrak{S}_n} \rho(w)\mathrm{R}^{\mathbf{G}}_{\mathbf{T}_w}(1)$$

(see Example 15.2 above). We choose an actual parametrization of $\mathrm{Irr}(\mathfrak{S}_n)$ by partitions of n, $\lambda \mapsto [\lambda]$ such that $[\lambda]$ is present in both $\mathrm{Ind}_{\mathfrak{S}_{\lambda^*}}^{\mathfrak{S}_n} 1$ and $\mathrm{Ind}_{\mathfrak{S}_\lambda}^{\mathfrak{S}_n}\epsilon$ where λ^* is the transpose partition of λ and ϵ is the signature character — note that this is the transpose of the convention in [JaKe] Sect. 2. We then get a parametrization $\lambda \mapsto \chi_{[\lambda]}$ of $\mathcal{E}(G,1)$ by the set $\mathfrak{P}_n$ of partitions of n.

The unitriangularity of the decomposition matrix of $B_\ell(G,1)$ then implies that we get in turn a parametrization $\lambda \mapsto \phi_\lambda$ of $\mathrm{IBr}(B_\ell(G,1))$ by $\mathfrak{P}_n$.

Assume then that $\lambda \vdash n$.

Theorem 19.1. *ϕ_λ represents a cuspidal kG-module if and only if $n = e(q,\ell)\ell^a$ for an integer $a \geq 0$ and $\lambda = (n)$. In that case $\phi_{(n)}$ is the reduction modulo ℓ of some $\pm\mathrm{R}^{\mathbf{G}}_{\mathbf{T}}(\theta)$ where $\theta = \theta_\ell$ is a regular character of $\mathbf{T}^F$ (the stabilizer of θ in $\mathrm{N}_{\mathbf{G}}(\mathbf{T})^F$ is $\mathbf{T}^F$) of multiplicative order a power of ℓ and $\mathbf{T}^F$ is cyclic of order q^n-1.*

The following description of Harish Chandra series (see 13.B above) is due to Dipper-Du (see [DipDu93], [CaE] 19.20).

We abbreviate $e(q,\ell) = e$. Let us decompose each λ_i in $\lambda = (\lambda_1 \geq \lambda_2 \geq \cdots) \vdash n$ as $\lambda_i = \lambda_i^{(-1)} + e(\sum_{j\geq 0}\ell^j\lambda_i^{(j)})$ with $0 \leq \lambda_i^{(-1)} < e$, and $0 \leq \lambda_i^{(j)} < \ell$ for any $j \geq 0$. Denoting $m_j = \sum_i \lambda_i^{(j)}$ for any $j \geq -1$, we associate now $\overline{\lambda} := (1^{m_{-1}}, (e^{(m_0)}, \ldots, (e\ell^j)^{m_j}) \vdash n$ (the partition with 1 repeated m_{-1} times and $e\ell^j$ repeated m_j times).

Theorem 19.2. *Let* $\mathrm{GL}_n(q) \supseteq L_{\overline{\lambda}} \cong \mathrm{GL}_1(q)^{m_{-1}} \times \Pi_{j\geq 0}\mathrm{GL}_{e\ell^j}(q)^{m_j}$ *a standard Levi subgroup endowed with the cuspidal Brauer character* ρ *defined in Theorem 19.1 for each component. Then* ϕ_λ *is in the Harish Chandra series associated with the cuspidal pair* $(L_{\overline{\lambda}}, \rho)$.

19.B. Other types

For other finite reductive groups $\mathbf{G}^F$, less complete results are known regarding indexation of $\mathrm{Irr}(k\mathbf{G}^F)$ - a problem related with triangularity of the decomposition matrix in characteristic ℓ - and Harish Chandra series over the field k of characteristic ℓ. See [GeHiMa96] and its references for the state of the art at the time of that paper. Most modules involved are produced by inducing Gelfand Graev representations of Levi subgroups, or their generalization due to Kawanaka.

Some new methods were used recently by Dudas to produce other modules using ℓ-adic cohomology (see [Du13]), thus giving decomposition numbers unknown until then, especially for groups of small rank.

19.C. Supercuspidality

We have defined in Definition 13.6 above the notion of supercuspidality, a strengthening of the notion of cuspidality. This leaves open the question of existence and disjointness of series modeled on the case of cuspidality (see (13.7) above). A simple $k\mathbf{G}^F$-module M would define a pair (L, ρ) where L is a standard Levi subgroup of $\mathbf{G}^F$, ρ is a supercuspidal element of $\mathrm{Irr}(kL)$ and $[\mathrm{R}_L^G\rho, M] \neq \{0\}$. The existence of such a pair was seen in Remark 13.10, its uniqueness (i.e. that the union in Remark 13.10 is disjoint) is less clear.

However, we have

Theorem 19.3. *(Hiss [Hi96]) Uniqueness above is ensured for* $G = \mathrm{GL}_n(q)$.

The proof goes roughly as follows. Let us recall first the relation between Lusztig series and Harish Chandra series of *ordinary* characters in this case of $\mathbf{G}^F = \mathrm{GL}_n(q)$. If $s \in \mathbf{G}^{*F} = \mathrm{GL}_n(q)$, there is a single $\mathbf{G}^F$-class of pairs $(\mathbf{L}_s, \zeta_s)$ such that $\mathbf{L}_s$ is a split Levi subgroup of an F-stable parabolic of $\mathbf{G}$, $\zeta_s \in \mathrm{Irr}(\mathbf{L}_s^F)$ is cuspidal and $\mathrm{R}_{\mathbf{L}_s}^{\mathbf{G}}\zeta$ has components in $\mathcal{E}(\mathbf{G}^F, s)$. Indeed, $\mathbf{L}_s$ is dual to the smallest split Levi subgroup $\mathbf{L}_s^*$ of $\mathbf{G}^*$ containing the centralizer of s (which implies that s is regular in $\mathbf{L}^*$) and $\zeta_s \in \mathcal{E}(\mathbf{L}_s^F, s)$ is the character corresponding with the trivial character of the torus $\mathrm{C}_{\mathbf{L}_s^*}(s)^F$ (of type the

Coxeter element of $\mathbf{L}_s^*$) through Jordan decomposition of characters (see Theorem 15.3).

Now, if $(\mathbf{L}^F, \rho)$ is a supercuspidal pair as mentioned before and $\mathrm{R}_\mathbf{L}^\mathbf{G}\rho$ intersects the sum of blocks $B_\ell(\mathbf{G}^F, s)$ for some ℓ-regular semi-simple element of $\mathbf{G}^{*F}$ (see Theorem 16.2 above), one shows that $\mathbf{L} = \mathbf{L}_s$ and ρ is the restriction of ζ_s to p-regular elements. This proves the claimed uniqueness (disjunction) by the fact that $\mathrm{R}_\mathbf{L}^\mathbf{G}$ functor preserves series s (itself an easy consequence of the transitivity of those functors).

First ρ belongs to the sum of blocks $B_\ell(\mathbf{L}^F, s')$ for some ℓ-regular semi-simple $s' \in \mathbf{L}^{*F}$. Since ρ is cuspidal, by the classification recalled in Theorem 19.1 above (and suitably generalized to arbitrary series, see [Dip85b] 3.5) its Brauer character is the restriction to ℓ-regular elements of some $\pm\mathrm{R}_\mathbf{T}^\mathbf{L}\theta$ where θ corresponds by duality to some semi-simple element $s't \in \mathbf{T}^{*F} \leq \mathbf{L}^{*F}$ with $t = (s't)_\ell$ and $s't$ is regular in $\mathbf{L}^*$. On the other hand the projective cover $\mathcal{P}_\rho$, seen as an $\mathcal{O}\mathbf{L}^F$-module, has a character in $\mathbb{Z}\mathcal{E}_\ell(\mathbf{L}^F, s')$. By the preservation of series by $\mathrm{R}_\mathbf{L}^\mathbf{G}$ functors, $\mathrm{R}_\mathbf{L}^\mathbf{G}\mathcal{P}_\rho$ is in the sum of blocks $B_\ell(\mathbf{G}^F, s)$, so we can assume $s' = s$. Moreover, thanks to Theorem 16.4 above, the character of $\mathcal{P}_\rho$ has non-zero projection on $\mathcal{E}(\mathbf{L}^F, s)$. By supercuspidality, those components have to be cuspidal as ordinary characters. Then $\mathcal{E}(\mathbf{L}^F, s)$ contains a cuspidal ordinary character, which readily implies that $\mathbf{L} = \mathbf{L}_s$ in the notation above and ζ_s is a component of the character of $\mathcal{P}_\rho$. On the other hand $\zeta_s = \pm\mathrm{R}_\mathbf{T}^\mathbf{L}\theta'$ where θ' correspond with s by duality. Now $(st)_{\ell'} = s$ implies $\theta'_{\ell'} = \theta$ as a multiplicative character of $\mathbf{T}^F$. Then $\pm\mathrm{R}_\mathbf{T}^\mathbf{L}\theta$ has same restriction to ℓ-regular elements of $\mathbf{L}^F$ as ζ_s and this gives our claim about ρ. ∎

In [H96], all supercuspidal pairs (L, ρ) with ρ in the "unipotent block" $B_\ell(L, 1)$ are determined in groups $\mathbf{G}^F$ of classical type and under mild restrictions in exceptional types.

Acknowledgments

This survey is based on four lectures given at IMS of National University of Singapore in April 2013. The author thanks Professors Wee Teck Gan and Kai Meng Tan for the invitation and the kind atmosphere during the whole stay.

References

[AJS94] H. H. Andersen, J. C. Jantzen, and W. Soergel. Representations of quantum groups at a p-th root of unity and of semisimple groups in

characteristic p: independence of p. *Astérisque*, **220**, 1994.
[Be] D. Benson. *Representations and Cohomology I: Basic Representation Theory of finite Groups and associative Algebras*, Cambridge, 1991.
[Bn] C. Bonnafé. *Representations of* $\mathrm{SL}_2(\mathbb{F}_q)$. Springer, 2011.
[BnMi11] C. Bonnafé and J. Michel. Computational proof of the Mackey formula for $q > 2$, *J. Algebra*, **327** (2011), 506–526.
[BnRo03] C. Bonnafé and R. Rouquier. Catégories dérivées et variétés de Deligne-Lusztig, *Publ. Math. Inst. Hautes Études Sci.*, **97** (2003), 1–59.
[Bo] A. Borel. *Linear algebraic groups*. Springer, 1991.
[Bou] N. Bourbaki. *Groupes et algèbres de Lie, IV, V, VI.* Masson, 1968.
[Br90] M. Broué. Isométries parfaites, types de blocs, catégories dérivées. *Astérisque*, **181–182** (1990), 61–92.
[BrM92] M. Broué and G. Malle. Théorèmes de Sylow génériques pour les groupes réductifs sur les corps finis. *Math. Ann.*, **292** (1992), 241–262.
[BrMaM93] M. Broué, G. Malle, and J. Michel. Generic blocks of finite reductive groups. *Astérisque*, **212** (1993), 7–92.
[BrMi89] M. Broué and J. Michel. Blocs et séries de Lusztig dans un groupe réductif fini, *J. reine angew. Math.*, **395** (1989), 56–67.
[Ca09] M. Cabanes. On Okuyama's theorems about Alvis-Curtis duality. *Nagoya Math. J.* **195** (2009), 1–19.
[CaE99] M. Cabanes and M. Enguehard. On blocks of finite reductive groups and twisted induction. *Adv. Math.* **145** (1999), 189–229.
[CaE] M. Cabanes and M. Enguehard. *Representation theory of finite reductive groups*, Cambridge, 2004.
[Cart1] R. W. Carter. *Simple groups of Lie type.* Wiley, 1972.
[Cart2] R. W. Carter. *Finite groups of Lie type, Conjugacy classes and complex characters.* Wiley, 1985.
[CartL76] R.W. Carter and G. Lusztig. Modular representations of finite groups of Lie type. *Proc. London Math. Soc.* **32** (1976), 347–384.
[Ch55] C. Chevalley. Sur certains groupes simples, *Tohoku Math. J.*, (2) **7**, (1955), 1–66.
[ChR08] J. Chuang and R. Rouquier. Derived equivalences for symmetric groups and $\mathfrak{sl}_2$-categorification, *Ann. of Math.*, **167** (2008), 245–298.
[Cr12] D. Craven. Perverse equivalences and Broué's conjecture II: The cyclic case. *Preprint*, (2012).
[DeLu76] P. Deligne and G. Lusztig. Representations of reductive groups over finite fields, *Ann. of Math.*, **103** (1976), 103–161.
[DiMi] F. Digne and J. Michel. *Representations of finite groups of Lie type*, Cambridge, 1991.
[Dip85a] R. Dipper. On the decomposition numbers of the finite general linear groups, *Trans. Amer. Soc.*, **290** (1985), 315–344.
[Dip85b] R. Dipper. On the decomposition numbers of the finite general linear groups II, *Trans. Amer. Soc.*, **292** (1985), 123–133.
[DipDu93] R. Dipper and J. Du. Harish-Chandra vertices, *J. reine angew. Math.*, **437** (1993), 101–130.
[DipJam89] R. Dipper and G. James. The q-Schur algebra, *Proc. London Math.*

Soc., **59** (1989), 23–50.
[Du13] O. Dudas. A note on decomposition numbers for groups of Lie type of small rank, *J. Algebra*, **388** (2013), 364–373.
[EW12] B. Elias and G. Williamson. The Hodge theory of Soergel bimodules. *preprint*, ArXiv:1212.0791.
[F12] P. Fiebig. An upper bound on the exceptional characteristics for Lusztig's character formula. *J. reine angew. Math.*, **673** (2012), 1–31.
[FoSe74] P. Fong and G.M. Seitz. Groups with a (B,N)-pair of rank 2, I. *Invent. Math.,* **21** (1973), 1–57; II, ibid., **24** (1974), 191–239.
[FoSr82] P. Fong and B. Srinivasan. The blocks of finite general and unitary groups, *Invent. Math.*, **69** (1982), 109–153.
[FoSr86] P. Fong and B. Srinivasan. Generalized Harish-Chandra theory for unipotent characters of finite classical groups, *J. of Algebra*, **104** (1986), 301–309.
[FoSr89] P. Fong and B. Srinivasan. The blocks of finite classical groups, *J. reine angew. Math.*, **396** (1989), 122–191.
[FoSr90] P. Fong and B. Srinivasan. Brauer trees in classical groups, *J. Algebra*, **131** (1990), 179–225.
[GeHi91] M. Geck and G. Hiss. Basic sets of Brauer characters of finite groups of Lie type. *J. reine angew. Math.*, **418** (1991), 173–188.
[GeHi97] M. Geck and G. Hiss. Modular representations of finite groups of Lie type in non-defining characteristic, in *Finite reductive groups,* (M. Cabanes ed.), *Prog. Math.*, **141** (1997), Birkhäuser, pp 195–249.
[GeHiMa94] M. Geck, G. Hiss and G. Malle. Cuspidal unipotent Brauer characters, *J. Algebra*, **168** (1994), 182–220.
[GeHiMa96] M. Geck, G. Hiss and G. Malle. Towards a classification of the irreducible representations in non-describing characteristic of a finite group of Lie type, *Math. Z.*, **221** (1996), 353–386.
[GeMa] M. Geck and G. Malle. *Reflection groups.* (Handbook of algebra. Vol. **4**, 337–383) Elsevier/North-Holland, 2006.
[GePf] M. Geck and G. Pfeiffer. *Characters of finite Coxeter groups and Iwahori-Hecke algebras.* Oxford, 2000.
[GLS] D. Gorenstein, R. Lyons, and R. Solomon. *The classification of the finite simple groups. Number 3. Part I. Chapter A.* AMS, 1998.
[GraL96] J.J. Graham and G.I. Lehrer. Cellular algebras. *Invent. Math.* **123** (1996), 1–34.
[Gre78] J.A. Green. On a theorem of Sawada, *J. London Math. Soc.*, **18** (1978), 247–252.
[GrHi97] J. Gruber and G. Hiss. Decomposition numbers of finite classical groups for linear primes, *J. reine angew. Math.*, **485** (1997), 55–91.
[GyUn89] A. Gyoja and K. Uno. On the semisimplicity of Hecke algebras. *J. Math. Soc. Japan*, **41-1** (1989), 75–79.
[Ha85] M.E. Harris. On the p-deficiency class of a finite group. *J. Algebra*, **94** (1985), 411–424.
[HeKaSe72] C. Hering, W. Kantor, and G.M. Seitz. Finite groups with a split BN-pair of rank 1, I. *J. Algebra*, **20** (1972), 435–475.

[Hi96] G. Hiss. Supercuspidal representations of finite reductive groups. *J. Algebra*, **184** (1996), 839–851.

[HowL80] B. Howlett and G. Lehrer. Induced cuspidal representations and generalized Hecke rings, *Invent. Math.*, **58** (1980), 37–64.

[HowL94] B. Howlett and G. Lehrer. On Harish-Chandra induction for modules of Levi subgroups, *J. Algebra*, **165** (1994), 172–183.

[Hum1] J. E. Humphreys. *Introduction to Lie algebras and representation theory.* Springer, 1972.

[Hum2] J. E. Humphreys. *Modular representations of finite groups of Lie type*, Cambridge, 2006.

[Isa] I. M. Isaacs. *Character theory of finite groups.* Academic, 1976.

[Ja90] G. James. The decomposition matrix of $\mathrm{GL}_n(q)$ for $n \leq 10$, *Proc. London Math. Soc.*, **60** (1990), 225–265.

[JaKe] G. James and A. Kerber. *The representation theory of the symmetric group.* Addison-Wesley, 1981.

[Jn] J.C. Jantzen. *Representations of algebraic groups. Second edition.* Mathematical Surveys and Monographs, **107**. AMS, 2003.

[KL79] D. Kazhdan and G. Lusztig. Representations of Coxeter groups and Hecke algebras. *Invent. Math.*, **53** (1979), 165–184.

[KM13] R. Kessar and G. Malle. Quasi-isolated blocks and Brauer's height zero conjecture. *Ann. of Math.* (2) **178** (2013), 321–384.

[Lus76] G. Lusztig. Irreducible representations of finite classical groups, *Invent. Math.*, **58** (1976), 37–64.

[Lus80] G. Lusztig. Some problems in the representation theory of finite Chevalley groups, *Proc. Sympos. Pure Math.*, **37** (1980), 313–317.

[Lus84] G. Lusztig. Characters of reductive groups over a finite field, *Ann. Math. Studies*, **107**, Princeton, 1984.

[Lus03] G. Lusztig. *Hecke algebras with unequal parameters*, CRM Monograph Series **18**, AMS, 2003.

[MT] G. Malle, D. Testerman. *Linear algebraic groups and finite groups of Lie type.* Cambridge, 2011.

[N] G. Navarro. *Characters and blocks of finite groups.* Cambridge, 1998.

[Sch] P. Schneider. *Modular representation theory of finite groups.* Springer, 2013.

[Spr] T. A. Springer. *Linear algebraic groups*, Birkhäuser, second edition, 1998.

[Sri] B. Srinivasan. *Representations of finite Chevalley groups.* Springer Lecture Notes in pure Mathematics, Springer, 1979.

[St63] R. Steinberg. Representations of algebraic groups. *Nagoya Math. J.*, **22** (1963), 33–56.

[W13] G. Williamson. Schubert calculus and torsion. (Sep 2013) arXiv: 1309.5055.

ℓ-MODULAR REPRESENTATIONS OF p-ADIC GROUPS ($\ell \neq p$)

Vincent Sécherre
Université de Versailles Saint-Quentin-en-Yvelines
Laboratoire de Mathématiques de Versailles
45 avenue des États-Unis
78035 Versailles cedex, France
vincent.secherre@math.uvsq.fr

In these notes, we give an overview of the representation theory of p-adic reductive groups with coefficients in fields of characteristic different from p. We emphasize the case of $\mathrm{GL}(n)$ and its inner forms.

Contents

Introduction

The representation theory of reductive p-adic groups with coefficients in the field of complex numbers has been developed since the 1960's (see Cartier's introduction [11]). It has inherited certain techniques coming from harmonic analysis on reductive groups over Archimedean fields, making a large use of the fact that the representations have complex coefficients (Harish-Chandra [16], Langlands' classification [25]). Then in the 1980's, Bernstein and Zelevinski developed a fully algebraic approach [2, 3, 4].

The arithmetic of modular forms has required to develop the representation theory of reductive p-adic groups with coefficients in fields — or even rings — other than complex numbers. Provided that p is invertible in the coefficient ring, a large part of Bernstein-Zelevinski's algebraic approach can be reproduced (Vignéras [27, 28]). In these lectures, I will assume that the coefficient ring is an algebraically closed field with characteristic different from p. For bibliographic references, see Vignéras [27], Blondel [5].

In Lecture 1, I define parabolic induction and restriction, and the notions of cuspidal representation and cuspidal support. An important aspect of the theory of ℓ-modular representations is that there is a difference between the two notions of cuspidal and supercuspidal representations (Example 1.11). This leads to the notion of supercuspidal support, then to the problem of classifying irreducible representations with a given supercuspidal support.

In Lecture 2, I discuss the case of the group GL_n and its inner forms. I explain how, thanks to the theory of types developed by Bushnell and Kutzko, one can prove the uniqueness of supercuspidal support for irreducible representations of these groups (Theorem 2.1).

In Lecture 3, I present the classification of all irreducible representations of $\mathrm{GL}_n(\mathrm{F})$ in terms of multisegments, generalizing Zelevinski's classification of complex irreducible representations. First, the theory of types allows one to reduce to the classification of all unipotent irreducible representations. Then one defines a map:

$$\mathrm{Z} : \mathfrak{m} \mapsto \mathrm{Z}(\mathfrak{m})$$

that associates a unipotent irreducible representation to any multisegment. The definition of this map and the proof that it is injective relies on the theory of generic representations of $\mathrm{GL}_n(\mathrm{F})$ (see also Remark 3.11 for inner forms). Proof of surjectivity requires a counting argument that relies on results of [1, 12] on the classification of simple modules over an affine Hecke algebra of type A at a root of unity.

In Lecture 4, I introduce the operation of reducing mod ℓ an irreducible ℓ-adic representation of G having a stable lattice. Then I present the properties of the map Z and of the local Langlands correspondence with respect to reduction mod ℓ.

1. Lecture 1

1.1. *Notation and preliminaries*

In all these lectures, we fix a locally compact non-Archimedean field F of residue characteristic denoted p; we write $\mathcal{O}$ for its ring of integers, $\mathfrak{p}$ for the maximal ideal of $\mathcal{O}$ and q for the cardinality of its residue field. We also fix an algebraically closed field R of characteristic not dividing q.

Let **G** be a connected reductive group defined over F, and let $\mathrm{G} = \mathbf{G}(\mathrm{F})$ be the group of its F-points. When endowed with the topology coming from that of F, the group G is locally compact, and its neutral element 1 has a basis of neighborhoods made of compact open pro-p-subgroups (that is, all of whose open subgroups have index of the form p^r, $r \geqslant 0$).

Example 1.1: If $\mathbf{G} = \mathrm{GL}_n$, then G is the group $\mathrm{GL}_n(\mathrm{F})$. The identity matrix has a basis of neighborhoods made of the compact open pro-p-subgroups $\mathrm{K}_i = 1 + \mathrm{M}_n(\mathfrak{p}^i)$ for $i \geqslant 1$.

Definition 1.2: A *smooth* R*-representation of* G is a pair (π, V) made of a vector space V over R together with a group homomorphism:

$$\pi : \mathrm{G} \mapsto \mathrm{GL}(\mathrm{V})$$

such that, for all $v \in \mathrm{V}$, there is a compact open subgroup of G fixing v.

In these lectures, all representations will be smooth R-representations. Therefore we will often write *representation* for *smooth* R*-representation.*

Given two smooth R-representations (π, V) and (σ, W) of G, a homomorphism from (π, V) to (σ, W) is an R-linear map $f : \mathrm{V} \to \mathrm{W}$ such that $f \circ \pi(g) = \sigma(g) \circ f$ for all $g \in \mathrm{G}$. The space of all such maps will be denoted $\mathrm{Hom}_{\mathrm{G}}(\pi, \sigma)$.

This defines the abelian category, denoted $\mathscr{R}_{\mathrm{R}}(\mathrm{G})$, of smooth R-representations of G.

Definition 1.3: A smooth R-representation (π, V) of G is said to be *admissible* if, for any open subgroup H of G, the space V^{H} of H-fixed vectors of V is finite-dimensional.

A smooth R-character (or character for short) of G is a group homomorphism from G to $R^\times$ with open kernel.

Given a representation π and a character χ of G, we write $\pi\chi$ for the twisted representation $g \mapsto \pi(g)\chi(g)$.

A first very important fact is that one can define R-valued Haar measures on G (see [27], I.2). Such a measure is a nonzero R-linear form on:

$$\mathcal{C}_c^\infty(G, R),$$

the space of compactly supported and locally constant R-valued functions on G, which is invariant under right translations (it is then automatically invariant under left translations). Such a measure is unique up to a nonzero scalar. To define a Haar measure, let us fix a compact open pro-p-subgroup $H \subseteq G$ and define the measure of any compact open subgroup K of G by:

$$\mathrm{meas}(K) = \frac{(K : K \cap H)}{(H : K \cap H)} \in R,$$

which is well defined since the denominator is a p-power and p is invertible in R.

We have here our first important difference between the complex theory (when R is the field **C** of complex numbers) and the modular theory (when R has characteristic $\ell > 0$). Unlike the complex case, meas(K) may be 0 in the modular case. Smooth R-representations of K need not be semi-simple, and the functor $V \mapsto V^K$ of K-fixed vectors need not be exact.

1.2. *Parabolic functors*

Let **P** be a parabolic subgroup of **G**, together with a Levi decomposition $\mathbf{P} = \mathbf{MN}$ defined over F, where **N** is the unipotent radical of **P**. Write $P = \mathbf{P}(F)$, $M = \mathbf{M}(F)$ and $N = \mathbf{N}(F)$.

Attached to the parabolic subgroup P there is a complex character δ_P of M defined by:

$$\delta_P(m) = (mKm^{-1} : mKm^{-1} \cap K)/(K : mKm^{-1} \cap K)$$

for all $m \in M$, where K is an arbitrary compact open pro-p-subgroup of N (the complex number $\delta_P(m)$ does not depend on the choice of K). These values are integer powers of q. Thus for $m \in M$, there is a $v(m) \in \mathbf{Z}$ such that $\delta_P(m) = q^{v(m)}$. Let us make a choice of a square root of q in R, denoted $\sqrt{q}$, and write:

$$\sqrt{\delta_P} : m \mapsto (\sqrt{q})^{v(m)} \in R^\times.$$

Let (σ, W) be a smooth representation of M. Write $\boldsymbol{i}_{\mathrm{M,P}}^{\mathrm{G}}(\mathrm{W})$ for the space of locally constant functions $f : \mathrm{G} \to \mathrm{W}$ such that:

$$f(mng) = \sqrt{\delta_{\mathrm{P}}}(m)\sigma(m)f(g), \quad m \in \mathrm{M}, n \in \mathrm{N}, g \in \mathrm{G}.$$

The representation of G on this space by right translation is smooth, denoted $\boldsymbol{i}_{\mathrm{M,P}}^{\mathrm{G}}(\sigma)$ and called the *parabolic induction of* (σ, W) to G along P. The functor:

$$\boldsymbol{i}_{\mathrm{M,P}}^{\mathrm{G}} : \mathscr{R}_{\mathrm{R}}(\mathrm{M}) \to \mathscr{R}_{\mathrm{R}}(\mathrm{G})$$

has a left adjoint $\boldsymbol{r}_{\mathrm{M,P}}^{\mathrm{G}}$, the *parabolic restriction* functor from G to M along P. Given (π, V) a smooth representation of G, write V(N) for the subspace of V spanned by $\pi(n)v - v$ for all $n \in \mathrm{N}$ and $v \in \mathrm{V}$. Then $\boldsymbol{r}_{\mathrm{M,P}}^{\mathrm{G}}(\pi)$ is the natural representation of M on the quotient V/V(N) twisted by the inverse of the character $\sqrt{\delta_{\mathrm{P}}}$.

Remark 1.4:

(1) The functors $\boldsymbol{i}_{\mathrm{M,P}}^{\mathrm{G}}$, $\boldsymbol{r}_{\mathrm{M,P}}^{\mathrm{G}}$ are exact, $\boldsymbol{r}_{\mathrm{M,P}}^{\mathrm{G}}$ preserves the property of being of finite type, and $\boldsymbol{i}_{\mathrm{M,P}}^{\mathrm{G}}$ preserves admissibility (see [27], II.2.1).
(2) More difficult: the functors $\boldsymbol{i}_{\mathrm{M,P}}^{\mathrm{G}}$ and $\boldsymbol{r}_{\mathrm{M,P}}^{\mathrm{G}}$ both preserve the property of having finite length ([27], II.5.13).

We have the following theorem (see [27], II.2.18 for more details), known as the Geometric Lemma.

Theorem 1.5: *Given parabolic subgroups* $\mathrm{P} = \mathrm{MN}$ *and* $\mathrm{Q} = \mathrm{LU}$ *of* G, *there is a formula describing the functor* $\boldsymbol{r}_{\mathrm{L,Q}}^{\mathrm{G}} \circ \boldsymbol{i}_{\mathrm{M,P}}^{\mathrm{G}}$.

If R is the field of complex numbers, the functor $\boldsymbol{i}_{\mathrm{M,P}}^{\mathrm{G}}$ has a right adjoint, which is $\boldsymbol{r}_{\mathrm{M,P^-}}^{\mathrm{G}}$ with P^- the parabolic subgroup of G opposite to P with respect to M (this is known as the second adjointness property; see [7]). In the modular case, this is not known in general, but partial results can be found in [13]. However, there is a version for admissible representations (see [27], II.3.8).

1.3. *Cuspidal representations*

Definition 1.6: A representation (π, V) of G is *cuspidal* if the following equivalent conditions are satisfied:

(1) The space $\boldsymbol{r}_{\mathrm{M,P}}^{\mathrm{G}}(\mathrm{V})$ is zero for all proper parabolic subgroups $\mathrm{P} = \mathrm{MN} \subsetneq \mathrm{G}$.

(2) The space $\mathrm{Hom}_G(V, \boldsymbol{i}_{M,P}^G(W))$ is zero for all smooth R-representations (σ, W) of M and all proper parabolic subgroups $P = MN \subsetneq G$.

Theorem 1.7: *(See [27, 22]) Given* (π, V) *an irreducible representation of* G*, there are a parabolic subgroup* $P = MN \subseteq G$ *and an irreducible cuspidal representation* (σ, W) *of* M *such that* π *embeds in* $\boldsymbol{i}_{M,P}^G(\sigma)$*. Moreover, the cuspidal pair* (M, σ) *is unique up to* G*-conjugacy.*

Definition 1.8: The G-conjugacy class of (M, σ) is called the *cuspidal support* of (π, V), denoted $\mathrm{cusp}(\pi, V)$.

The cuspidal support depends on the choice of $\sqrt{q} \in R^{\times}$ that we have made.

Problem 1: Classify all irreducible representations of G having given cuspidal support.

There ia another characterization of cuspidality. Write Z for the centre of G.

Proposition 1.9: *An irreducible representation* (π, V) *is cuspidal if and only if, for all* $v \in V$ *and all smooth linear form* $\xi : V \to R$*, the function* $g \mapsto \xi(\pi(g)v)$ *has support whose image in* G/Z *is compact.*

Corollary 1.10: *(See [27]) All irreducible representation of* G *are admissible and have a central character.*

When $R = \mathbf{C}$, Proposition 1.9 is crucial. It is one of the key properties used for the Bernstein decomposition of the category $\mathscr{R}_{\mathbf{C}}(G)$ into blocks with respect to the notion of (inertial) cuspidal support. This is related to the fact that an irreducible cuspidal representation of G with central character ω is projective in the full subcategory of $\mathscr{R}_{\mathbf{C}}(G)$ made of smooth R-representations having central character ω.

As observed by Vignéras, this is no longer true in the modular case, since irreducible cuspidal representations of G may occur as subquotients of proper parabolically induced representations.

Example 1.11: Assume $G = GL_2(\mathbf{Q}_5)$ and $\ell = 3$ (where $\mathbf{Q}_p$ is the field of p-adic numbers). Let B be the subgroup of upper triangular matrices. The representation of G on the space V of R-valued locally constant functions on $B\backslash G$ is indecomposable and has length 3. Its unique irreducible subrepresentation is the trivial character of G, and its unique irreducible quotient is $g \mapsto |\det g|$, where $x \mapsto |x|$ denotes the absolute value of F giving value

q^{-1} to any uniformizer. The remaining (infinite-dimensional) subquotient is cuspidal.

Definition 1.12: An irreducible representation of G is *supercuspidal* if for all proper $\mathrm{P} = \mathrm{MN}$ and all irreducible representation (σ, W) of M, it does not occur as a subquotient of $\boldsymbol{i}_{\mathrm{M,P}}^{\mathrm{G}}(\sigma)$.

All supercuspidal representations of G are cuspidal, but the converse need not be true (see Example 1.11).

Remark 1.13: When $\mathrm{R} = \mathbf{C}$, any cuspidal representation is supercuspidal.

Remark 1.14:

(1) Assume (π, V) is a supercuspidal irreducible representation of G having a projective cover P_π of finite type. Then P_π is cuspidal (by Frobenius reciprocity). This implies that π does not occur as a subquotient of $\boldsymbol{i}_{\mathrm{M,P}}^{\mathrm{G}}(\sigma)$ for any smooth (σ, W).
(2) It is not known in general whether or not a supercuspidal irreducible representation of G has a projective cover of finite type. This is known for $\mathrm{G} = \mathrm{GL}_n(\mathrm{F})$, $n \geqslant 1$ ([14]).

Proposition 1.15: *For all irreducible representation* (π, V) *of* G, *there are a parabolic subgroup* $\mathrm{P} = \mathrm{MN}$ *and a supercuspidal irreducible representation* (σ, W) *of* M *such that* π *occurs as a subquotient of* $\boldsymbol{i}_{\mathrm{M,P}}^{\mathrm{G}}(\sigma)$.

Let us denote by $\mathrm{scusp}(\pi, \mathrm{V})$ the set of all possible such pairs (M, σ), called the *supercuspidal support* of (π, V).

Problem 2: Given an irreducible representation (π, V) of G, is $\mathrm{scusp}(\pi, \mathrm{V})$ made of a single G-conjugacy class?

The answer to Problem 2 is known only for the groups $\mathrm{GL}_n(\mathrm{F})$, $n \geqslant 1$ and their inner forms (see Lecture 2). See also [19] for the unitary group $\mathrm{U}(2, 1)$ with respect to an unramified quadratic extension.

2. Lecture 2

From now on, we will assume that $\mathbf{G}$ is GL_n, $n \geqslant 1$ or possibly one of its inner form, so that G is of the form $\mathrm{GL}_m(\mathrm{D})$ with D a central division F-algebra of degree d^2, with $n = md$.

Let $\alpha = (m_1, \ldots, m_r)$ be a family of positive integers of sum m. Denote by M_α the subgroup of $\mathrm{GL}_m(\mathrm{D})$ of invertible matrices which are diagonal by blocks of size $m_1, \dots, m_r$ respectively (it is isomorphic to $\mathrm{GL}_{m_1}(\mathrm{D}) \times \cdots \times \mathrm{GL}_{m_r}(\mathrm{D})$) and by P_α the subgroup of $\mathrm{GL}_m(\mathrm{D})$ generated by M_α and the upper triangular matrices.

Write $\boldsymbol{i}_\alpha$ for the functor of parabolic induction associated with M_α and P_α, and $\boldsymbol{r}_\alpha$ for its left adjoint. If $\pi_1, \dots, \pi_r$ are smooth representations of $\mathrm{GL}_{m_1}(\mathrm{D}), \dots, \mathrm{GL}_{m_r}(\mathrm{D})$ respectively, write:

$$\pi_1 \times \pi_2 \times \cdots \times \pi_r = \boldsymbol{i}_\alpha(\pi_1 \otimes \pi_2 \otimes \cdots \otimes \pi_r).$$

If (M, σ) is a cuspidal pair, then up to conjugacy we have $\mathrm{M} = \mathrm{M}_\alpha$ for some α as above and σ has the form $\sigma_1 \otimes \cdots \otimes \sigma_r$ where σ_i is a cuspidal irreducible representation of $\mathrm{GL}_{m_i}(\mathrm{D})$. Therefore the $\mathrm{GL}_m(\mathrm{D})$-conjugacy class of (M, σ) will be identified with the formal sum $\sigma_1 + \cdots + \sigma_r$.

2.1. *Supercuspidal support*

Theorem 2.1: *(See [28, 22]) Assume* G *is* $\mathrm{GL}_n(\mathrm{F})$ *or one of its inner forms. For all irreducible representations* (π, V) *of* G*, the set* $\mathrm{scusp}(\pi, \mathrm{V})$ *is a single* G*-conjugacy class.*

We need to introduce Bushnell-Kutzko's theory of types [9]. This is a monumental machinery, initially developed by Bushnell and Kutzko in order to prove that any complex irreducible cuspidal representation of $\mathrm{GL}_n(\mathrm{F})$ is compactly induced from an irreducible representation of a compact mod centre, open subgroup of $\mathrm{GL}_n(\mathrm{F})$. More precisely:

Theorem 2.2: *(See [9, 27, 23]) Assume* G *is* $\mathrm{GL}_n(\mathrm{F})$ *or one of its inner forms. Then there is a family of pairs* (J, λ) *made of a compact open subgroup* $\mathrm{J} \subseteq \mathrm{G}$ *and an irreducible representation* λ *of* J *with the following properties:*

(1) For any irreducible cuspidal representation ρ *of* G*, there is a pair* (J, λ)*, unique up to* G*-conjugacy, such that* λ *embeds in the restriction of* ρ *to* J.

(2) If ρ *is an irreducible representation of* G *containing a pair* (J, λ)*, then it is cuspidal.*

(3) If two irreducible cuspidal representations ρ, ρ' *both contain* (J, λ)*, then there is an unramified character* $\chi : \mathrm{G} \to \mathrm{R}^\times$ *(that is, trivial on all compact subgroups of* G*) such that* $\rho' \simeq \rho\chi$.

(4) Given (J, λ)*, any irreducible subquotient of* $\mathrm{ind}_\mathrm{J}^\mathrm{G}(\lambda)$*, the compact induction of* λ *to* G*, is isomorphic to a quotient of* $\mathrm{ind}_\mathrm{J}^\mathrm{G}(\lambda)$.

(5) *Given* (J, λ), *the representation* λ *extends to its* G*-normalizer* $\widehat{\mathrm{J}}$; *compact induction from* $\widehat{\mathrm{J}}$ *to* G *induces a bijection between the set of representations of* $\widehat{\mathrm{J}}$ *extending* λ *and that of isomorphism classes of irreducible cuspidal representations of* G *containing* λ.

Example 2.3: Write $\mathrm{GL}_n(q)$ for the group of invertible $n \times n$ matrices with entries in the residue field of F, and let σ be an irreducible cuspidal representation of $\mathrm{GL}_n(q)$. Inflate σ into an irreducible representation of $\mathrm{K} = \mathrm{GL}_n(\mathcal{O})$ that is trivial on $\mathrm{K}_1 = 1 + \mathscr{M}_n(\mathfrak{p})$, still denoted σ. Then the pairs of the form (K, σ) obtained this way fullfil all the properties 1 to 5 for irreducible cuspidal representations of $\mathrm{G} = \mathrm{GL}_n(\mathrm{F})$ having nonzero K_1-fixed vectors (such irreducible representations are said to have level 0).

Moreover, if $\widehat{\sigma}$ is a representation of KZ (where Z is the centre of G) extending σ and if ρ is the representation of G compactly induced form $\widehat{\sigma}$, then the space ρ^{K_1} of K_1-fixed vectors of ρ – which is naturally a representation of $\mathrm{K}/\mathrm{K}_1 \simeq \mathrm{GL}_n(q)$ – is isomorphic to σ.

Moreover, if ρ is an irreducible representation of G containing (K, σ), then it is supercuspidal if and only if σ is supercuspidal as a representation of $\mathrm{GL}_n(q)$.

Example 2.4: We give a positive level example for $\mathrm{G} = \mathrm{GL}_n(\mathrm{F})$. Let us fix a uniformizer ϖ of F, a character $\psi : \mathrm{F} \to \mathrm{R}^\times$ trivial on $\mathfrak{p}$ but not on $\mathcal{O}$, and a character $\omega : \mathrm{F}^\times \to \mathrm{R}^\times$ trivial on $1 + \mathfrak{p}$. Define a compact open subgroup:

$$\mathrm{I}_1 = 1 + \begin{pmatrix} \mathfrak{p} & \mathcal{O} & \cdots & \mathcal{O} \\ \vdots & \ddots & \ddots & \vdots \\ \vdots & & \ddots & \mathcal{O} \\ \mathfrak{p} & \cdots & \cdots & \mathfrak{p} \end{pmatrix}$$

of G. Given $t \in \mathcal{O}^\times$, define a character $\theta = \theta_t$ of I_1 by:

$$\theta(1 + x) = \psi(x_{1,2} + \cdots + x_{n-1,n} + t\varpi^{-1}x_{n,1}), \quad 1 + x \in \mathrm{I}_1.$$

Write $\mathrm{J} = \mathcal{O}^\times \mathrm{I}_1$ and let $\lambda = \lambda_t$ be the character of J defined by $\lambda(xg) = \omega(x)\theta(g)$ for all $x \in \mathcal{O}^\times$ and $g \in \mathrm{I}_1$. The F-algebra E generated by:

$$\beta = \begin{pmatrix} 0 & t \cdot \mathrm{id}_{n-1} \\ \varpi & 0 \end{pmatrix} \in \mathscr{M}_n(\mathrm{F})$$

is a totally ramified extension of degree n and uniformizer β. The group $\mathrm{E}^\times$ normalizes the pair (J, λ), and the G-normalizer of the latter is $\widehat{\mathrm{J}} = \mathrm{E}^\times \mathrm{J}$.

Given $z \in \mathrm{R}^\times$, there is a unique character ${}_z\lambda = {}_z\lambda_t$ of $\widehat{\mathrm{J}}$ extending λ such that:

$$ {}_z\lambda(\beta) = z $$

and its compact induction ${}_z\rho = {}_z\rho_t$ is a supercuspidal irreducible representation (of level $1/n$) of G. It is obtained from ${}_1\rho$ by twisting by the unramified character $g \mapsto z^{\mathrm{val}(\det(g))}$, where val denotes the valuation of F giving value 1 to any uniformizer.

Two pairs (J, λ_t), (J, λ_u) with $t, u \in \mathcal{O}^\times$ are G-conjugate if and only if $tu^{-1} \in 1 + \mathfrak{p}$.

Remark 2.5: This construction shows that, given ℓ and $n \geqslant 1$, there is an irreducible mod ℓ supercuspidal representation of $\mathrm{GL}_n(\mathrm{F})$. Note that it may happen that all level 0 mod ℓ cuspidal representations of $\mathrm{GL}_n(\mathrm{F})$ are non-supercuspidal (this is the case, for instance, if $n = q = 2$ and $\ell = 3$).

The pairs (J, λ) appearing in Theorem 2.2 are called the *maximal simple types* of G. The proof of Theorem 2.1 requires a larger family of types, called the *semisimple types* of G (see [10, 23] for a precise definition). We will only give the following crucial fact about these types.

Let $\alpha = (n_1, \dots, n_r)$ be a family of positive integers of sum n. For $i \in \{1, \dots, r\}$, let $(\mathrm{J}_i, \lambda_i)$ be a maximal simple type of $\mathrm{GL}_{n_i}(\mathrm{F})$. Then there exists a semisimple type $(\mathbf{J}, \boldsymbol{\lambda})$ of $\mathrm{G} = \mathrm{GL}_n(\mathrm{F})$ such that the compact induction $\mathrm{ind}_{\mathbf{J}}^{\mathrm{G}}(\boldsymbol{\lambda})$ is isomorphic to the parabolic induction:

$$ \mathrm{ind}_{\mathrm{J}_1}^{\mathrm{GL}_{n_1}(\mathrm{F})}(\lambda_1) \times \cdots \times \mathrm{ind}_{\mathrm{J}_r}^{\mathrm{GL}_{n_r}(\mathrm{F})}(\lambda_r). \tag{2.1} $$

For instance, if $m_1 = \cdots = m_r = 1$ and if λ_i is the trivial character of $\mathrm{J}_i = \mathcal{O}^\times$ for all i, then one can choose for $\boldsymbol{\lambda}$ the trivial character of the standard Iwahori subgroup I of G.

Let us sketch the proof of Theorem 2.1 for the group $\mathrm{G} = \mathrm{GL}_n(\mathrm{F})$. Let (π, V) be an irreducible representation of G.

Step 1. — We first reduce to the case where π is cuspidal, by applying an appropriate parabolic restriction functor $\boldsymbol{r}_\alpha$ (such that $\boldsymbol{r}_\alpha(\pi)$ is cuspidal) and by using the Geometric Lemma 1.5.

Step 2. — We prove uniqueness up to inertia: there are a unique positive integer r dividing n and an irreducible supercuspidal representation ρ of $\mathrm{GL}_{n/r}(\mathrm{F})$ such that any pair in $\mathrm{scusp}(\pi, \mathrm{V})$ is G-conjugate to a pair of the form:

$$ (\mathrm{GL}_{n/r}(\mathrm{F}) \times \cdots \times \mathrm{GL}_{n/r}(\mathrm{F}), \rho\chi_1 \otimes \cdots \otimes \rho\chi_r) $$

where the χ_i's are unramified characters of $\mathrm{GL}_{n/r}(\mathrm{F})$. This step requires type theory and uniqueness of supercuspidal support for irreducible representations of the groups $\mathrm{GL}_m(q^u)$, where mu divides n.

Step 3. — We finally prove that any pair in scusp(π, V) is G-conjugate to a pair of the form:

$$(\mathrm{GL}_{n/r}(\mathrm{F}) \times \cdots \times \mathrm{GL}_{n/r}(\mathrm{F}), \rho \otimes \rho|\det| \otimes \cdots \otimes \rho|\det|^{r-1})$$

where $x \mapsto |x|$ denotes the absolute value of F (normalized so that $|\varpi| = q^{-1}$) and ρ is uniquely determined up to a twist by some power of $|\det|$ (note that $\rho|\det|^r$ is isomorphic to ρ). This step requires the theory of Whittaker models [27] , which allows one to prove the following crucial fact: if $\chi_1, \ldots, \chi_r$ are unramified characters of $\mathrm{GL}_{n/r}(\mathrm{F})$ such that $\rho\chi_1 \times \cdots \times \rho\chi_r$ has a cuspidal subquotient, there is an $i \in \{1, \ldots, r\}$ such that, for all $k \in \mathbf{Z}$, there is a $j \in \{1, \ldots, r\}$ such that $\rho\chi_i|\det|^k \simeq \rho\chi_j$ (see [22], 8.2).

This proof uses strong results that are known so far for $\mathrm{GL}_n(\mathrm{F})$ and its inner forms only. They are of two kinds: (a) results from the theory of types, and (b) results from the representation theory of finite reductive groups.

We give more details on Step 2 in the case where π is a level 0 cuspidal representation.

We first introduce a useful tool. Given a smooth R-representation (σ, W) of G, let us form the space $\mathrm{W}^{\mathrm{K}_1}$ of K_1-fixed vectors of W. Since K_1 is normal in $\mathrm{K} = \mathrm{GL}_n(\mathcal{O})$, there is a representation of K on $\mathrm{W}^{\mathrm{K}_1}$, and K_1 acts trivially. This gives us a representation of the quotient K/K_1, which naturally identifies with $\mathrm{GL}_n(q)$. This defines an exact functor from $\mathscr{R}_{\mathrm{R}}(\mathrm{G})$ to the category of R-representations of $\mathrm{GL}_n(q)$.

Let $\mathrm{P} = \mathrm{MN}$ be a standard parabolic subgroup (that is, containing all upper triangular matrices) of G. We write $\mathrm{M}(q)$ for the standard Levi subgroup of $\mathrm{GL}_n(q)$ that corresponds to M. By restricting functions from G to K, one can see from the Iwasawa decomposition $\mathrm{G} = \mathrm{PK}$ that the functor $\mathrm{W} \mapsto \mathrm{W}^{\mathrm{K}_1}$ transforms the parabolic induction functor $\boldsymbol{i}_{\mathrm{M,P}}^{\mathrm{G}}$ into the Harish-Chandra induction functor from $\mathrm{M}(q)$ to $\mathrm{GL}_n(q)$, denoted R.

Assume that π occurs as an irreducible subquotient of $\rho_1 \times \cdots \times \rho_r$ where ρ_i is an irreducible supercuspidal representation of $\mathrm{GL}_{n_i}(\mathrm{F})$, and with $n_1 + \cdots + n_r = n$. Then π^{K_1}, which is nonzero, is a subquotient of:

$$(\rho_1 \times \cdots \times \rho_r)^{\mathrm{K}_1} \simeq \mathsf{R}\big(\rho_1^{\mathrm{K}_1(n_1)} \otimes \cdots \otimes \rho_r^{\mathrm{K}_1(n_r)}\big)$$

where $\mathrm{K}_1(n)$ stands for the K_1-group of $\mathrm{GL}_n(\mathrm{F})$, $n \geqslant 1$. This implies that all the ρ_i's have level 0. Following Example 2.3, write:

$$\pi \simeq \mathrm{ind}_{\mathrm{KF}^\times}^{\mathrm{G}}(\widehat{\sigma}), \quad \rho_i \simeq \mathrm{ind}_{\mathrm{K}(n_i)\mathrm{F}^\times}^{\mathrm{GL}_{n_i}(\mathrm{F})}(\widehat{\sigma}_i),\ i \in \{1, 2, \ldots, r\},$$

where $\mathrm{K}(n) = \mathrm{GL}_n(\mathcal{O})$, $n \geqslant 1$. By taking the K_1-fixed vectors, we get that σ is a subquotient of the Harish-Chandra induction of $\sigma_1 \otimes \cdots \otimes \sigma_r$. This implies (see Marc Cabanes's lectures) that $\sigma_1 \simeq \sigma_2 \simeq \cdots \simeq \sigma_r$ (denoted σ_0) and $n_1 = n_2 = \cdots = n_r$ (denoted n_0). We get the result by choosing $r = n/n_0$ and $\rho = \rho_1$.

Remark 2.6: The level 0 case is easy because it is the smallest possible level for an irreducible representation, and because the compatibility of the functor of K_1-fixed vectors with parabolic induction follows from the Iwasawa decomposition. The positive level case is more difficult, and requires the use of endo-classes [8, 6].

2.2. *Decomposition of* $\mathscr{R}_{\mathrm{R}}(\mathrm{G})$

A (super)cuspidal pair of G is a pair (M, ρ) made of a Levi subgroup $\mathrm{M} \subseteq \mathrm{G}$ and an irreducible (super)cuspidal representation ρ of M.

Definition 2.7: Two cuspidal pairs (M, ρ) and (M', ρ') in G are *inertially equivalent* if there is an unramified character χ of M such that (M', ρ') is G-conjugate to $(\mathrm{M}, \rho\chi)$.

Let $\mathcal{S}(\mathrm{G})$ denote the set of all inertial classes of supercuspidal pairs of G. Let us fix $\Omega \in \mathcal{S}(\mathrm{G})$ and choose $(\mathrm{M}, \rho) \in \Omega$ with:

$$\mathrm{M} = \mathrm{GL}_{n_1}(\mathrm{F}) \times \cdots \times \mathrm{GL}_{n_r}(\mathrm{F}),$$
$$\rho = \rho_1 \otimes \cdots \otimes \rho_r,$$

where ρ_i is a supercuspidal irreducible representation of $\mathrm{GL}_{n_i}(\mathrm{F})$. For each $i = 1, \ldots, r$ choose a pair $(\mathrm{J}_i, \lambda_i)$ for ρ_i as in Theorem 2.2 and write:

$$\mathscr{U}(\Omega) = \mathrm{ind}_{\mathrm{J}_1}^{\mathrm{GL}_{n_1}(\mathrm{F})}(\lambda_1) \times \cdots \times \mathrm{ind}_{\mathrm{J}_r}^{\mathrm{GL}_{n_r}(\mathrm{F})}(\lambda_r).$$

According to (2.1), this representation can be described as the compact induction of a semisimple type. For instance, if $n_1 = \cdots = n_r = 1$ and all ρ_i are the trivial character of $\mathrm{F}^\times$, then $\mathscr{U}(\Omega)$ is the compact induction $\mathrm{ind}_{\mathrm{I}}^{\mathrm{G}}(1)$ of the trivial character of an Iwahori subgroup I of G.

Theorem 2.8: *(See [24])*

(1) For all irreducible representations (π, V) *of* G *and all* $\Omega \in \mathcal{S}(\mathrm{G})$, *one has:*

$$\mathrm{scusp}(\pi, \mathrm{V}) \in \Omega \quad \Leftrightarrow \quad \pi \text{ is a subquotient of } \mathscr{U}(\Omega).$$

(2) For all smooth representations (π, V) *of* G *and all* $\Omega \in \mathcal{S}(\mathrm{G})$, *let* $\mathrm{V}(\Omega)$ *denote the maximal subrepresentation of* V *all of whose irreducible subquotients have supercuspidal support in* Ω. *Then:*

$$\mathrm{V} = \bigoplus_{\Omega \in \mathcal{S}(\mathrm{G})} \mathrm{V}(\Omega).$$

(3) If (π, V),(σ, W) *are smooth representations of* G, *then* $\mathrm{Hom}_{\mathrm{G}}(\mathrm{V}, \mathrm{W})$ *decomposes canonically as the product of all* $\mathrm{Hom}_{\mathrm{G}}(\mathrm{V}(\Omega), \mathrm{W}(\Omega))$*'s for* $\Omega \in \mathcal{S}(\mathrm{G})$.

(4) The full subcategory $\mathscr{R}_{\mathrm{R}}(\Omega)$ *made of all smooth representations* (π, V) *of* G *such that* $\mathrm{V} = \mathrm{V}(\Omega)$ *is indecomposable.*

The strategy of the proof is very different from Bernstein's proof for complex representations. It uses type theory as well as a decomposition theorem with respect to the supercuspidal support for representations of $\mathrm{GL}_n(q)$.

Example 2.9: Let (π, V) be a smooth level zero R-representation of G, that is, V is generated by $\mathrm{V}^{\mathrm{K}_1}$. As a representation of $\mathrm{GL}_n(q)$, $\mathrm{V}^{\mathrm{K}_1}$ decomposes as a direct sum:

$$\bigoplus_{[\mathrm{L},\sigma]} \mathrm{V}^{\mathrm{K}_1}([\mathrm{L}, \sigma])$$

where $[\mathrm{L}, \sigma]$ ranges over all possible supercuspidal supports of $\mathrm{GL}_n(q)$ and where $\mathrm{V}^{\mathrm{K}_1}([\mathrm{L}, \sigma])$ is the maximal subrepresentation of $\mathrm{V}^{\mathrm{K}_1}$ all of whose irreducible subquotients have supercuspidal support $[\mathrm{L}, \sigma]$. Write $\mathrm{V}[\mathrm{L}, \sigma]$ for the subrepresentation of V generated by $\mathrm{V}^{\mathrm{K}_1}([\mathrm{L}, \sigma])$. Then V decomposes as the direct sum of the $\mathrm{V}[\mathrm{L}, \sigma]$'s. Now write:

$$\mathrm{L} = \mathrm{GL}_{n_1}(q) \times \cdots \times \mathrm{GL}_{n_r}(q),$$
$$\sigma = \sigma_1 \otimes \cdots \otimes \sigma_r,$$

where σ_i is a supercuspidal irreducible representation of $\mathrm{GL}_{n_i}(q)$. For each $i = 1, \ldots, r$, inflate σ_i to an irreducible representation of $\mathrm{J}_i = \mathrm{K}(n_i) = \mathrm{GL}_{n_i}(\mathcal{O})$, still denoted σ_i. By Example 2.3, the pair (J_i, σ_i) is a level 0 maximal simple type. Choose a supercuspidal irreducible representation ρ_i

of $\mathrm{GL}_{n_i}(\mathrm{F})$ containing the pair (J_i, σ_i). The representation $\rho = \rho_1 \otimes \cdots \otimes \rho_r$ is a supercuspidal irreducible representation of a standard Levi subgroup M of G. Let Ω be the inertial class of the supercuspidal pair (M, ρ). The process $[\mathrm{L}, \sigma] \mapsto \Omega$ is well defined and induces a bijection between supercuspidal supports of $\mathrm{GL}_n(q)$ and inertial class of level 0 supercuspidal pairs of G. Moreover, one has $\mathrm{V}(\Omega) = \mathrm{V}[\mathrm{L}, \sigma]$.

By type theory, one can prove that the endomorphism algebra of $\mathscr{U}(\Omega)$ is a finite tensor product of affine Hecke algebras of type A. Together with part (2) of Theorem 2.8, this implies that $\mathscr{R}_{\mathrm{R}}(\Omega)$ is indecomposable.

Problem 3: What is the structure of $\mathscr{R}_{\mathrm{R}}(\Omega)$ for $\Omega \in \mathcal{S}(\mathrm{G})$? Find a progenerator P_Ω in $\mathscr{R}_{\mathrm{R}}(\Omega)$ and compute $\mathrm{End}_{\mathrm{G}}(\mathrm{P}_\Omega)$.

For $\mathrm{G} = \mathrm{GL}_m(\mathrm{D})$, type theory provides candidates for the P_Ω's.

Theorem 2.10: *(See [14]) Let ρ be an irreducible supercuspidal representation of $\mathrm{G} = \mathrm{GL}_n(\mathrm{F})$. Let Ω be its inertial class. Write $n(\rho)$ for the number of unramified characters χ of G such that $\rho\chi \simeq \rho$ and v for the ℓ-adic valuation of $q^{n(\rho)} - 1$. There is a progenerator P_Ω of $\mathscr{R}_{\mathrm{R}}(\Omega)$, with:*

$$\mathrm{End}_{\mathrm{G}}(\mathrm{P}_\Omega) \simeq \mathrm{R}[\mathrm{X}, \mathrm{X}^{-1}, \mathrm{T}]/(\mathrm{T}^{\ell^v}).$$

When Ω has level 0, Guiraud [15] proved that there is a progenerator P_Ω, and that the computation of its endomorphism algebra reduces to the case where $n_1 = \cdots = n_r$ and $\rho_1 = \cdots = \rho_r$ (with the notation of the beginning of the paragraph).

3. Lecture 3

In this lecture, our goal is the classification of all irreducible representations of $\mathrm{G} = \mathrm{GL}_m(\mathrm{D})$ having given (super)cuspidal support. For complex representations, this has been done by Zelevinski [32] for $\mathrm{GL}_n(\mathrm{F})$, and by Tadić [26] for its inner forms.

3.1. *Reduction to the (super)unipotent case*

Assume $\mathrm{G} = \mathrm{GL}_n(\mathrm{F})$ for simplicity. The indecomposable block corresponding to the inertial class Ω_1 of the pair $(\mathrm{F}^{\times n}, 1_{\mathrm{F}^\times}^{\otimes n})$ is called the *unipotent block*.

An irreducible representation (π, V) of G is *unipotent* if it has supercuspidal support in Ω_1, that is, if π is a subquotient of $\mathscr{U}(\Omega_1) \simeq \mathrm{ind}_{\mathrm{I}}^{\mathrm{G}}(1)$ where I denotes the standard Iwahori subgroup of G.

An irreducible representation (π, V) of G is *superunipotent* if $\mathrm{cusp}(\pi) \in \Omega_1$, that is, if V^{I} is nonzero.

Theorem 3.1: *(See [23]) Let $\mathfrak{s}$ be a cuspidal support in* G. *It writes uniquely as $\mathfrak{s}_1 + \cdots + \mathfrak{s}_t$ such that (a) $\mathfrak{s}_j, \mathfrak{s}_k$ have inertially equivalent terms in common if and only if $j = k$, and (b) any two terms in $\mathfrak{s}_j$ are inertially equivalent, for all j. Then:*

(1) The map $(\pi_1, \ldots, \pi_t) \to \pi_1 \times \cdots \times \pi_t$ induces a bijection:

$$\mathrm{cusp}^{-1}(\mathfrak{s}_1) \times \cdots \times \mathrm{cusp}^{-1}(\mathfrak{s}_t) \to \mathrm{cusp}^{-1}(\mathfrak{s}).$$

(2) Assume that the terms of $\mathfrak{s}$ are supercuspidal. Then the same map induces a bijection:

$$\mathrm{scusp}^{-1}(\mathfrak{s}_1) \times \cdots \times \mathrm{scusp}^{-1}(\mathfrak{s}_t) \to \mathrm{scusp}^{-1}(\mathfrak{s}).$$

We are thus reduced to describe $\mathrm{cusp}^{-1}(\Omega_\rho)$ and $\mathrm{scusp}^{-1}(\Omega_\rho)$ where Ω_ρ is the inertial class of $(\mathrm{GL}_{n/r}(\mathrm{F})^r, \rho^{\otimes r})$ and ρ a (super)cuspidal irreducible representation of $\mathrm{GL}_{n/r}(\mathrm{F})$, r dividing n.

Theorem 3.2: *(See [23]) Let ρ be a cuspidal irreducible representation of* $\mathrm{GL}_{n/r}(\mathrm{F})$.

(1) There are a finite extension F'/F of degree dividing n and a bijective map from $\mathrm{cusp}^{-1}(\Omega_\rho)$ to the set of superunipotent representations of $\mathrm{GL}_r(\mathrm{F}')$.

(2) Assume that ρ is supercuspidal. Then there is a bijective map from $\mathrm{scusp}^{-1}(\Omega_\rho)$ to the set of unipotent representations of $\mathrm{GL}_r(\mathrm{F}')$.

Moreover, these bijections preserve the (super)cuspidal support in the following sense. Given an unramified character χ of G, it writes $\chi = \xi \circ \det$ where ξ is an unramified character of $\mathrm{F}^\times$. Then write χ' for the unramified character $\xi' \circ \det$ of $\mathrm{G}' = \mathrm{GL}_r(\mathrm{F}')$, where ξ' is the unramified character of $\mathrm{F}'^\times$ whose value at a uniformizer of F' is the same as that of ξ at a uniformizer of F. We get a bijective map $\chi \mapsto \chi'$ between unramified characters of G and unramified characters of G'. Then an irreducible representation $\pi \in \mathrm{cusp}^{-1}(\Omega_\rho)$ has cuspidal support $\rho\chi_1 + \cdots + \rho\chi_r$ if and only if it corresponds to a superunipotent representation π' of G' having cuspidal support $\chi_1' + \cdots + \chi_r'$. There is a similar statement for the supercuspidal support in the case where ρ is supercuspidal.

Therefore we are reduced to classify (super)unipotent representations of $\mathrm{GL}_n(\mathrm{F})$.

3.2. *Classification of (super)unipotent representations, I*

We write $\mathcal{H}_R(G, I)$ for the Hecke-Iwahori R-algebra, that is the space of functions $f : G \to R$ with compact support and such that $f(xgx') = f(g)$ for all $g \in G$, $x, x' \in I$, endowed with the convolution product with respect to the Haar measure on G giving measure 1 to I.

For all smooth representation (π, V), the space:

$$V^I \simeq \mathrm{Hom}_G(\mathcal{C}_c^\infty(I\backslash G, R), V)$$

is made into a right module over the Iwahori-Hecke algebra $\mathcal{H}_R(G, I)$ by the formula:

$$v * f = \sum f(g)\pi(g^{-1})v$$

for $v \in V^I$ and $f \in \mathcal{H}_R(G, I)$, where g ranges over a set of representatives of $I\backslash G$ in G.

If R has characteristic ℓ different from $0, p$, the functor $V \mapsto V^I$ from smooth representations of G to right modules over $\mathcal{H}_R(G, I)$ need not be exact (more precisely, this functor is exact if and only if ℓ does not divide $q - 1$). Thus $\mathcal{C}_c^\infty(I\backslash G, R)$ need not be projective as a representation of G, but it has the following crucial property.

Theorem 3.3: *(See [28])*

(1) The representation $\mathcal{C}_c^\infty(I\backslash G, R)$ is quasi-projective, that is, for any surjective G-homomorphism $\mathcal{C}_c^\infty(I\backslash G, R) \to V$, the restriction $\mathcal{H}_R(G, I) \to V^I$ is also surjective.

(2) The functor $V \to V^I$ induces a bijection between the isomorphism classes of superunipotent representations of G and the isomorphism classes of simple right modules over $\mathcal{H}_R(G, I)$.

But this functor kills all unipotent non-superunipotent representations. In order to deal with these representations as well as the superunipotent ones, Vignéras has introduced the following affine Schur algebra. Write:

$$V = \bigoplus_{\mathcal{P}} \mathcal{C}_c^\infty(\mathcal{P}\backslash G, R)$$

where $\mathcal{P}$ ranges over all standard parahoric subgroups (that is, $I \subseteq \mathcal{P} \subseteq K$) and their conjugates by the G-normalizer of I.

The endomorphism algebra $\mathcal{S}_R(G, I) = \mathrm{End}_G(V)$ is called the affine Schur algebra (see [30]).

Fix a Haar measure μ on G and let $\mathcal{H}_R(G, \mu)$ denote the space $\mathcal{C}_c^\infty(G, R)$ endowed with the convolution product with respect to μ. Any smooth R-representation (π, V) is given a structure of left $\mathcal{H}_R(G, \mu)$-module by:

$$f \cdot v = \int_G f(g)\pi(g)v \, d\mu(g).$$

Let $\mathscr{J} = \mathscr{J}_R(G)$ be the ideal of $\mathcal{H}_R(G, \mu)$ that annihilates the representation $\mathcal{C}_c^\infty(I\backslash G, R)$.

Theorem 3.4: *(See [30])*

(1) There is a functor from $\mathscr{R}_R(G)$ to the category of right $\mathcal{S}_R(G, I)$-modules inducing an equivalence between the full subcategory of $\mathscr{R}_R(\Omega_1)$ made of all representations that are killed by $\mathscr{J}$ and the category of right $\mathcal{S}_R(G, I)$-modules.

(2) This induces a bijection between the isomorphism classes of unipotent representations of G and the isomorphism classes of simple modules over $\mathcal{S}_R(G, I)$.

(3) There is an integer $N \geqslant 1$ such that $\mathscr{J}^N$ annihilates the whole block $\mathscr{R}_R(\Omega_1)$.

3.3. *Classification of (super)unipotent representations, II*

We now classify unipotent representations of $GL_n(F)$ by multisegments.

Definition 3.5:

(1) A *segment* is a pair $(\xi, n) \in R^\times \times \mathbf{Z}_{>0}$.
(2) A *multisegment* is a formal finite sum of segments.
(3) The integer n is called the *length* of the segment (ξ, n), and the length of a multisegment is the sum of the lengths of its segments.

Given a segment (ξ, n), write $Z(\xi, n)$ for the character:

$$g \mapsto \xi^{-\mathrm{val}(\det(g))}$$

of the group $GL_n(F)$. For a multisegment $\mathfrak{m} = (\xi_1, n_1) + \cdots + (\xi_r, n_r)$ of length $n \geqslant 1$, we want to define an irreducible subquotient $Z(\mathfrak{m})$ of:

$$I(\mathfrak{m}) = Z(\xi_1, n_1) \times \cdots \times Z(\xi_r, n_r).$$

Let U be the subgroup of upper triangular unipotent matrices of $GL_n(F)$. Let us fix a smooth nontrivial character ψ_F of F, and set $\psi(u) = \psi_F(u_{1,2} + \cdots + u_{n-1,n})$ for all $u \in U$.

Definition 3.6: An irreducible representation (π, V) of G is *generic* if the space $\mathrm{Hom}_{\mathrm{U}}(\pi, \psi)$ is nonzero. (This notion does not depend on the choice of ψ_{F}.)

Given a multisegment $\mathfrak{m}$ of length $n \geqslant 1$ as above, write $\mu_{\mathfrak{m}}$ for the partition of n conjugate to $(n_1 \geqslant n_2 \geqslant \cdots \geqslant n_r)$.

Proposition 3.7:

(1) $\mu_{\mathfrak{m}}$ *is the unique maximal partition (for the dominance order) of* n *such that the parabolic restriction* $\boldsymbol{r}_{\mu_{\mathfrak{m}}}(\mathrm{I}(\mathfrak{m}))$ *contains a generic irreducible subquotient.*

(2) Such a generic irreducible subquotient is unique, and occurs with multiplicity 1.

Write $\mathrm{Z}(\mathfrak{m})$ for the unique irreducible subquotient of $\mathrm{I}(\mathfrak{m})$ such that the parabolic restriction $\boldsymbol{r}_{\mu_{\mathfrak{m}}}(\mathrm{Z}(\mathfrak{m}))$ contains a generic irreducible subquotient.

Theorem 3.8: *(See* [22]*) The map* $\mathfrak{m} \mapsto \mathrm{Z}(\mathfrak{m})$ *is a bijection between multisegments of length* n *and isomorphism classes of unipotent representations of* $\mathrm{GL}_n(\mathrm{F})$.

This bijection does not depend on the choice of $\sqrt{q} \in \mathrm{R}^{\times}$.

Proof: For injectivity, $\mu_{\mathfrak{m}}$ can be recovered by the uniqueness property in Proposition 3.7(1). Then $\mathfrak{m}$ can be recovered by looking at the generic irreducible subquotient in $\boldsymbol{r}_{\mu_{\mathfrak{m}}}(\mathrm{Z}(\mathfrak{m}))$.

For surjectivity, one uses a counting argument that is based on the classification of all simple modules over $\mathcal{H}_{\mathrm{R}}(\mathrm{G}, \mathrm{I})$ by aperiodic multisegments [1, 12]. □

Now write ℓ for the characteristic of R and define an integer $e \geqslant 0$ by:

$$e = \begin{cases} 0 & \text{if } \ell = 0, \\ \text{the smallest } k \geqslant 2 \text{ such that } 1 + q + \cdots + q^{k-1} = 0 & \text{if } \ell > 0. \end{cases}$$

Then the multisegment $\mathfrak{m}$ writes uniquely as:

$$\mathfrak{m} = \mathfrak{a} + \sum_{u \geqslant 0} \sum_{n \geqslant 1} \sum_{\xi} c(\xi, n, u) \cdot (\xi, n)_u$$

where ξ ranges over a set of representatives of $\mathrm{R}^{\times}/q^{\mathbf{Z}}$ in $\mathrm{R}^{\times}$ and:

(1) $(\xi, n)_u$ is the multisegment:

$$(\xi, n) + (\xi q, n) + \cdots + (\xi q^{e\ell^u - 1}, n)$$

where e is as above, and $c(\xi, n, u)$ is an integer in $\{0, 1, \ldots, \ell - 1\}$;

(2) $\mathfrak{a}$ is q-aperiodic, which means that for all segments (ζ, d) and all $v \geqslant 0$, the multisegment $(\zeta, d)_v$ does not occur in $\mathfrak{a}$ (when $\ell = 0$, any multisegment is q-aperiodic thus $\mathfrak{a} = \mathfrak{m}$).

Given an integer $u \geqslant 0$, the induced representation $1 \times \nu \times \cdots \times \nu^{e\ell^u - 1}$ (where ν is the absolute value $x \mapsto |x|$) possesses a unique cuspidal irreducible subquotient, denoted ρ_u. Given $\xi \in \mathrm{R}^\times$, we also write $\chi_{\xi,u}$ for the unramified character $\mathrm{Z}(\xi, e\ell^u)$ of $\mathrm{GL}_{e\ell^u}(\mathrm{F})$.

Theorem 3.9: *(See [22]) The cuspidal support of* $\mathrm{Z}(\mathfrak{m})$ *is:*

$$\mathrm{scusp}(\mathrm{Z}(\mathfrak{a})) + \sum_{u \geqslant 0} \sum_{n \geqslant 1} \sum_{\xi} n \cdot c(\xi, n, u) \cdot \rho_u \chi_{\xi,u}.$$

We finally have the following decomposition theorem.

Theorem 3.10: *(See [22]) Let* $\mathfrak{m}$ *be a multisegment. Then the semisimplification of* $\mathrm{I}(\mathfrak{m})$ *writes:*

$$\mathrm{Z}(\mathfrak{m}) + \sum_{\mathfrak{n}} d_{\mathfrak{m},\mathfrak{n}} \cdot \mathrm{Z}(\mathfrak{n})$$

where $\mathfrak{n}$ *ranges over all multisegments and* $d_{\mathfrak{m},\mathfrak{n}} \in \mathbf{Z}_{\geqslant 0}$*, with the following property: if* $d_{\mathfrak{m},\mathfrak{n}} \neq 0$*, then* $\mu_{\mathfrak{n}}$ *is smaller than* $\mu_{\mathfrak{m}}$ *(for the dominance order).*

Remark 3.11: When G is a non-split inner form of $\mathrm{GL}_n(\mathrm{F})$, there is no theory of generic representations for G. In order to define the irreducible representation $\mathrm{Z}(\mathfrak{m})$ of G, we introduce the notion of residually generic representation (see [22]) by using the functor $\mathrm{W} \mapsto \mathrm{W}^{\mathrm{K}_1}$ defined in Paragraph 2.1 (and more general functors coming from type theory to deal with positive level representations).

3.4. *Comments*

When R is the field of complex numbers, the map $\mathfrak{m} \mapsto \mathrm{Z}(\mathfrak{m})$ gives Zelevinski's classification of all irreducible representations having nonzero Iwahori-fixed vectors. When the segments of $\mathfrak{m}$ are put in a suitable order, the representation $\mathrm{Z}(\mathfrak{m})$ can be characterized as the unique irreducible subrepresentation of $\mathrm{I}(\mathfrak{m})$.

There is also a Langlands classification $\mathfrak{m} \mapsto \mathrm{L}(\mathfrak{m})$ where $\mathrm{L}(\mathfrak{m})$ is uniquely determined as the unique irreducible quotient of:

$$\mathrm{J}(\mathfrak{m}) = \mathrm{L}(\xi_1, n_1) \times \cdots \times \mathrm{L}(\xi_r, n_r)$$

when the segments are put in a suitable order, and where $\mathrm{L}(\xi, n)$ is the unique generic irreducible representation with the same cuspidal support as $\mathrm{Z}(\xi, n)$. These two classifications are exchanged by the Zelevinski involution.

4. Lecture 4

In this section, G is the group $\mathrm{GL}_n(\mathrm{F})$.

4.1. *Reduction mod* ℓ

Let us fix a prime number $\ell \neq p$ and an algebraic closure $\overline{\mathbf{Q}}_\ell$ of the field of ℓ-adic numbers. The residue field $\overline{\mathbf{F}}_\ell$ of its ring of integers $\overline{\mathbf{Z}}_\ell$ is an algebraic closure of a finite field of characteristic ℓ.

Definition 4.1: An irreducible $\overline{\mathbf{Q}}_\ell$-representation (π, V) of G is said to be *integral* if V has a G-stable lattice L, that is a free $\overline{\mathbf{Z}}_\ell$-module generated by a $\overline{\mathbf{Q}}_\ell$-basis of V.

Then $\mathrm{L} \otimes \overline{\mathbf{F}}_\ell$ is a smooth $\overline{\mathbf{F}}_\ell$-representation of finite length of G, and its semisimplification does not depend on the choice of L; it is denoted $\mathbf{r}_\ell(\pi)$ (see [31] and [27], II.5.11).

We fix a square root of q in $\overline{\mathbb{Z}}_\ell^\times \subseteq \overline{\mathbb{Q}}_\ell^\times$. By reducing mod the maximal ideal of $\overline{\mathbf{Z}}_\ell$, it gives us a square root of q in $\overline{\mathbb{F}}_\ell^\times$.

We write MS(R) for the semigroup of multisegments made of segments (ξ, n) with $\xi \in \mathrm{R}^\times$ and Z_R for the bijection from MS(R) to the set of all isomorphism classes of unipotent representations.

A multisegment $\mathfrak{m} = (\xi_1, n_1) + \cdots + (\xi_r, n_r) \in \mathrm{MS}(\overline{\mathbf{Q}}_\ell)$ is *integral* if the ξ_i's belong to $\overline{\mathbb{Z}}_\ell^\times$. If $\mathfrak{m}$ as above is integral, define:

$$\mathbf{r}_\ell(\mathfrak{m}) = (\overline{\xi}_1, n_1) + \cdots + (\overline{\xi}_r, n_r) \in \mathrm{MS}(\overline{\mathbf{F}}_\ell)$$

where $\overline{\xi}$ denotes the image of $\xi \in \overline{\mathbb{Z}}_\ell^\times$ in $\overline{\mathbb{F}}_\ell^\times$.

Lemma 4.2:

(1) Let $\mathfrak{m} \in \mathrm{MS}(\overline{\mathbf{Q}}_\ell)$. *Then* $\mathrm{Z}(\mathfrak{m})$ *is integral if and only if* $\mathfrak{m}$ *is integral.*

(2) Let π be an integral unipotent $\overline{\mathbf{Q}}_\ell$-representation of G*. Then all irreducible subquotients of $\mathbf{r}_\ell(\pi)$ are unipotent.*

Proof: For 2, let $\mathfrak{m} = (\xi_1, n_1) + \cdots + (\xi_r, n_r) \in \mathrm{MS}(\overline{\mathbf{Q}}_\ell)$ be an integral multisegment such that π is isomorphic to $\mathrm{Z}(\mathfrak{m})$. By construction, π is an irreducible subquotient of:

$$\mathrm{I}(\mathfrak{m}) = \mathrm{Z}_{\overline{\mathbf{Q}}_\ell}(\xi_1, n_1) \times \cdots \times \mathrm{Z}_{\overline{\mathbf{Q}}_\ell}(\xi_r, n_r).$$

Since reduction mod ℓ commutes with parabolic induction, all irreducible subquotients of $\mathbf{r}_\ell(\pi)$ are subquotients of:

$$\mathbf{r}_\ell(\mathrm{I}(\mathfrak{m})) = \mathrm{Z}_{\overline{\mathbf{Q}}_\ell}(\overline{\xi}_1, n_1) \times \cdots \times \mathrm{Z}_{\overline{\mathbf{Q}}_\ell}(\overline{\xi}_r, n_r) = \mathrm{I}(\mathbf{r}_\ell(\mathfrak{m})).$$

The result follows. □

Theorem 4.3: *Let $\mathfrak{m} \in \mathrm{MS}(\overline{\mathbf{Q}}_\ell)$ be an integral multisegment. Then:*

$$\mathbf{r}_\ell(\mathrm{Z}_{\overline{\mathbf{Q}}_\ell}(\mathfrak{m})) = \mathrm{Z}_{\overline{\mathbf{F}}_\ell}(\mathbf{r}_\ell(\mathfrak{m})) + \sum_{\mathfrak{n}} a(\mathfrak{m}, \mathfrak{n}) \cdot \mathrm{Z}_{\overline{\mathbf{F}}_\ell}(\mathfrak{n})$$

where $\mathfrak{n}$ ranges over all multisegments and $a(\mathfrak{m}, \mathfrak{n}) \in \mathbf{Z}_{\geqslant 0}$, with the property: if $a(\mathfrak{m}, \mathfrak{n})$ is nonzero, then $\mu_{\mathfrak{n}}$ is smaller than $\mu_{\mathfrak{m}}$ (for the dominance order).

Example 4.4: Assume $\mathrm{G} = \mathrm{GL}_2(\mathbf{Q}_5)$ and $\ell = 3$ (see Example 1.11). Thus we have $q = 5$.

The unipotent $\overline{\mathbf{Q}}_\ell$-representation corresponding to $(1, 2) \in \mathrm{MS}(\overline{\mathbf{Q}}_\ell)$ is the trivial $\overline{\mathbf{Q}}_\ell$-character of G, whose reduction mod ℓ is the trivial $\overline{\mathbf{F}}_\ell$-character of G, that corresponds to the multisegment $(\overline{1}, 2) \in \mathrm{MS}(\overline{\mathbf{F}}_\ell)$.

Now consider $\mathfrak{m} = (1, 1) + (q, 1) \in \mathrm{MS}(\overline{\mathbf{Q}}_\ell)$. We write St for the Steinberg $\overline{\mathbf{Q}}_\ell$-representation of G and π for the cuspidal subquotient of the $\overline{\mathbf{F}}_\ell$-representation V of Example 1.11. We have the following diagram:

$$\begin{array}{ccc}
(1,1) + (q,1) & \xrightarrow{\ \mathrm{Z}_{\overline{\mathbf{Q}}_\ell}\ } & \mathrm{St}\mathrm{Z}_{\overline{\mathbf{Q}}_\ell}(\sqrt{q}, 2) \\
\Big\downarrow {\scriptstyle \mathbf{r}_\ell} & & \Big\downarrow {\scriptstyle \mathbf{r}_\ell} \\
(1,1) + (-1,1) & \xrightarrow[\ \mathrm{Z}_{\overline{\mathbf{F}}_\ell}\ ]{} & \pi + \mathrm{Z}_{\overline{\mathbf{F}}_\ell}(-1, 2)
\end{array}$$

and $\pi = \mathrm{Z}_{\overline{\mathbf{F}}_\ell}((1, 1) + (-1, 1))$ is unipotent but not superunipotent.

4.2. *The local Langlands correspondence*

Let us fix a separable closure $\overline{\mathrm{F}}$ of F. Its residue field $\overline{\mathfrak{k}}$ is a separable closure of the residue field $\mathfrak{k}$ of F. The Galois group $\Gamma_{\mathrm{F}} = \mathrm{Gal}(\overline{\mathrm{F}}/\mathrm{F})$ acts on $\overline{\mathfrak{k}}$ such that there is a surjective group homomorphism:

$$\gamma_{\mathrm{F}} : \Gamma_{\mathrm{F}} \to \mathrm{Gal}(\overline{\mathfrak{k}}/\mathfrak{k}).$$

The group Γ_{F} is profinite, and $\mathrm{I}_{\mathrm{F}} = \mathrm{Ker}(\gamma_{\mathrm{F}})$ is a closed subgroup of Γ_{F}. Let $\mathrm{Frob}_{\mathrm{F}} \in \mathrm{Gal}(\overline{\mathfrak{k}}/\mathfrak{k})$ be the Frobenius automorphism $x \mapsto x^q$. Then write:

$$\mathrm{W}_{\mathrm{F}} = \{g \in \Gamma_{\mathrm{F}} \mid \gamma_{\mathrm{F}}(g) \in \mathrm{Frob}_{\mathrm{F}}^{\mathbf{Z}}\}.$$

It is called the Weil group of F. There is a unique topology on W_{F} such that:

(1) the topology of I_{F} induced by W_{F} and that induced by Γ_{F} coincide;
(2) the subgroup I_{F} is open in W_{F}.

Note that this is not the topology on W_{F} induced by Γ_{F} (for which (2) is not satisfied).

Remark 4.5: The group W_{F} is locally compact, and its neutral element has a basis of neighborhoods made of compact open pro-p-subgroups. There is a notion of smooth R-representation of W_{F}, just as for the group G. There is also a notion of reduction mod ℓ for (integral) finite-dimensional $\overline{\mathbf{Q}}_\ell$-representations of W_{F}.

Now write:

$$\mathscr{G}_n^0(\mathrm{F})$$

for the set of isomorphism classes of n-dimensional irreducible **C**-representations of W_{F}, and:

$$\mathscr{A}_n^0(\mathrm{F})$$

for the set of isomorphism classes of irreducible cuspidal **C**-representations of $\mathrm{GL}_n(\mathrm{F})$.

The local Langlands correspondence [20, 17, 18] asserts that there exists a unique family of bijections:

$$\pi_n^0 : \mathscr{G}_n^0(\mathrm{F}) \to \mathscr{A}_n^0(\mathrm{F}), \quad n \geqslant 1$$

satisfying certain specific conditions that we do not give explicitly here.

Now fix an isomorphism of fields $\alpha : \mathbf{C} \to \overline{\mathbf{Q}}_\ell$. By extension of scalars, any smooth complex representation of $\mathrm{W_F}, \mathrm{G}$ gives rise to a smooth $\overline{\mathbf{Q}}_\ell$-representation of $\mathrm{W_F}, \mathrm{G}$. It also gives bijections:

$$_\alpha\pi_n^0 : \mathscr{G}_n^0(\mathrm{F}, \overline{\mathbf{Q}}_\ell) \to \mathscr{A}_n^0(\mathrm{F}, \overline{\mathbf{Q}}_\ell), \quad n \geqslant 1$$

depending on α, because π_n^0 for n even depends on the choice of $\sqrt{q} \in \mathbf{C}^\times$.

Theorem 4.6: *(See [29])*

(1) For any $\rho \in \mathscr{A}_n^0(\mathrm{F}, \overline{\mathbf{Q}}_\ell)$, the reduction $\mathbf{r}_\ell(\rho)$ is irreducible and cuspidal.

(2) The map:

$$\mathbf{r}_\ell : \mathscr{A}_n^0(\mathrm{F}, \overline{\mathbf{Q}}_\ell)^{\mathrm{int}} \to \mathscr{A}_n^0(\mathrm{F}, \overline{\mathbf{F}}_\ell)$$

is surjective, where $\mathscr{A}_n^0(\mathrm{F}, \overline{\mathbf{Q}}_\ell)^{\mathrm{int}}$ denotes the subset of integral representations in $\mathscr{A}_n^0(\mathrm{F}, \overline{\mathbf{Q}}_\ell)$.

(3) A representation $\sigma \in \mathscr{G}_n^0(\mathrm{F}, \overline{\mathbf{Q}}_\ell)$ is integral if and only if $_\alpha\pi_n^0(\sigma)$ is integral.

(4) Assume $\sigma, \sigma' \in \mathscr{G}_n^0(\mathrm{F}, \overline{\mathbf{Q}}_\ell)$ are integral. Then:

$$\mathbf{r}_\ell(\sigma) = \mathbf{r}_\ell(\sigma') \quad \Leftrightarrow \quad \mathbf{r}_\ell({}_\alpha\pi_n^0(\sigma)) = \mathbf{r}_\ell({}_\alpha\pi_n^0(\sigma')).$$

(5) $\mathbf{r}_\ell(\sigma)$ is irreducible if and only if $\mathbf{r}_\ell({}_\alpha\pi_n^0(\sigma))$ is supercuspidal.

(6) This induces a bijection $_\alpha\overline{\pi}_n^0$ between isomorphism classes of irreducible n-dimensional $\overline{\mathbf{F}}_\ell$-representationsof $\mathrm{W_F}$ and isomorphism classes of irreducible supercuspidal $\overline{\mathbf{F}}_\ell$-representations of G.

Acknowledgments

The content of these notes is based on four lectures given at IMS of National University of Singapore in April 2013. I thank the organizers and IMS for providing such an opportunity.

References

1. S. Ariki – *On the decomposition numbers of the Hecke algebra of $G(m, 1, n)$*, J. Math. Kyoto Univ. **36** (1996), no. 4, p. 789-808.
2. J. Bernstein – *Le centre de Bernstein*, in *Representations of reductive groups over a local field*, Travaux en Cours, p. 1-32. Hermann, Paris, 1984. Written by P. Deligne.
3. J. Bernstein – *Representations of p-adic groups*, Lectures at Harvard University, 1992. Written by K. E. Rumelhart.
4. J. Bernstein & A. Zelevinski – *Induced representations of reductive p-adic groups. I*, Ann. Sci. École Norm. Sup. (4) **10** (1977), no. 4, p. 441-472.

5. C. Blondel – *Basic representation theory of reductive p-adic groups*, in *p-adic representations, theta correspondence and the Langlands-Shahidi method*, 2013, Science Press.
6. P. Broussous, V. Sécherre & S. Stevens – *Smooth representations of* $\mathrm{GL}(m, D)$*, V: endo-classes*, Documenta Math. **17** (2012), p. 23-77.
7. C. J. Bushnell – *Representations of reductive p-adic groups: localization of Hecke algebras and applications*, J. London Math. Soc. (2) **63** (2001), no. 2, p. 364-386.
8. C. J. Bushnell & G. Henniart – *Local tame lifting for* $\mathrm{GL}(N)$*. I. Simple characters*, Inst. Hautes Études Sci. Publ. Math. **83** (1996), p. 105-233.
9. C. J. Bushnell & P. C. Kutzko – *The admissible dual of* $\mathrm{GL}(N)$ *via compact open subgroups*, Princeton University Press, Princeton, NJ, 1993.
10. C. J. Bushnell & P. C. Kutzko – *Semisimple types in* GL_n, Compositio Math. **119** (1999), no. 1, p. 53-97.
11. P. Cartier – *Representations of* $\mathfrak{p}$*-adic groups: a survey*, Proceedings of Symposia in Pure Mathematics, vol. 33 (1979), part 1, p. 111-155.
12. N. Chriss & V. Ginzburg – *Representation theory and complex geometry*, Birkhäuser Boston Inc., Boston, MA, 1997.
13. J.-F. Dat – *Finitude pour les représentations lisses de groupes p-adiques*, J. Inst. Math. Jussieu **8** (2009), no. 2, p. 261-333.
14. J.-F. Dat – *Théorie de Lubin-Tate non abélienne* ℓ*-entière*, Duke Math. J. **161** (2012), no. 6, p. 951-1010.
15. D. Guiraud – *On semisimple* ℓ*-modular Bernstein-blocks of a p-adic general linear group*, J. Number Theory **133** (2013), 3524-3548.
16. Harish-Chandra – *Harmonic analysis on reductive p-adic groups*, Notes by G. van Dijk. Lecture Notes in Mathematics, vol. 162. Springer-Verlag, Berlin-New York, 1970.
17. M. Harris & R. Taylor – *The geometry and cohomology of some simple Shimura varieties*, Annals of Mathematics Studies **151**, Princeton University Press. With an appendix by Vladimir G. Berkovich.
18. G. Henniart – *Une preuve simple des conjectures de Langlands pour* $\mathrm{GL}(n)$ *sur un corps p-adique*, Invent. Math. **139** (2000), no. 2, p. 439-455.
19. R. Kurinczuk – ℓ*-modular representations of unramified p-adic* $\mathrm{U}(2,1)$, Algebra Number Theory **8** (2014), No. 8, 1801-1838.
20. G. Laumon, M. Rapoport & U. Stuhler – $\mathcal{D}$*-elliptic sheaves and the Langlands correspondence*, Invent. Math. **113** (1993), no. 2, p. 217-338.
21. A. Mínguez & V. Sécherre – *Unramified* ℓ*-modular representations of* $\mathrm{GL}_n(\mathrm{F})$ *and its inner forms*, Int. Math. Res. Notices **8** (2014), p. 2090-2128.
22. A. Mínguez & V. Sécherre – *Représentations lisses modulo* ℓ *de* $\mathrm{GL}_m(\mathrm{D})$, Duke Math. J. **163** (2014), p. 795-887.
23. A. Mínguez & V. Sécherre – *Types modulo* ℓ *pour les formes intérieures de* GL_n *sur un corps local non archimédien*, to appear in Proc. Lond. Math. Soc. (2014). With an appendix by V. Sécherre and S. Stevens. Preprint available at `http://lmv.math.cnrs.fr/annuaire/vincent-secherre/`.
24. V. Sécherre & S. Stevens – *Blocks of the category of* ℓ*-modular smooth representations of* $\mathrm{GL}_m(\mathrm{D})$. Preprint available at:

`http://lmv.math.cnrs.fr/annuaire/vincent-secherre/`.

25. A. Silberger – *The Langlands quotient theorem for p-adic groups*, Math. Ann. **236** (1978), no. 2, p. 95-104.
26. M. Tadić – *Induced representations of* $\mathrm{GL}(n, A)$ *for p-adic division algebras* A, J. Reine Angew. Math. **405** (1990), p. 48-77.
27. M.-F. Vignéras – *Représentations l-modulaires d'un groupe réductif p-adique avec* $l \neq p$, Progress in Mathematics, vol. 137, Birkhäuser Boston Inc., Boston, MA, 1996.
28. M.-F. Vignéras – *Induced R-representations of p-adic reductive groups*, Selecta Math. (N.S.) **4** (1998), no. 4, p. 549-623. With an appendix by Alberto Arabia.
29. M.-F. Vignéras – *Correspondance de Langlands semi-simple pour* $\mathrm{GL}(n, F)$ *modulo* $\ell \neq p$, Invent. Math. **144** (2001), no. 1, p. 177-223.
30. M.-F. Vignéras – *Schur algebras of reductive p-adic groups I*, Duke Math. J. **116** (2003), no. 1, p. 35-75.
31. M.-F. Vignéras – "On highest Whittaker models and integral structures, in *Contributions to Automorphic forms, Geometry and Number theory: Shalikafest 2002*, John Hopkins Univ. Press, 2004, p. 773-801.
32. A. Zelevinski – *Induced representations of reductive* $\mathfrak{p}$*-adic groups. II. On irreducible representations of* $\mathrm{GL}(n)$, Ann. Sci. École Norm. Sup. (4) **13** (1980), no. 2, p. 165-210.

p-MODULAR REPRESENTATIONS OF p-ADIC GROUPS*

Florian Herzig
Department of Mathematics
University of Toronto
40 St. George Street, #6290
Toronto, ON, M5S 2E4 Canada
herzig@math.toronto.edu

These notes are an introduction to the p-modular (or "mod-p") representation theory of p-adic reductive groups. We will focus on the group $\mathrm{GL}_2(\mathbb{Q}_p)$, but we try to provide statements that generalize to an arbitrary p-adic reductive group G (for example, $\mathrm{GL}_n(\mathbb{Q}_p)$).[a]

1. Motivation

We fix two primes p and ℓ. The motivation for studying the representation theory of p-adic groups comes from Local Langlands Conjectures.

1.1. *The case $\ell \neq p$*

In this case, we are in the setting of the classical Local Langlands Correspondence, which can be stated (roughly) as follows: Let $n \geq 1$. We then have an injective map

$$\left\{\begin{array}{c}\text{continuous representations of}\\ \mathrm{Gal}(\overline{\mathbb{Q}}_p/\mathbb{Q}_p)\text{ on }n\text{-dimensional}\\ \overline{\mathbb{Q}}_\ell\text{-vector spaces, up to}\\ \text{isomorphism}\end{array}\right\} \hookrightarrow \left\{\begin{array}{c}\text{irreducible, admissible}\\ \text{representations of }\mathrm{GL}_n(\mathbb{Q}_p)\\ \text{on }\overline{\mathbb{Q}}_\ell\text{-vector spaces,}\\ \text{up to isomorphism}\end{array}\right\} \tag{1.1}$$

The statement above is a bit imprecise; one needs to impose that Frobenius acts semisimply on the left-hand side. Usually, the left-hand side is

*Notes transcribed by Karol Kozioł
[a]We thank Vytas Paškūnas for his help with Section 10.

enlarged by replacing it by the set of Frobenius-semisimple Weil-Deligne representations, so that one obtains a bijection. This correspondence can be uniquely characterized by a list of properties (equivalence of L- and ε-factors of pairs, compatibility with contragredients, etc.). In particular, for $n = 1$, the correspondence reduces to local class field theory.

The correspondence was first established by Harris-Taylor ([13]) and Henniart ([14]) in 2000, and more recently by Scholze ([23]).

1.2. *The case $\ell = p$*

In this case, we would like to have a p-adic analog of (1.1), to be dubbed the "p-adic Local Langlands Correspondence." Additionally, we would like to have a "mod-p version;" that is, a "mod-p Local Langlands Correspondence," which would be compatible with the p-adic correspondence by reduction of lattices on both sides.

When $\ell = p$, we have "more" Galois representations, due to the fact that $\mathrm{Gal}(\overline{\mathbb{Q}}_p/\mathbb{Q}_p)$ and $\mathrm{GL}_n(\overline{\mathbb{Q}}_p)$ have compatible topologies. Breuil proposed to replace the right-hand side above with certain (not necessarily irreducible) Banach space representations of $\mathrm{GL}_n(\mathbb{Q}_p)$ over $\overline{\mathbb{Q}}_p$ (or, more precisely, a fixed finite extension of $\mathbb{Q}_p$). For the mod-p correspondence, one takes admissible representations of $\mathrm{GL}_n(\mathbb{Q}_p)$ over $\overline{\mathbb{F}}_p$.

When $n = 2$, such a correspondence has been made precise and proven by work of Breuil, Colmez, Paškūnas and others (see [7], [10], [12], [19], [22], and the references therein). However, for $n > 2$, there are fewer tools one has to attack this problem. For an overview of what is known, see [8].

The goal of these notes will be to describe the irreducible representations of $\mathrm{GL}_2(\mathbb{Q}_p)$ over $\overline{\mathbb{F}}_p$, or more generally, an algebraically closed field E of characteristic p. The classification was first obtained by Barthel-Livné ([3], [4]), and completed by Breuil ([6]). One of the main differences between the mod-p theory and the classical theory (i.e., the theory of complex representations) is the absence of an $\overline{\mathbb{F}}_p$-valued Haar measure, which limits the techniques at one's disposal.

2. p-Adic Groups

We begin with the basics. Let p be a prime number, and let $\mathbb{Z}_p$ denote the ring of p-adic integers; it is a discrete valuation ring with uniformizer p. We have

$$\mathbb{Z}_p/p^r\mathbb{Z}_p \cong \mathbb{Z}/p^r\mathbb{Z},$$

so in particular, the residue field of $\mathbb{Z}_p$ is $\mathbb{F}_p$. We let $\mathbb{Q}_p = \mathrm{Frac}(\mathbb{Z}_p)$ be the field of p-adic numbers. The topology of $\mathbb{Q}_p$ is defined by a fundamental system of neighborhoods of 0:

$$\mathbb{Z}_p \supset p\mathbb{Z}_p \supset p^2\mathbb{Z}_p \supset \ldots \supset p^r\mathbb{Z}_p \supset \ldots$$

The basis for the topology is then given by the cosets of these neighborhoods. With this topology, $\mathbb{Z}_p$ and $\mathbb{Q}_p$ are topological rings. We note that the sets $p^r\mathbb{Z}_p$ are all compact and open.

We take $n \geq 2$, and set $G = \mathrm{GL}_n(\mathbb{Q}_p)$. This is again a topological group, and a fundamental system of neighborhoods of 1 is given by

$$\mathrm{GL}_n(\mathbb{Z}_p) \supset 1 + p\mathrm{M}_n(\mathbb{Z}_p) \supset 1 + p^2\mathrm{M}_n(\mathbb{Z}_p) \supset \ldots \supset 1 + p^r\mathrm{M}_n(\mathbb{Z}_p) \supset \ldots.$$

We again note that the subgroups $1+p^r\mathrm{M}_n(\mathbb{Z}_p)$ are compact and open. We define

$$K := \mathrm{GL}_n(\mathbb{Z}_p),$$
$$K(r) := 1 + p^r\mathrm{M}_n(\mathbb{Z}_p).$$

The group K is a maximal compact subgroup of $\mathrm{GL}_n(\mathbb{Q}_p)$, and we have

$$K/K(r) \cong \mathrm{GL}_n(\mathbb{Z}_p/p^r\mathbb{Z}_p).$$

In particular, $K/K(1) \cong \mathrm{GL}_n(\mathbb{F}_p)$.

When $n = 2$, we define the following closed subgroups of $\mathrm{GL}_2(\mathbb{Q}_p)$:

$$B := \begin{pmatrix} * & * \\ 0 & * \end{pmatrix}, \qquad T := \begin{pmatrix} * & 0 \\ 0 & * \end{pmatrix}, \qquad U := \begin{pmatrix} 1 & * \\ 0 & 1 \end{pmatrix},$$

$$\overline{B} := \begin{pmatrix} * & 0 \\ * & * \end{pmatrix}, \qquad \overline{U} := \begin{pmatrix} 1 & 0 \\ * & 1 \end{pmatrix},$$

where a $*$ indicates an arbitrary entry. The group T is a maximal torus of $\mathrm{GL}_2(\mathbb{Q}_p)$, B is a Borel subgroup, and U its unipotent radical. We have factorizations as follows:

$$B = T \ltimes U, \quad \overline{B} = T \ltimes \overline{U}.$$

3. Smooth Representations

In what follows, we let Γ be a closed subgroup of $\mathrm{GL}_n(\mathbb{Q}_p)$ with the subspace topology, or a finite group with the discrete topology. These assumptions ensure that the identity element of Γ has a neighborhood basis of compact open subgroups. We take E to be an algebraically closed field of characteristic p (e.g., $\overline{\mathbb{F}}_p$). This will serve as the coefficient field for all representations we consider. Given a representation π of Γ, we will often identify π with its underlying vector space. Moreover, given a vector $v \in \pi$, we denote the action of $\gamma \in \Gamma$ on v by $\gamma.v$. If S is a subset of π and H is a submonoid of Γ, we let $\langle H.S \rangle_E$ be the smallest subspace of π containing S and stable by the action of H.

Definition 3.1: A representation π of Γ on an E-vector space is said to be *smooth* if

$$\pi = \bigcup_{\text{open subgroups } W} \pi^W,$$

where π^W denotes the subspace of vectors fixed by a subgroup W. Equivalently, π is smooth if the above equality holds with W running over compact open subgroups.

We remark that the smoothness condition is equivalent to saying that every vector is fixed by an open subgroup. This, in turn, is equivalent to saying that the action map

$$\begin{aligned} \Gamma \times \pi &\longrightarrow \pi \\ (\gamma, v) &\longmapsto \gamma.v \end{aligned}$$

is continuous, where π is given the discrete topology.

Definition 3.2: If π is any representation of Γ (not necessarily smooth), we define

$$\pi^\infty := \bigcup_{\text{open subgroups } W} \pi^W \subset \pi$$

to be the largest smooth subrepresentation of π.

We now proceed to define two types of induction functors.

Assume first that H is a closed subgroup of Γ, and let σ be a smooth representation of H. We define

$$\mathrm{Ind}_H^\Gamma(\sigma) := \{f : \Gamma \longrightarrow \sigma : f(h\gamma) = h.f(\gamma) \text{ for all } h \in H, \gamma \in \Gamma\}^\infty, \quad (3.1)$$

where the action of Γ is given by

$$(g.f)(\gamma) = f(\gamma g)$$

for every $\gamma, g \in \Gamma$. This procedure gives a smooth representation of Γ, and we call this functor *induction*.

Assume now that W is an open subgroup of Γ. Since Γ is a topological group, this implies that W is also closed in Γ. Let τ be a smooth representation of W, and define

$$\mathrm{ind}_W^\Gamma(\tau) := \left\{ f : \Gamma \longrightarrow \tau : \begin{array}{l} \diamond\ f(w\gamma) = w.f(\gamma) \text{ for all } w \in W, \gamma \in \Gamma \\ \diamond\ W\backslash\mathrm{supp}(f) \text{ is compact} \end{array} \right\}. \tag{3.2}$$

We note that $\mathrm{supp}(f)$ will be a union of W-cosets, so the quotient $W\backslash\mathrm{supp}(f)$ makes sense. Additionally, since W is open, the condition of $W\backslash\mathrm{supp}(f)$ being compact is equivalent to $W\backslash\mathrm{supp}(f)$ being finite. We define an action of Γ on $\mathrm{ind}_W^\Gamma(\tau)$ as above, and again obtain a smooth representation (without having to take smooth vectors!). We call this functor *compact induction.*

For $\gamma \in \Gamma, x \in \tau$, we will denote by $[\gamma, x] \in \mathrm{ind}_W^\Gamma(\tau)$ the function satisfying $\mathrm{supp}([\gamma, x]) = W\gamma^{-1}$ and $[\gamma, x](\gamma^{-1}) = x$. We have

$$[\gamma w, x] = [\gamma, w.x]$$

for $w \in W$, and moreover, the action of $g \in \Gamma$ on $[\gamma, x]$ is given by

$$g.[\gamma, x] = [g\gamma, x].$$

Remark 3.3: If the index $(\Gamma : H)$ is finite, then the subgroup H is open if and only if it is closed. In this case, the functors $\mathrm{Ind}_H^\Gamma(-)$ and $\mathrm{ind}_H^\Gamma(-)$ coincide.

Remark 3.4: Since the elements of the compact induction $\mathrm{ind}_W^\Gamma(\tau)$ have finite support modulo W, we have a Γ-equivariant isomorphism

$$\mathrm{ind}_W^\Gamma(\tau) \cong E[\Gamma] \otimes_{E[W]} \tau,$$

as in the representation theory of finite groups.

The following result is central in representation theory.

Theorem 3.5: (Frobenius Reciprocity) *Let H be a closed subgroup of Γ and W an open subgroup of Γ. Assume that π is a smooth representations of Γ, σ a smooth representation of H, and τ a smooth representation of W. We then have natural isomorphisms:*

(1) $\mathrm{Hom}_\Gamma(\pi, \mathrm{Ind}_H^\Gamma(\sigma)) \cong \mathrm{Hom}_H(\pi|_H, \sigma)$.
(2) $\mathrm{Hom}_\Gamma(\mathrm{ind}_W^\Gamma(\tau), \pi) \cong \mathrm{Hom}_W(\tau, \pi|_W)$.

Proof: (1) Given a homomorphism $\varphi \in \mathrm{Hom}_\Gamma(\pi, \mathrm{Ind}_H^\Gamma(\sigma))$, we obtain an element of $\mathrm{Hom}_H(\pi|_H, \sigma)$ by post-composing φ with the natural H-equivariant evaluation map

$$\begin{array}{rcl} \mathrm{Ind}_H^\Gamma(\sigma) & \longrightarrow & \sigma \\ f & \longmapsto & f(1). \end{array}$$

(2) Again, given a homomorphism $\varphi \in \mathrm{Hom}_\Gamma(\mathrm{ind}_W^\Gamma(\tau), \pi)$, we obtain an element of $\mathrm{Hom}_W(\tau, \pi|_W)$ by pre-composing φ with the W-equivariant map θ, given by

$$\begin{array}{rcl} \tau & \xrightarrow{\theta} & \mathrm{ind}_W^\Gamma(\tau) \\ x & \longmapsto & [1, x]. \end{array}$$

The remaining details are left as an exercise. □

Remark 3.6: Note that any element of the compact induction $\mathrm{ind}_W^\Gamma(\tau)$ can be written as

$$\sum_{i=1}^r [\gamma_i, x_i] = \sum_{i=1}^r \gamma_i.[1, x_i] = \sum_{i=1}^r \gamma_i.\theta(x_i),$$

where $\gamma_i \in \Gamma, x_i \in \tau$. This shows that $\theta(\tau)$ generates $\mathrm{ind}_W^\Gamma(\tau)$ as a Γ-representation.

Example 3.7: Let $n = 2$, and $G = \mathrm{GL}_2(\mathbb{Q}_p)$. We let $\chi_1, \chi_2 : \mathbb{Q}_p^\times \longrightarrow E^\times$ be two smooth characters (i.e., smooth, one-dimensional representations) of $\mathbb{Q}_p^\times$, and denote by

$$\begin{array}{rrcl} \chi_1 \otimes \chi_2 : & \overline{B} & \longrightarrow & E^\times \\ & \begin{pmatrix} \alpha & 0 \\ \gamma & \delta \end{pmatrix} & \longmapsto & \chi_1(\alpha)\chi_2(\delta) \end{array}$$

the character of $\overline{B}$ obtained by inflation from T. Taking the induction of $\chi_1 \otimes \chi_2$ from $\overline{B}$ to G, we obtain a smooth G-representation $\mathrm{Ind}_{\overline{B}}^G(\chi_1 \otimes \chi_2)$, called a *principal series representation.* Note that this representation is infinite-dimensional: it is not hard to verify that we have a vector space isomorphism

$$\begin{aligned}\{f \in \mathrm{Ind}_{\overline{B}}^{G}(\chi_1 \otimes \chi_2) : \mathrm{supp}(f) \subset \overline{B}U\} &\xrightarrow{\sim} C_c^{\infty}(\mathbb{Q}_p, E) \\ f &\longmapsto \left(x \longmapsto f\begin{pmatrix}1 & x\\ 0 & 1\end{pmatrix}\right),\end{aligned}$$

where $C_c^{\infty}(\mathbb{Q}_p, E)$ denotes the space of locally constant, compactly supported, E-valued functions on $\mathbb{Q}_p$. We also remark that this vector space isomorphism can actually be upgraded to an isomorphism of B-representations, as the left-hand side is B-stable.

We now proceed to discuss some distinguished features of representation theory in characteristic p.

Definition 3.8: A *pro-p group* is a topological group which is compact and Hausdorff, and possesses a fundamental system of neighborhoods of the identity consisting of normal subgroups of p-power index.

Example 3.9:

(1) A finite p-group with the discrete topology is obviously a pro-p group.
(2) The ring of p-adic integers $\mathbb{Z}_p$ is a pro-p group (using that $\mathbb{Z}_p/p^r\mathbb{Z}_p \cong \mathbb{Z}/p^r\mathbb{Z}$).
(3) Using the fact that $K(r)/K(r+1) \cong \mathrm{M}_n(\mathbb{F}_p)$ for $r > 0$ (exercise), one sees that $K(1)$ is a pro-p group. This implies in particular that any E-valued Haar measure μ of G vanishes on $K(r)$ for all $r > 0$, hence $\mu = 0$.
(4) Let $I(1)$ denote the preimage in $K = \mathrm{GL}_n(\mathbb{Z}_p)$ of the unipotent (p-Sylow) subgroup

$$\begin{pmatrix}1 & * & *\\ 0 & \ddots & *\\ 0 & 0 & 1\end{pmatrix} \subset \mathrm{GL}_n(\mathbb{F}_p).$$

This subgroup is pro-p by the same argument as in (3), and is called the *pro-p-Iwahori subgroup* of K.

The following lemma is very useful in mod-p representation theory.

Lemma 3.10: *Any nonzero smooth representation τ of a pro-p group H has H-fixed vectors; that is, $\tau^H \neq 0$.*

Proof: Forgetting the E-vector-space structure on τ, we can reduce to the case where τ is an $\mathbb{F}_p$-linear representation of H. Fix a vector $x \in \tau - \{0\}$;

since the action of H on τ is smooth, $\mathrm{Stab}_H(x)$ is open, and hence of finite index in H (by compactness). The orbit of x under H is then finite, and therefore $\langle H.x\rangle_{\mathbb{F}_p}$ is a finite dimensional $\mathbb{F}_p$-vector-space (isomorphic to $\mathbb{F}_p^d$, say). Now, since H is pro-p, the image of $H \longrightarrow \mathrm{GL}_d(\mathbb{F}_p)$ will be a p-group. As the p-Sylow subgroups of $\mathrm{GL}_d(\mathbb{F}_p)$ are all conjugate to

$$\begin{pmatrix} 1 & * & * \\ 0 & \ddots & * \\ 0 & 0 & 1 \end{pmatrix},$$

the image of H (after conjugation) will fix the first basis vector. □

We may now begin to analyze the mod-p representations of G.

Definition 3.11: We shall call the irreducible mod-p representations of $\mathrm{GL}_n(\mathbb{F}_p)$ *weights.*

Corollary 3.12:

(1) We have a canonical bijection

$$\{\textit{irreducible smooth representations of } K \textit{ over } E\}$$

$$\stackrel{1:1}{\longleftrightarrow} \{\textit{irreducible representations of } \mathrm{GL}_n(\mathbb{F}_p) \textit{ over } E\}.$$

(2) If π is a nonzero smooth representation of $G = \mathrm{GL}_n(\mathbb{Q}_p)$, then $\pi|_K$ contains a weight.

Proof: (1) Since $K(1)$ is open and normal in K and $K/K(1) \cong \mathrm{GL}_n(\mathbb{F}_p)$, we obtain irreducible smooth representations of K by inflation from $K/K(1)$. Conversely, let V be a smooth irreducible representation of K. As $K(1)$ is pro-p, we obtain $V^{K(1)} \neq 0$ by Lemma 3.10, and $V^{K(1)}$ is K-stable (again using normality of $K(1)$ in K). Hence, $V = V^{K(1)}$, and V is a representation of $K/K(1)$.

(2) Choosing $x \in \pi^{K(1)} - \{0\}$, we have (by abuse of notation)

$$\langle K.x\rangle_E = \langle \mathrm{GL}_n(\mathbb{F}_p).x\rangle_E.$$

As this space is finite dimensional, it contains an irreducible subrepresentation of K. □

Example 3.13: Using Corollary 3.12 above and the decomposition $\mathbb{Q}_p^\times = \mathbb{Z}_p^\times \times p^{\mathbb{Z}}$, we obtain a bijection

$$\begin{array}{ccc}\{\text{smooth characters } \chi : \mathbb{Q}_p^\times \longrightarrow E^\times\} & \stackrel{1:1}{\longleftrightarrow} & \mathrm{Hom}(\mathbb{F}_p^\times, E^\times) \times E^\times \\ \chi & \longmapsto & (\chi|_{\mathbb{Z}_p^\times}, \chi(p)).\end{array}$$

4. Weights

From now on, we focus on the case $n = 2$. This will allow us to do computations very explicitly. We first describe the weights of $K = \mathrm{GL}_2(\mathbb{Z}_p)$.

For any group X defined over $\mathbb{Q}_p$, we denote by X_p denote the analogous group over $\mathbb{F}_p$; for example

$$G_p := \mathrm{GL}_2(\mathbb{F}_p), \quad B_p := \begin{pmatrix} \mathbb{F}_p^\times & \mathbb{F}_p \\ 0 & \mathbb{F}_p^\times \end{pmatrix},$$

$$T_p := \begin{pmatrix} \mathbb{F}_p^\times & 0 \\ 0 & \mathbb{F}_p^\times \end{pmatrix}, \quad U_p := \begin{pmatrix} 1 & \mathbb{F}_p \\ 0 & 1 \end{pmatrix}, \quad \text{etc.}$$

Proposition 4.1: *Up to isomorphism, the weights of $G_p = \mathrm{GL}_2(\mathbb{F}_p)$ are precisely the following:*

$$F(a,b) := \mathrm{Sym}^{a-b}(E^2) \otimes \det{}^b,$$

where $0 \leq a - b \leq p-1,\ 0 \leq b < p-1$.

The action of G_p on E^2 is the standard one, given by the inclusion $G_p \hookrightarrow \mathrm{GL}_2(E)$.

We may describe the action of G_p on $F(a,b)$ explicitly as follows. We have an isomorphism $F(a,b) \cong E[X,Y]_{(a-b)}$, where $E[X,Y]_{(a-b)}$ is the space of homogeneous polynomials of degree $a-b$, with the action of $\left(\begin{smallmatrix} \alpha & \beta \\ \gamma & \delta \end{smallmatrix}\right) \in G_p$ on $f \in E[X,Y]_{(a-b)}$ given by

$$\left(\begin{pmatrix} \alpha & \beta \\ \gamma & \delta \end{pmatrix} \cdot f\right)(X,Y) = f(\alpha X + \gamma Y, \beta X + \delta Y)(\alpha\delta - \beta\gamma)^b.$$

We remark that this definition makes sense for $\left(\begin{smallmatrix} \alpha & \beta \\ \gamma & \delta \end{smallmatrix}\right) \in \mathrm{GL}_2(E)$, so that $E[X,Y]_{(a-b)}$ becomes a representation of the algebraic group $\mathrm{GL}_2(E)$. For $\mathrm{GL}_n(\mathbb{F}_p)$ with $n > 2$, there is no simple known explicit way to describe the irreducible mod-p representations. However, they still arise by restricting certain irreducible representations of the algebraic group $\mathrm{GL}_n(E)$ to $\mathrm{GL}_n(\mathbb{F}_p)$.

Proof: (Idea of proof (Proposition 4.1)) The proof of irreducibility relies on several observations:

(1) The space $F(a,b)^{U_p}$ is one-dimensional, and

$$F(a,b)^{U_p} = EX^{a-b}. \tag{4.1}$$

The group T_p acts on this space by the character $\eta_a \otimes \eta_b$, where $\eta_a(x) = x^a$ for $x \in \mathbb{F}_p^\times$.

(2) We have

$$\langle G_p.X^{a-b}\rangle_E = F(a,b).$$

To show this, it suffices to consider the action of $\overline{U}_p$, and then the claim follows from a computation with a Vandermonde determinant (that is, one checks that the elements $\left(\begin{smallmatrix}1&0\\\gamma&1\end{smallmatrix}\right).X^{a-b}$ for $0 \le \gamma \le a-b$ span $F(a,b)$).

The first observation along with Lemma 3.10 implies that the U_p-invariants of any nonzero subrepresentation of $F(a,b)$ must contain X^{a-b}. Thus, this subrepresentation must be all of $F(a,b)$ by the second observation. This shows that the representations $F(a,b)$ are all irreducible.

By considering the action of T_p on $F(a,b)^{U_p}$, we see that the representations $F(a,b)$ are all pairwise inequivalent, except for possibly the representations $F(b,b)$ and $F(p-1+b,b)$. Since $\dim_E F(b,b) = 1$ and $\dim_E F(p-1+b,b) = p$, we conclude that all of the $F(a,b)$ are indeed distinct.

Now, the number of p-modular representations of the finite group G_p is equal to the number of p-regular conjugacy classes in G_p. Using rational canonical form, it follows that the latter number is exactly $p(p-1)$, which shows that the representations $F(a,b)$ with $0 \le a-b \le p-1$ and $0 \le b < p-1$ form a full system of representatives for the p-modular representations of G_p. □

Lemma 4.2: *We have*

$$F(a,b)_{\overline{U}_p} \cong \eta_a \otimes \eta_b$$

as T_p-representations, where $V_{\overline{U}_p}$ denotes the $\overline{U}_p$-coinvariants of a G_p-representation V.

Proof: This follows from observation (1) of the previous proof, and the identity

$$(V_{\overline{U}_p})^* \cong (V^*)^{\overline{U}_p}$$

(note that we have $F(a,b)^* \cong F(a,b) \otimes \det^{-a-b}$ and, by conjugation, $F(a,b)^{\overline{U}_p} = EY^{a-b}$). Concretely, we have

$$F(a,b)_{\overline{U}_p} = F(a,b)/\langle EX^iY^{a-b-i} : 0 \le i < a - b\rangle. \qquad \square$$

Corollary 4.3: *The natural map obtained by composing*

$$F(a,b)^{U_p} \longhookrightarrow F(a,b) \longrightarrow\!\!\!\!\!\to F(a,b)_{\overline{U}_p}$$

is a T_p-linear isomorphism.

Using the above results, we can classify the weights appearing in principal series representations.

Proposition 4.4: *Fix two smooth characters $\chi_1, \chi_2 : \mathbb{Q}_p^\times \longrightarrow E^\times$. Then*

$$\dim_E \operatorname{Hom}_K(V, \operatorname{Ind}_{\overline{B}}^G(\chi_1 \otimes \chi_2)|_K) \le 1$$

for all weights V.

If $\chi_1|_{\mathbb{Z}_p^\times} \neq \chi_2|_{\mathbb{Z}_p^\times}$, then there is precisely one V such that equality holds, and $\dim_E V > 1$. If $\chi_1|_{\mathbb{Z}_p^\times} = \chi_2|_{\mathbb{Z}_p^\times}$, then two choices of V such that equality holds, and either $\dim_E V = 1$ or $\dim_E V = p$.

Proof: The proof essentially follows from Frobenius Reciprocity. Note first that, by Corollary 3.12, the characters $\chi_i|_{\mathbb{Z}_p^\times} : \mathbb{Z}_p^\times \longrightarrow E^\times$ must factor as

$$
\begin{array}{ccc}
\mathbb{Z}_p^\times & \xrightarrow{\;\chi_i|_{\mathbb{Z}_p^\times}\;} & E^\times \\
& \searrow \qquad \nearrow {\scriptstyle \eta_i} & \\
& \mathbb{F}_p^\times &
\end{array}
$$

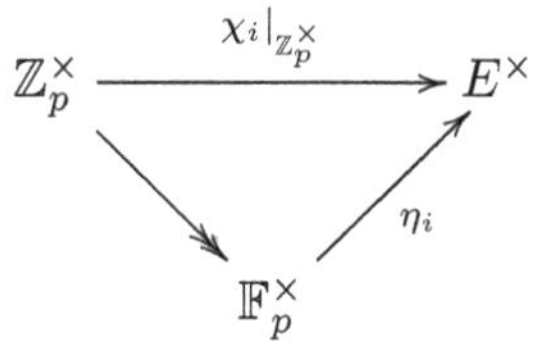

for some $\eta_i : \mathbb{F}_p^\times \longrightarrow E^\times$.

Now, the Iwasawa Decomposition says that we have a factorization of the form

$$G = \overline{B}K,$$

from which it follows that

$$\operatorname{Ind}_{\overline{B}}^G(\chi_1 \otimes \chi_2)|_K \cong \operatorname{Ind}_{\overline{B} \cap K}^K(\chi_1|_{\mathbb{Z}_p^\times} \otimes \chi_2|_{\mathbb{Z}_p^\times})$$

(this is a special case of the Mackey Decomposition). Both of these facts are left as an exercise for $GL_2(\mathbb{Q}_p)$. Hence, we obtain

$$\operatorname{Hom}_K(V, \operatorname{Ind}_{\overline{B}}^G(\chi_1 \otimes \chi_2)|_K) \overset{\text{Mackey}}{\cong} \operatorname{Hom}_K(V, \operatorname{Ind}_{\overline{B} \cap K}^K(\chi_1|_{\mathbb{Z}_p^\times} \otimes \chi_2|_{\mathbb{Z}_p^\times}))$$

$$\overset{\text{Frobenius}}{\cong} \quad \mathrm{Hom}_{\overline{B}\cap K}(V|_{\overline{B}\cap K},\ \chi_1|_{\mathbb{Z}_p^\times} \otimes \chi_2|_{\mathbb{Z}_p^\times})$$

$$\overset{\substack{\text{action factors}\\ \text{mod } p}}{\cong} \quad \mathrm{Hom}_{\overline{B}_p}(V|_{\overline{B}_p},\ \eta_1 \otimes \eta_2)$$

$$\overset{\substack{\overline{U}_p \text{ acts}\\ \text{trivially}}}{\cong} \quad \mathrm{Hom}_{T_p}(V_{\overline{U}_p},\ \eta_1 \otimes \eta_2).$$

Lemma 4.2 shows that $\mathrm{Hom}_K(V, \mathrm{Ind}_{\overline{B}}^G(\chi_1 \otimes \chi_2)|_K)$ is one-dimensional if $V_{\overline{U}_p} \cong \eta_1 \otimes \eta_2$, and zero-dimensional otherwise. A quick computation verifies that if $\eta_1 \neq \eta_2$, there is one such weight V (satisfying $\dim_E V > 1$), and two otherwise (satisfying $\dim_E V = 1$ or p). □

5. Hecke Algebras

Throughout this discussion, fix a weight V, and π a smooth representation of $G = \mathrm{GL}_2(\mathbb{Q}_p)$. Since the representation $\pi|_K$ contains a weight (cf. Corollary 3.12), it is natural to consider the multiplicity space

$$\mathrm{Hom}_K(V, \pi|_K)$$

of the weight V in π. By Frobenius Reciprocity, we have

$$\mathrm{Hom}_K(V, \pi|_K) \cong \mathrm{Hom}_G(\mathrm{ind}_K^G(V), \pi),$$

and the latter space comes equipped with a right action of

$$\mathcal{H}(V) := \mathrm{End}_G(\mathrm{ind}_K^G(V))$$

by pre-composition. The following proposition gives a more concrete description of the algebra $\mathcal{H}(V)$.

Proposition 5.1: *We have*

$$\mathcal{H}(V) \cong \left\{ \varphi : G \longrightarrow \mathrm{End}_E(V) : \begin{array}{c} \diamond\ \varphi(k_1 g k_2) = k_1 \circ \varphi(g) \circ k_2 \\ \textit{for all } k_1, k_2 \in K, g \in G \\ \diamond\ K\backslash \mathrm{supp}(\varphi)/K \textit{ is finite} \end{array} \right\}.$$

The composition product on the left-hand side becomes the convolution product on the right-hand side: for $\varphi_1, \varphi_2 : G \longrightarrow \mathrm{End}_E(V)$, we have

$$(\varphi_1 * \varphi_2)(g) = \sum_{\gamma \in G/K} \varphi_1(g\gamma)\varphi_2(\gamma^{-1}).$$

Proof: (Idea of proof) By Frobenius Reciprocity, we have

$$\mathcal{H}(V) \cong \mathrm{Hom}_K(V, \mathrm{ind}_K^G(V)|_K)$$

$$\subset \mathrm{Map}(V, \mathrm{Map}(G, V)) \cong \mathrm{Map}(G, \mathrm{Map}(V, V)),$$

where the last isomorphism is the natural map. Since the right-hand side of Proposition 5.1 is naturally a subspace of

$$\mathrm{Map}(G, \mathrm{Map}(V, V)),$$

one simply needs to check that the conditions on $\mathcal{H}(V)$ cut out the same subspace (see, for example, [18], Proposition 12 for more details). It is then straightforward to verify the final statement. □

Remark 5.2: Taking $V = \mathbf{1}_K$, the trivial representation of K, we obtain

$$\mathcal{H}(\mathbf{1}_K) \cong C_c(K\backslash G/K, E),$$

the usual double coset algebra of compactly supported, K-biinvariant, E-valued functions on G.

Remark 5.3: If $\varphi \in \mathcal{H}(V), f \in \mathrm{Hom}_K(V, \pi|_K)$, and $v \in V$, then the right action of $\mathcal{H}(V)$ on $\mathrm{Hom}_K(V, \pi|_K)$ is given by

$$(f * \varphi)(v) = \sum_{g \in K\backslash G} g^{-1}.f(\varphi(g).v). \tag{5.1}$$

To further analyze the structure of the algebra $\mathcal{H}(V)$, we will need to know about the double coset space $K\backslash G/K$:

Lemma 5.4: (Cartan Decomposition)

$$G = \bigsqcup_{\substack{r,s\in\mathbb{Z} \\ r\leq s}} K \begin{pmatrix} p^r & 0 \\ 0 & p^s \end{pmatrix} K.$$

This decomposition holds more generally (cf. [25], Section 3.3.3); in the case $G = \mathrm{GL}_2(\mathbb{Q}_p)$, it can be proven using the theory of elementary divisors.

Theorem 5.5:

(1) For every $r, s \in \mathbb{Z}$ with $r \leq s$, there exists a unique function $\mathrm{T}_{r,s} \in \mathcal{H}(V)$ such that $\mathrm{supp}(\mathrm{T}_{r,s}) = K\begin{pmatrix} p^r & 0 \\ 0 & p^s \end{pmatrix}K$ and such that $\mathrm{T}_{r,s}\begin{pmatrix} p^r & 0 \\ 0 & p^s \end{pmatrix} \in \mathrm{End}_E(V)$ is a linear projection.

(2) The set $\{\mathrm{T}_{r,s}\}_{r,s\in\mathbb{Z},r\leq s}$ forms a basis for $\mathcal{H}(V)$.

(3) We have an algebra isomorphism

$$\mathcal{H}(V) \cong E[\mathrm{T}_1, \mathrm{T}_2^{\pm 1}],$$

where $\mathrm{T}_1 := \mathrm{T}_{0,1}$ *and* $\mathrm{T}_2 := \mathrm{T}_{1,1}$. *In particular,* $\mathcal{H}(V)$ *is commutative.*

Proof: (Idea of proof) (1) As an example, consider the following identity in G:

$$\begin{pmatrix} \alpha & \beta \\ p\gamma & \delta \end{pmatrix} \begin{pmatrix} 1 & 0 \\ 0 & p \end{pmatrix} = \begin{pmatrix} 1 & 0 \\ 0 & p \end{pmatrix} \begin{pmatrix} \alpha & p\beta \\ \gamma & \delta \end{pmatrix},$$

where $\alpha, \delta \in \mathbb{Z}_p^\times, \beta, \gamma \in \mathbb{Z}_p$. Since the action of G on V factors through $\mathrm{GL}_2(\mathbb{F}_p)$, we have

$$\begin{pmatrix} \alpha & \beta \\ 0 & \delta \end{pmatrix} \circ \varphi \begin{pmatrix} 1 & 0 \\ 0 & p \end{pmatrix} = \varphi \begin{pmatrix} 1 & 0 \\ 0 & p \end{pmatrix} \circ \begin{pmatrix} \alpha & 0 \\ \gamma & \delta \end{pmatrix}$$

for any $\varphi \in \mathcal{H}(V)$. Taking $\alpha = \delta = 1, \gamma = 0$ shows that the image of the operator $\varphi \left(\begin{smallmatrix} 1 & 0 \\ 0 & p \end{smallmatrix}\right)$ is contained in V^{U_p}. Likewise, taking $\alpha = \delta = 1, \beta = 0$ shows that $\varphi \left(\begin{smallmatrix} 1 & 0 \\ 0 & p \end{smallmatrix}\right)$ factors through $V_{\overline{U}_p}$. Hence, the map $\varphi \left(\begin{smallmatrix} 1 & 0 \\ 0 & p \end{smallmatrix}\right) : V \longrightarrow V$ factors as

$$\begin{array}{ccc} V & \xrightarrow{\varphi\left(\begin{smallmatrix} 1 & 0 \\ 0 & p \end{smallmatrix}\right)} & V \\ \downarrow & & \uparrow \\ V_{\overline{U}_p} & \dashrightarrow & V^{U_p} \end{array}$$

Taking $\beta = \gamma = 0$, we see that the dotted arrow is T_p-linear. Thus, by Corollary 4.3, we see that there is a one-dimensional space of such maps. We take $\mathrm{T}_{0,1}$ to be the function supported on $K \left(\begin{smallmatrix} 1 & 0 \\ 0 & p \end{smallmatrix}\right) K$, such that $\mathrm{T}_{0,1} \left(\begin{smallmatrix} 1 & 0 \\ 0 & p \end{smallmatrix}\right)$ is (the map associated to) the inverse of the obvious map $V^{U_p} \xrightarrow{\sim} V_{\overline{U}_p}$ of Corollary 4.3. It follows that $\mathrm{T}_{0,1} \left(\begin{smallmatrix} 1 & 0 \\ 0 & p \end{smallmatrix}\right)$ is a projection.

More generally, for $r < s$, the argument is the same as above. For $r = s$, it is more elementary, and left as an exercise.
(2) This follows from the Cartan decomposition and the above proof of (1).
(3) It is clear that the elements $\{\mathrm{T}_{r,s}\}_{r,s \in \mathbb{Z}, r \leq s}$ form a basis, and one can show that this basis is related to $\{\mathrm{T}_1^i \mathrm{T}_2^j\}_{i \geq 0, j \in \mathbb{Z}}$ by a unitriangular change of coordinates (for a suitable ordering of both sides). □

To see this presentation of the Hecke algebra more naturally, we may use the mod-p Satake transform. We let

$$\mathcal{H}_T(V_{\overline{U}_p}) := \mathrm{End}_T(\mathrm{ind}_{T\cap K}^T(V_{\overline{U}_p})),$$

and note that Proposition 5.1 remains valid for $\mathcal{H}_T(V_{\overline{U}_p})$ if (G, K, V) is replaced with $(T, T\cap K, V_{\overline{U}_p})$. The Satake transform $\mathcal{S}_G : \mathcal{H}(V) \longrightarrow \mathcal{H}_T(V_{\overline{U}_p})$ is given explicitly by

$$\mathcal{S}_G(\varphi)(t) := \mathrm{pr}_{\overline{U}} \circ \left(\sum_{\overline{u}\in\overline{U}\cap K\backslash\overline{U}} \varphi(\overline{u}t) \right),$$

where $\varphi \in \mathcal{H}(V), t \in T$, and $\mathrm{pr}_{\overline{U}} : V \twoheadrightarrow V_{\overline{U}_p}$ is the natural map. It is straightforward to check that with this definition $\mathcal{S}_G$ is well-defined and an algebra homomorphism. With more work one shows that $\mathcal{S}_G$ is injective and that

$$\mathrm{im}\ \mathcal{S}_G = \left\{ \psi \in \mathcal{H}_T(V_{\overline{U}_p}) : \psi \begin{pmatrix} x & 0 \\ 0 & y \end{pmatrix} = 0 \text{ if } yx^{-1} \notin \mathbb{Z}_p \right\}. \qquad (5.2)$$

As T is abelian, one easily obtains the following analog of Theorem 5.5: $\mathcal{H}_T(V_{\overline{U}_p})$ has basis $\{\tau_{r,s} = \tau_1^{s-r}\tau_2^r\}_{r,s\in\mathbb{Z}}$, where $\tau_{r,s}$ is determined by

$$\mathrm{supp}(\tau_{r,s}) = (T\cap K)\begin{pmatrix} p^r & 0 \\ 0 & p^s \end{pmatrix}, \qquad \tau_{r,s}\begin{pmatrix} p^r & 0 \\ 0 & p^s \end{pmatrix} = 1,$$

and $\tau_1 := \tau_{0,1}, \tau_2 := \tau_{1,1}$. From (5.2), we get

$$\mathcal{H}(V) \cong \mathrm{im}\ \mathcal{S}_G = E[\tau_1, \tau_2^{\pm 1}].$$

In fact, it is not hard to verify that $\mathcal{S}_G(\mathrm{T}_1) = \tau_1$ and $\mathcal{S}_G(\mathrm{T}_2) = \tau_2$. We remark that the Satake transform exists more generally for any p-adic reductive group (see [16] and [15]).

Proposition 5.6: *Suppose* $\chi_1, \chi_2 : \mathbb{Q}_p^\times \longrightarrow E^\times$ *are two smooth characters, and let* $f : V \hookrightarrow \mathrm{Ind}_{\overline{B}}^G(\chi_1 \otimes \chi_2)|_K$ *be a nonzero* K*-linear map. Then*

$$f * \mathrm{T}_1 = \chi_2(p)^{-1} f$$
$$f * \mathrm{T}_2 = \chi_1(p)^{-1}\chi_2(p)^{-1} f.$$

Proof: This is a calculation using equation (5.1). Alternatively, the map f induces a T-linear map

$$\overline{f} : \mathrm{ind}_{T\cap K}^T(V_{\overline{U}_p}) \twoheadrightarrow \chi_1 \otimes \chi_2$$

(see the proof of Proposition 4.4). The map f then factors as $\mathrm{Ind}_{\overline{B}}^G(\overline{f}) \circ F_0$, where

$$F_0 : \mathrm{ind}_K^G(V) \longrightarrow \mathrm{Ind}_{\overline{B}}^G(\mathrm{ind}_{T\cap K}^T(V_{\overline{U}_p}))$$

is the map introduced in the proof of Theorem 6.1. Since

$$F_0 \circ \varphi = \mathrm{Ind}_{\overline{B}}^G(\mathcal{S}_G(\varphi)) \circ F_0,$$

one reduces the proof to the analogous, but much simpler, problem for T:

$$\overline{f} \circ \tau_1 = \chi_2(p)^{-1}\overline{f}$$
$$\overline{f} \circ \tau_2 = \chi_1(p)^{-1}\chi_2(p)^{-1}\overline{f}. \qquad \square$$

6. Comparison Isomorphisms

Using the results of the previous section, we can now describe comparison isomorphisms between compact and parabolic induction.

We fix two smooth characters $\chi_1, \chi_2 : \mathbb{Q}_p^\times \longrightarrow E^\times$, let V be a weight, and suppose

$$f : V \hookrightarrow \mathrm{Ind}_{\overline{B}}^G(\chi_1 \otimes \chi_2)|_K$$

is a nonzero K-linear homomorphism. Frobenius Reciprocity provides us with a nonzero G-linear map

$$\widetilde{f} : \mathrm{ind}_K^G(V) \longrightarrow \mathrm{Ind}_{\overline{B}}^G(\chi_1 \otimes \chi_2).$$

By Proposition 5.6, if $\varphi \in \mathcal{H}(V)$, we have

$$\widetilde{f} \circ \varphi = \chi'(\varphi)\widetilde{f},$$

where $\chi' : \mathcal{H}(V) \longrightarrow E$ is the algebra homomorphism defined by

$$\chi'(\mathrm{T}_1) = \chi_2(p)^{-1}$$
$$\chi'(\mathrm{T}_2) = \chi_1(p)^{-1}\chi_2(p)^{-1}.$$

The universal property of tensor products implies that we have a G-linear map

$$\overline{f} : \mathrm{ind}_K^G(V) \otimes_{\mathcal{H}(V)} \chi' \longrightarrow \mathrm{Ind}_{\overline{B}}^G(\chi_1 \otimes \chi_2).$$

Theorem 6.1: *The map* $\overline{f}$ *is an isomorphism if* $\dim_E V > 1$.

Proof: (Idea of proof) One can even prove a "universal" version of this theorem: consider the G-linear map

$$F_0 : \mathrm{ind}_K^G(V) \longrightarrow \mathrm{Ind}_{\overline{B}}^G(\mathrm{ind}_{T\cap K}^T(V_{\overline{U}_p}))$$

obtained as in the proof of Proposition 4.4 from the $T \cap K$-linear map $V_{\overline{U}_p} \longrightarrow \mathrm{ind}_{T\cap K}^T(V_{\overline{U}_p})$, which corresponds to the identity map in

$\mathrm{End}_T(\mathrm{ind}_{T\cap K}^T(V_{\overline{U}_p}))$ under Frobenius Reciprocity. A calculation shows that F_0 is $\mathcal{H}(V)$-linear with respect to $\mathcal{S}_G$, i.e., that $F_0 \circ \varphi = \mathrm{Ind}_{\overline{B}}^G(\mathcal{S}_G(\varphi)) \circ F_0$ for all $\varphi \in \mathcal{H}(V)$. As $\mathcal{S}_G(\mathrm{T}_1) = \tau_1$ is invertible in $\mathcal{H}_T(V_{\overline{U}_p})$ and the actions of G and $\mathcal{H}(V)$ commute, F_0 induces a G-linear and $\mathcal{H}(V)[\mathrm{T}_1^{-1}]$-linear map

$$F : \mathrm{ind}_K^G(V)[\mathrm{T}_1^{-1}] \longrightarrow \mathrm{Ind}_{\overline{B}}^G(\mathrm{ind}_{T\cap K}^T(V_{\overline{U}_p})). \qquad (6.1)$$

It now suffices to show F is an isomorphism, and then specialize T_1 and T_2 (that is, we apply the functor $-\otimes_{\mathcal{H}(V)[\mathrm{T}_1^{-1}]}\chi'$ to both sides of the isomorphism (6.1) above to recover the map $\overline{f}$).

To show F is surjective, one shows that the elements $F(\mathrm{T}_1^{-n}([1,x]))$, where $n \geq 0$ and $x \in V^{U_p} - \{0\}$, generate the right-hand side of (6.1) as a G-representation (this argument uses the Bruhat Decomposition and the fact that $\dim_E V > 1$). For the injectivity of F, one can use an elegant argument of Abe which uses the Satake transform (and which works even if $\dim_E V = 1$). See [18], Theorems 16 and 32 for more details. □

Remark 6.2: Notice that any pair (V, χ') subject to $\dim_E V > 1$ and $\chi'(\mathrm{T}_1) \neq 0$ arises in this theorem. Hence, for any such pair, $\mathrm{ind}_K^G(V) \otimes_{\mathcal{H}(V)} \chi'$ is a principal series representation.

Corollary 6.3: *If* $\dim_E V > 1$, *the weight* $f(V)$ *generates* $\mathrm{Ind}_{\overline{B}}^G(\chi_1 \otimes \chi_2)$ *as a* G*-representation.*

Proof: We have already seen that the image of V inside $\mathrm{ind}_K^G(V)$ generates the compact induction as a G-representation. By Theorem 6.1, we have

$$\mathrm{ind}_K^G(V) \twoheadrightarrow \mathrm{ind}_K^G(V) \otimes_{\mathcal{H}(V)} \chi' \cong \mathrm{Ind}_{\overline{B}}^G(\chi_1 \otimes \chi_2),$$

and the claim follows. □

Corollary 6.4: *Let* $\chi_1, \chi_2 : \mathbb{Q}_p^\times \longrightarrow E^\times$ *be two smooth characters, and suppose* $\chi_1|_{\mathbb{Z}_p^\times} \neq \chi_2|_{\mathbb{Z}_p^\times}$. *Then*

$$\pi = \mathrm{Ind}_{\overline{B}}^G(\chi_1 \otimes \chi_2)$$

is an irreducible G*-representation.*

Proof: By Proposition 4.4, π contains a unique weight V and it satisfies $\dim_E V > 1$. Moreover, V generates π as a G-representation by Corollary 6.3. Suppose σ is a nonzero subrepresentation of π; then σ contains the unique weight V, and thus has to be the whole of π. □

7. The Steinberg Representation

In this section we discuss the Steinberg representation, which arises when one takes $\chi_1 = \chi_2 = \mathbf{1}$ (the trivial character) in the definition of principal series.

Definition 7.1: We define the *Steinberg representation* St by the following short exact sequence:

$$0 \longrightarrow \mathbf{1}_G \longrightarrow \operatorname{Ind}_{\overline{B}}^G(\mathbf{1} \otimes \mathbf{1}) \longrightarrow \mathrm{St} \longrightarrow 0,$$

where the first map arises from Frobenius Reciprocity. More concretely, using (3.1), we can identify St as the quotient

$$\{\text{locally constant functions on } \overline{B}\backslash G \cong \mathbb{P}^1(\mathbb{Q}_p)\}/\{\text{constant functions}\}.$$

Theorem 7.2: *The representation* St *is irreducible.*

Proof: (Idea of proof) Recall that the subgroup $I(1)$, defined in Example 3.9(4), is a pro-p subgroup of G. One can prove in a straightforward way that $\dim_E \mathrm{St}^{I(1)} = 1$ (one needs to show that any $I(1)$-invariant function in the quotient is the image of an $I(1)$-invariant function in $\operatorname{Ind}_{\overline{B}}^G(\mathbf{1} \otimes \mathbf{1})$; see [4], Lemma 27 for more details). This implies that St contains a unique weight, which is of multiplicity 1.

Let $V = F(p-1, 0) = \mathrm{Sym}^{p-1}(E^2)$, and note that $\dim_E V = p$. We thus obtain the exact sequence

$$0 \longrightarrow \operatorname{Hom}_K(V, \mathbf{1}_G|_K) \longrightarrow \operatorname{Hom}_K(V, \operatorname{Ind}_{\overline{B}}^G(\mathbf{1} \otimes \mathbf{1})|_K) \longrightarrow \operatorname{Hom}_K(V, \mathrm{St}|_K).$$

Since $\operatorname{Hom}_K(V, \mathbf{1}_G|_K) = 0$ and $\dim_E \operatorname{Hom}_K(V, \operatorname{Ind}_{\overline{B}}^G(\mathbf{1} \otimes \mathbf{1})|_K) = 1$ (by Proposition 4.4), we obtain $\dim_E \operatorname{Hom}_K(V, \mathrm{St}|_K) \geq 1$, which shows that V is the aforementioned unique weight of St. By Corollary 6.3, V generates $\operatorname{Ind}_{\overline{B}}^G(\mathbf{1} \otimes \mathbf{1})$ as a G-representation, and therefore generates St. The result now follows as in the proof of Corollary 6.4 above. □

Remark 7.3: The sequence defining St does not split. To see this, just note that the unique nontrivial weight of $\operatorname{Ind}_{\overline{B}}^G(\mathbf{1} \otimes \mathbf{1})$ generates $\operatorname{Ind}_{\overline{B}}^G(\mathbf{1} \otimes \mathbf{1})$ as a G-representation, by Corollary 6.3.

Remark 7.4: If $\chi : \mathbb{Q}_p^\times \longrightarrow E^\times$ is a smooth character, we can twist the exact sequence defining St to obtain

$$0 \longrightarrow \chi \circ \det \longrightarrow \operatorname{Ind}_{\overline{B}}^G(\chi \otimes \chi) \longrightarrow \mathrm{St} \otimes (\chi \circ \det) \longrightarrow 0.$$

By what we have proved for the representation St, we know that the representations $\chi \circ \det$ and $\mathrm{St} \otimes (\chi \circ \det)$ are irreducible, and contain unique weights (of dimensions 1 and p, respectively). Moreover, the Hecke eigenvalues on these representations are the same as those of $\mathrm{Ind}_{\overline{B}}^{G}(\chi \otimes \chi)$, and are given by

$$\mathrm{T}_1 \longmapsto \chi(p)^{-1}, \qquad \mathrm{T}_2 \longmapsto \chi(p)^{-2}.$$

8. Change of Weight

We now fix two weights V, V', and consider the module of intertwiners given by

$$\mathcal{H}(V, V') := \mathrm{Hom}_G(\mathrm{ind}_K^G(V), \mathrm{ind}_K^G(V')).$$

It naturally has the structure of an $(\mathcal{H}(V'), \mathcal{H}(V))$-bimodule by pre- and post-composition.

Proposition 8.1:

(1) We have

$$\mathcal{H}(V, V') \cong \left\{ \varphi : G \longrightarrow \mathrm{Hom}_E(V, V') : \begin{array}{c} \diamond\ \varphi(k_1 g k_2) = k_1 \circ \varphi(g) \circ k_2 \\ \textit{for all } k_1, k_2 \in K, g \in G \\ \diamond\ K \backslash \mathrm{supp}(\varphi)/K \textit{ is finite} \end{array} \right\}.$$

The bimodule structure on the right-hand side is given by convolution (as in Proposition 5.1).

(2) We have $\mathcal{H}(V, V') \neq 0$ if and only if $V_{\overline{U}_p} \cong V'_{\overline{U}_p}$ as T_p-representations.

(3) If $V \not\cong V'$ and $V_{\overline{U}_p} \cong V'_{\overline{U}_p}$, then there exists $\varphi \in \mathcal{H}(V, V')$ satisfying $\mathrm{supp}(\varphi) = K \begin{pmatrix} p^r & 0 \\ 0 & p^s \end{pmatrix} K$ if and only if $r < s$.

Proof: The proof is exactly the same as for Proposition 5.1. □

In case (3) of the above proposition, the only possible choices for V and V' are $V = F(b, b)$ and $V' = F(p-1+b, b)$ for $0 \leq b < p-1$ (or vice versa). Therefore, there exist G-linear maps

$$\mathrm{ind}_K^G(V) \underset{\varphi^-}{\overset{\varphi^+}{\rightleftarrows}} \mathrm{ind}_K^G(V')$$

satisfying $\mathrm{supp}(\varphi^-) = \mathrm{supp}(\varphi^+) = K \begin{pmatrix} 1 & 0 \\ 0 & p \end{pmatrix} K$. These maps are unique up to a scalar.

For the next proposition, it will be convenient to make the identifications

$$\mathcal{H}(V) \cong E[\mathrm{T}_1, \mathrm{T}_2^{\pm 1}] \cong \mathcal{H}(V'),$$

and call both algebras $\mathcal{H}$. (More naturally, both algebras are identified with the same subalgebra of $\mathcal{H}_T(V_{\overline{U}_p}) = \mathcal{H}_T(V'_{\overline{U}_p})$ by the Satake transform.)

Proposition 8.2:

(1) The maps φ^-, φ^+ are $\mathcal{H}$-linear and commute with each other.
(2) We have (up to a scalar)

$$\varphi^+ \circ \varphi^- = \mathrm{T}_1^2 - \mathrm{T}_2.$$

Proof: (Sketch of proof) The mod-p Satake transform gives an injective map $\mathcal{H}(V, V') \hookrightarrow \mathcal{H}_T(V_{\overline{U}_p}, V'_{\overline{U}_p})$ with the same formula as before. This map is compatible with convolution. Then (1) follows from the fact that $\mathcal{H}_T(V_{\overline{U}_p}) = \mathcal{H}_T(V'_{\overline{U}_p})$ is commutative, while (2) is a calculation. □

Corollary 8.3: *If $\chi' : \mathcal{H} \longrightarrow E$ is an algebra homomorphism such that $\chi'(\mathrm{T}_1^2 - \mathrm{T}_2) \neq 0$, then we obtain a G-linear isomorphism*

$$\mathrm{ind}_K^G(V) \otimes_{\mathcal{H}} \chi' \cong \mathrm{ind}_K^G(V') \otimes_{\mathcal{H}} \chi'.$$

Proof: By Proposition 8.2, part (1), the maps φ^-, φ^+ induce G-linear maps between the two representations. Their composite is $\chi'(\mathrm{T}_1^2 - \mathrm{T}_2) \neq 0$, so we obtain an isomorphism. □

Proposition 8.4: *Let $\chi_1, \chi_2 : \mathbb{Q}_p^\times \longrightarrow E^\times$ be two smooth characters, and suppose $\chi_1 \neq \chi_2$. Then*

$$\pi = \mathrm{Ind}_{\overline{B}}^G(\chi_1 \otimes \chi_2)$$

is an irreducible G-representation.

Proof: If $\chi_1|_{\mathbb{Z}_p^\times} \neq \chi_2|_{\mathbb{Z}_p^\times}$, then the claim follows from Corollary 6.4. We therefore may assume $\chi_1|_{\mathbb{Z}_p^\times} = \chi_2|_{\mathbb{Z}_p^\times}$ and

$$\chi_1(p) \neq \chi_2(p). \tag{8.1}$$

By Proposition 4.4, π contains two weights V, V' of the form $F(b, b)$ and $F(p-1+b, b)$ with $0 \leq b < p-1$ (or vice versa). Let us label these weights so that $V = F(b, b)$ and $V' = F(p-1+b, b)$. Corollary 6.3 then implies that V' generates π as a G-representation, since $\dim_E V' = p > 1$.

Now let σ be a nonzero G-subrepresentation of π. We claim that σ must contain V'. Indeed, if this were not the case, we would necessarily have a K-linear injection $V \hookrightarrow \sigma|_K$ (as $\sigma|_K$ has to contain a weight). This implies the existence of a nonzero G-linear map

$$\mathrm{ind}_K^G(V) \longrightarrow \sigma,$$

which descends to

$$\mathrm{ind}_K^G(V) \otimes_{\mathcal{H}} \chi' \longrightarrow \sigma$$

for some $\chi' : \mathcal{H} \longrightarrow E$, as $\mathrm{Hom}_K(V, \pi|_K)$ is one-dimensional. By Proposition 5.6, $\chi' : \mathcal{H} \longrightarrow E$ is the character given by

$$\chi'(\mathrm{T}_1) = \chi_2(p)^{-1}, \qquad \chi'(\mathrm{T}_2) = \chi_1(p)^{-1}\chi_2(p)^{-1}.$$

Hence, we have

$$\chi'(\mathrm{T}_1^2 - \mathrm{T}_2) = \chi_2(p)^{-1}(\chi_2(p)^{-1} - \chi_1(p)^{-1}) \neq 0,$$

by equation (8.1). Therefore, Corollary 8.3 shows that we have a nonzero G-linear map

$$\mathrm{ind}_K^G(V') \otimes_{\mathcal{H}} \chi' \longrightarrow \sigma,$$

which, by Frobenius Reciprocity, gives a nonzero map $V' \hookrightarrow \sigma|_K$.

Since σ contains V' and V' generates π as a G-representation, we obtain $\sigma = \pi$, and therefore π is irreducible. □

9. Classification of Representations

We are now in a position to give a classification of smooth irreducible representations of $G = \mathrm{GL}_2(\mathbb{Q}_p)$, at least under an admissibility assumption. We remark that this assumption is in fact unnecessary (see [5], [3], [6]).

For the following definition and proposition, suppose that Γ is a closed subgroup of G.

Definition 9.1: A smooth Γ-representation π is called *admissible* if $\dim_E \pi^W < \infty$ for all open subgroups W of Γ.

Proposition 9.2: *Let π be a smooth Γ-representation. Then π is admissible if and only if $\dim_E \pi^W < \infty$ for* one *open* pro-p *subgroup W of Γ.*

Proof: Let W be a fixed pro-p subgroup such that $\dim_E \pi^W < \infty$, and let W' be an arbitrary open subgroup. Since $\dim_E \pi^{W'} \leq \dim_E \pi^{W\cap W'}$, it is enough to show $\dim_E \pi^{W\cap W'} < \infty$. We may therefore assume that W' is an open subgroup of W (and hence of finite index). This gives

$$\pi^{W'} = \mathrm{Hom}_{W'}(\mathbf{1}_{W'}, \pi|_{W'}) \cong \mathrm{Hom}_W(\mathrm{ind}_{W'}^W(\mathbf{1}_{W'}), \pi|_W).$$

It thus suffices to show that $\mathrm{Hom}_W(M, \pi|_W)$ is finite-dimensional for any finite-dimensional smooth W-representation M. We argue by induction on $\dim_E M$. By Lemma 3.10, we have a short exact sequence

$$0 \longrightarrow \mathbf{1}_W \longrightarrow M \longrightarrow M/\mathbf{1}_W \longrightarrow 0.$$

Applying $\mathrm{Hom}_W(-, \pi|_W)$ yields

$$0 \longrightarrow \mathrm{Hom}_W(M/\mathbf{1}_W, \pi|_W) \longrightarrow \mathrm{Hom}_W(M, \pi|_W)$$
$$\longrightarrow \mathrm{Hom}_W(\mathbf{1}_W, \pi|_W) \cong \pi^W.$$

The last term is finite-dimensional by assumption and the first term is finite-dimensional by induction. Hence the middle term is finite-dimensional. □

The proof of the following proposition is left as an exercise (see also Lemmas 23 and 24 in [18]).

Proposition 9.3: *Let π be a smooth representation of G.*

(1) The representation π is admissible if and only if $\dim_E \mathrm{Hom}_K(V, \pi|_K) < \infty$ for any weight V.
(2) If π is admissible, then π possesses a central character.

Corollary 9.4: *All principal series representations and all representations of the form* $\mathrm{St} \otimes (\chi \circ \det)$, *with* $\chi : \mathbb{Q}_p^\times \longrightarrow E^\times$ *a smooth character, are admissible.*

Proof: This follows from part (2) of the previous proposition, Proposition 4.4, and (the proof of) Theorem 7.2. □

Definition 9.5: Let π be an irreducible, admissible G-representation. We say π is *supersingular* if for any weight V the action of T_1 on $\mathrm{Hom}_K(V, \pi|_K)$ is nilpotent (or, equivalently, if all eigenvalues of T_1 are zero).

The following gives a coarse classification of representations of G.

Theorem 9.6: (Barthel-Livné) *Every irreducible, admissible representation of $G = \mathrm{GL}_2(\mathbb{Q}_p)$ falls into one of the following four families:*

(1) the principal series $\mathrm{Ind}_{\overline{B}}^{G}(\chi_1 \otimes \chi_2)$ *with* $\chi_1 \neq \chi_2$,
(2) the characters $\chi \circ \det$,
(3) twists of the Steinberg representation $\mathrm{St} \otimes (\chi \circ \det)$,
(4) the supersingular representations.

The four families are disjoint, and the characters appearing in cases (1) to (3) are uniquely determined.

Proof: Let π be an irreducible, admissible representation, and let V be a weight of π. As π is admissible and $\mathcal{H}(V)$ is commutative, the (finite-dimensional) weight space $\mathrm{Hom}_K(V, \pi|_K)$ contains a common eigenvector $f : V \hookrightarrow \pi|_K$ for $\mathcal{H}(V)$, with eigenvalues given by an algebra homomorphism $\chi' : \mathcal{H}(V) \longrightarrow E$. If $\chi'(\mathrm{T}_1) = 0$ for all such pairs (V, χ'), then π is supersingular. We may therefore assume without loss of generality that $\chi'(\mathrm{T}_1) \neq 0$.

In this case, we get a nonzero (and in fact surjective) G-linear map

$$\mathrm{ind}_K^G(V) \otimes_{\mathcal{H}(V)} \chi' \twoheadrightarrow \pi.$$

We consider several possibilities:

- If $\dim_E V > 1$, then $\mathrm{ind}_K^G(V) \otimes_{\mathcal{H}(V)} \chi' \cong \mathrm{Ind}_{\overline{B}}^{G}(\chi_1 \otimes \chi_2)$ for some choice of χ_1, χ_2 (cf. Remark 6.2), and we obtain that π is either an irreducible principal series representation or a twist of a Steinberg representation.
- If $\dim_E V = 1$ and $\chi'(\mathrm{T}_1^2 - \mathrm{T}_2) \neq 0$, then by Corollary 8.3, we have $\mathrm{ind}_K^G(V) \otimes_{\mathcal{H}(V)} \chi' \cong \mathrm{ind}_K^G(V') \otimes_{\mathcal{H}(V)} \chi'$, for some p-dimensional weight V'. We may now proceed as in the previous case.
- If $\dim_E V = 1$ and $\chi'(\mathrm{T}_1^2 - \mathrm{T}_2) = 0$, then (after twisting π by a character of the form $\chi \circ \det$) we may assume $V = \mathbf{1}_K$ and $\chi'(\mathrm{T}_1) = \chi'(\mathrm{T}_2) = 1$. The claim now follows from the following fact (cf. [3], Theorem 30 or [1], Proposition 4.7):

$$(\mathrm{ind}_K^G(\mathbf{1}_K) \otimes_{\mathcal{H}(\mathbf{1}_K)} \chi')^{\mathrm{ss}} \cong (\mathrm{Ind}_{\overline{B}}^{G}(\mathbf{1} \otimes \mathbf{1}))^{\mathrm{ss}} \cong \mathbf{1}_G \oplus \mathrm{St}.$$

To show that the four families are disjoint, we analyze their weights and Hecke eigenvalues. Note firstly that if π is any subquotient of $\mathrm{Ind}_{\overline{B}}^{G}(\chi_1 \otimes \chi_2)$ and V is any weight of π, we have

$$V_{\overline{U}_p} \cong \chi_1|_{\mathbb{Z}_p^\times} \otimes \chi_2|_{\mathbb{Z}_p^\times}$$

as T_p-representations. This implies that the eigenvalues of $\mathcal{H}(V)$ on the space $\mathrm{Hom}_K(V, \pi|_K)$ are given by

$$\chi'(\mathrm{T}_1) = \chi_2(p)^{-1}, \tag{9.1}$$

$$\chi'(\mathrm{T}_2) = \chi_1(p)^{-1}\chi_2(p)^{-1}. \tag{9.2}$$

The condition $\chi'(\mathrm{T}_1) = 0$ distinguishes the supersingular representations (family (4)). The irreducible principal series representations (family (1)) are distinguished by the condition $1 < \dim_E V < p$ or $\chi'(\mathrm{T}_1^2 - \mathrm{T}_2) \neq 0$. Finally, the characters of G (family (2)) are determined by $\dim_E V = 1$ and $\chi'(\mathrm{T}_1^2 - \mathrm{T}_2) = 0$, while the twists of the Steinberg representation (family (3)) are determined by $\dim_E V = p$ and $\chi'(\mathrm{T}_1^2 - \mathrm{T}_2) = 0$. Moreover, equations (9.1) and (9.2) above (along with knowledge of $V_{\overline{U}_p}$) imply how to uniquely recover the characters χ_1, χ_2 (respectively, χ) from χ'. □

Definition 9.7: An irreducible, admissible G-representation is called *supercuspidal* if it is not a subquotient of a principal series representation.

Corollary 9.8: *If π is an irreducible, admissible representation of $G = \mathrm{GL}_2(\mathbb{Q}_p)$, then π is supercuspidal if and only if π is supersingular.*

Remark 9.9: The above description of irreducible, admissible representations in terms of parabolic induction, supersingular representations, and (generalized) Steinberg representations generalizes to arbitrary p-adic reductive groups. Also, Corollary 9.8 still holds. See [17] for the case of GL_n, [1] for the case of a general split, connected, reductive group, and [2] for the general case.

10. Supersingular Representations

Almost all of the arguments we have considered up to this point apply with little change to the group $\mathrm{GL}_2(F)$, where F is a finite extension of $\mathbb{Q}_p$. In this section, however, we will crucially use the fact that $F = \mathbb{Q}_p$ to give a more precise description of the classification of Theorem 9.6.

Our main goal will be to prove the following theorem, due to Breuil. We follow an argument due to Paškūnas [21] and Emerton [11].

Theorem 10.1: (Breuil) *The irreducible, admissible, supersingular representations of $G = \mathrm{GL}_2(\mathbb{Q}_p)$ are exactly*

$$\mathrm{ind}_K^G(V) \otimes_{\mathcal{H}(V)} \chi',$$

where V is a weight and $\chi'(\mathrm{T}_1) = 0$.

By Theorem 9.6 above, any irreducible, admissible, supersingular representation of G is a quotient of $\mathrm{ind}_K^G(V) \otimes_{\mathcal{H}(V)} \chi'$ with $\chi'(\mathrm{T}_1) = 0$, so it is enough to show that this quotient is irreducible and admissible. In order to do this, we shall use a slightly different model for $\mathrm{ind}_K^G(V) \otimes_{\mathcal{H}(V)} \chi'$, given as follows.

Let Z denote the center of G. We inflate V to a representation of KZ by decreeing that the element $\left(\begin{smallmatrix} p & 0 \\ 0 & p \end{smallmatrix}\right)$ acts trivially.

Proposition 10.2: *We have an isomorphism*

$$\mathrm{End}_G(\mathrm{ind}_{KZ}^G(V)) \cong E[\mathrm{T}],$$

where (via an isomorphism analogous to that of Proposition 5.1) T *is the function supported on the double coset* $KZ\left(\begin{smallmatrix} 1 & 0 \\ 0 & p \end{smallmatrix}\right)KZ$*, such that* $\mathrm{T}\left(\begin{smallmatrix} 1 & 0 \\ 0 & p \end{smallmatrix}\right) \in \mathrm{End}_E(V)$ *is a linear projection.*

Proof: The proof is analogous to the proof of Theorem 5.5. □

Now let $\eta : \mathbb{Q}_p^\times \longrightarrow E^\times$ be a (smooth) character which is trivial on $\mathbb{Z}_p^\times$, and which satisfies $\eta(p)^{-2} = \chi'(\mathrm{T}_2)$. Then one easily checks that we have a surjective G-linear map

$$\begin{aligned} \mathrm{ind}_K^G(V) &\twoheadrightarrow \mathrm{ind}_{KZ}^G(V) \otimes \eta \circ \det \\ [g, v]_K &\longmapsto [g, v]_{KZ} \otimes \eta(\det g), \end{aligned}$$

where the subscript denotes the group with respect to which the element $[g, v]$ is equivariant. Note that the action of T_1 on the left is compatible with the action of $\mathrm{T} \cdot \eta(p)^{-1}$ on the right, while the action of T_2 is compatible with the action of the scalar $\eta(p)^{-2} = \chi'(\mathrm{T}_2)$. It is not hard to check that this induces an isomorphism

$$\mathrm{ind}_K^G(V) \otimes_{\mathcal{H}(V)} \chi' \xrightarrow{\sim} \frac{\mathrm{ind}_{KZ}^G(V)}{(\mathrm{T} - \chi'(\mathrm{T}_1)\eta(p))} \otimes (\eta \circ \det).$$

Therefore, to prove Theorem 10.1 it is enough to show $\mathrm{ind}_{KZ}^G(V)/(\mathrm{T})$ is irreducible and admissible as a G-representation.

Proposition 10.3: *The representation* $\mathrm{ind}_{KZ}^G(V)/(\mathrm{T})$ *is nonzero.*

Proof: (Idea of proof) As the operator T is not invertible (by Proposition 10.2), it suffices to show that T is injective on $\mathrm{ind}_{KZ}^G(V)$. Consider the special case where $V = \mathbf{1}_K$. We then have

$$\mathrm{ind}_{KZ}^G(\mathbf{1}_K) = C_c(KZ\backslash G, E),$$

so that the compactly induced representation is equal to the space of finitely supported E-valued functions on $KZ\backslash G$. In this context, it is convenient to use the Bruhat-Tits tree $\mathfrak{X}$ of $\mathrm{GL}_2(\mathbb{Q}_p)$ (pictured below for $p = 2$; see [24] for an overview).

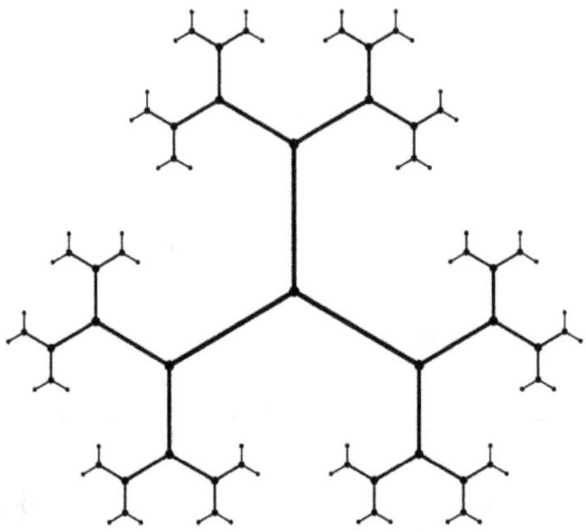

The vertices of $\mathfrak{X}$ correspond to homothety classes of $\mathbb{Z}_p$-lattices contained in $\mathbb{Q}_p^{\oplus 2}$. The group $\mathrm{GL}_2(\mathbb{Q}_p)$ acts transitively on the set of such lattices, and the stabilizer of the class of $\mathbb{Z}_p \oplus \mathbb{Z}_p$ is exactly KZ. Hence, the representation $\mathrm{ind}_{KZ}^G(\mathbf{1}_K)$ is exactly the set of compactly supported functions on the vertices of $\mathfrak{X}$. In this setup, the operator T is simply the "sum over neighbors" map; that is, $(\mathrm{T}f)(x) = \sum f(y)$, where y runs through the neighbors of x.

Now let $f \in \mathrm{ind}_{KZ}^G(\mathbf{1}_K)$ be an arbitrary nonzero function, and let $\mathfrak{T}$ be the convex hull of $\mathrm{supp}(f)$; it is a finite subtree of $\mathfrak{X}$. If we let x denote an extremal vertex of $\mathfrak{T}$ and $y \notin \mathfrak{T}$ a neighbor of x, then we have $(\mathrm{T}f)(y) \neq 0$, which shows T is injective. □

We introduce some additional notation for the proof of Theorem 10.1. Let

$$\mathrm{red} : K = \mathrm{GL}_2(\mathbb{Z}_p) \twoheadrightarrow G_p = \mathrm{GL}_2(\mathbb{F}_p)$$

denote the "reduction-modulo-p" map, and define

$$I := \mathrm{red}^{-1}(B_p), \qquad I(1) := \mathrm{red}^{-1}(U_p).$$

We let

$$\Pi := \begin{pmatrix} 0 & 1 \\ p & 0 \end{pmatrix};$$

we then have

$$\Pi I(1)\Pi^{-1} = I(1) \tag{10.1}$$

$$\Pi^2 = \begin{pmatrix} p & 0 \\ 0 & p \end{pmatrix}. \tag{10.2}$$

We also set

$$t := \begin{pmatrix} p & 0 \\ 0 & 1 \end{pmatrix}, \qquad s := \begin{pmatrix} 0 & 1 \\ 1 & 0 \end{pmatrix}, \qquad \mathcal{X} := \begin{pmatrix} 1 & 1 \\ 0 & 1 \end{pmatrix} - 1 \in E[I \cap U].$$

Note that $\Pi = st$.

We will also need the technique of weight cycling mentioned above. Let π denote the quotient $\mathrm{ind}_{KZ}^{G}(V)/(\mathrm{T})$, and write $V = F(a,b)$. We denote by f the composition of natural maps

$$V \longrightarrow \mathrm{ind}_{KZ}^{G}(V) \twoheadrightarrow \pi.$$

Note that f is injective, as V generates $\mathrm{ind}_{KZ}^{G}(V)$, and $f * \mathrm{T} = 0$, by definition of π.

The map f allows us to identify V with a K-subrepresentation of π. Let us fix a nonzero element $v \in V^{U_p} \subset \pi^{I(1)}$. By equation (4.1), we see that I acts on v by the character $\eta_a \otimes \eta_b$ (using the identification $I/I(1) \cong T_p$). By (10.1) above, the element $v' := \Pi.v$ also lies in $\pi^{I(1)}$, and I acts on v' by the character $\eta_b \otimes \eta_a$. Hence, by Frobenius Reciprocity, we get a nonzero K-linear map

$$\begin{array}{rcl} \mathrm{ind}_{I}^{K}(\eta_b \otimes \eta_a) & \stackrel{j}{\longrightarrow} & \pi \\ {[1,1]} & \longmapsto & v'. \end{array}$$

Additionally, using Lemma 4.2 and Frobenius Reciprocity, we have a complex

$$0 \longrightarrow V = F(a,b) \stackrel{i}{\longrightarrow} \mathrm{ind}_{I}^{K}(\eta_b \otimes \eta_a) \longrightarrow V' := F(b+p-1,a) \longrightarrow 0.$$

See also [9], Theorem 7.1 for an alternate derivation of this complex. Since $\dim_E F(a,b) = a-b+1$ and $\dim_E F(b+p-1,a) = p+b-a$, the complex above is exact.

Lemma 10.4: *We have $j \circ i = f * \mathrm{T}$ (up to a scalar).*

Proof: Calculation. □

The lemma holds equally well for any smooth representation π. In our situation we therefore have $j \circ i = 0$, so we see that the map j factors through the projection to V':

$$\begin{array}{ccc} \mathrm{ind}_I^K(\eta_b \otimes \eta_a) & \twoheadrightarrow & V' \\ {\scriptstyle j}\downarrow & \swarrow & \\ \pi & & \end{array}$$

This implies that the K-subrepresentation of π generated by v' is isomorphic to V'. The upshot of this discussion is that Π acts on $\pi^{I(1)}$, exchanging v and v' (by (10.2) above), and we have

$$\langle K.v\rangle_E \cong F(a,b), \qquad \langle K.v'\rangle_E \cong F(b+p-1,a).$$

Note that these two weights are nonisomorphic.

Lemma 10.5: (Iwahori Decomposition) *We have a factorization*

$$I = \underbrace{(I \cap U)}_{I^+}\underbrace{(I \cap T)}_{I^0}\underbrace{(I \cap \overline{U})}_{I^-} = \begin{pmatrix} 1 & \mathbb{Z}_p \\ 0 & 1 \end{pmatrix}\begin{pmatrix} \mathbb{Z}_p^\times & 0 \\ 0 & \mathbb{Z}_p^\times \end{pmatrix}\begin{pmatrix} 1 & 0 \\ p\mathbb{Z}_p & 1 \end{pmatrix},$$

where the factors may be taken in any order. In addition, we have

$$tI^+t^{-1} \subset I^+, \qquad tI^0t^{-1} = I^0, \qquad tI^-t^{-1} \supset I^-.$$

Proof: This decomposition is a general fact about reductive groups over local fields. For the case of $\mathrm{GL}_2(\mathbb{Q}_p)$, one can prove this directly: for example, one has (for $\alpha, \delta \in \mathbb{Z}_p^\times, \beta, \gamma \in \mathbb{Z}_p$)

$$\begin{pmatrix} \alpha & \beta \\ p\gamma & \delta \end{pmatrix} = \begin{pmatrix} 1 & \beta\delta^{-1} \\ 0 & 1 \end{pmatrix}\begin{pmatrix} \alpha - p\beta\gamma\delta^{-1} & 0 \\ 0 & \delta \end{pmatrix}\begin{pmatrix} 1 & 0 \\ p\gamma\delta^{-1} & 1 \end{pmatrix}.$$ □

Proposition 10.6:

(1) Let

$$M := \langle I^+t^{\mathbb{N}}.(Ev \oplus Ev')\rangle_E \subset \pi$$

be the subspace of π generated by the orbit of $I^+t^{\mathbb{N}}$ on the vectors v and v'. Then M is stable by and irreducible for the action of the monoid

$$I^0I^+t^{\mathbb{N}} = \begin{pmatrix} \mathbb{Z}_p - \{0\} & \mathbb{Z}_p \\ 0 & \mathbb{Z}_p^\times \end{pmatrix}.$$

(2) We have $V \subset M$.

Given this proposition, we may prove irreducibility of the representation $\pi = \mathrm{ind}_{KZ}^G(V)/(\mathrm{T})$.

Proof: (Proof of Theorem 10.1 (irreducibility), assuming Proposition 10.6) By Proposition 10.6, we have

$$M = \langle I^0 I^+ t^{\mathbb{N}}.V\rangle_E = \left\langle \begin{pmatrix} \mathbb{Z}_p - \{0\} & \mathbb{Z}_p \\ 0 & 1 \end{pmatrix}.V \right\rangle_E,$$

since I^0 normalizes I^+. Moreover, since $t.M \subset M$, we have an increasing chain of subspaces

$$M \subset t^{-1}.M \subset t^{-2}.M \subset \dots.$$

Using the Iwasawa decomposition (cf. proof of Proposition 4.4) and the fact that the center Z acts by scalars on π, we obtain

$$\begin{aligned} \bigcup_{n\geq 0} t^{-n}.M &= \left\langle \begin{pmatrix} \mathbb{Q}_p - \{0\} & \mathbb{Q}_p \\ 0 & 1 \end{pmatrix}.V \right\rangle_E \\ &\overset{Z \text{ acts by scalars}}{=} \langle B.V\rangle_E \\ &\overset{V \text{ is a weight}}{=} \langle BK.V\rangle_E \\ &\overset{\text{Iwasawa}}{=} \langle G.V\rangle_E \\ &= \pi. \end{aligned}$$

Now, let $\sigma \subset \pi|_B$ be a nonzero B-subrepresentation. Then, by the above computation, there exists an integer $n \geq 0$ and $m \in M - \{0\}$ such that

$$t^{-n}.m \in \sigma.$$

As σ is B-stable, we have

$$m \in t^n.\sigma = \sigma,$$

which implies $M \subset \sigma$ by Proposition 10.6. Hence, we obtain

$$\pi = \bigcup_{n\geq 0} t^{-n}.M \subset \sigma,$$

which shows that π is irreducible, *even as a B-representation.* □

Thus, we have shown that it suffices to prove Proposition 10.6 to show the irreducibility of π. We let

$$N := \langle I^+ t^{2\mathbb{N}}.v\rangle_E \subset M, \qquad N' := \langle I^+ t^{2\mathbb{N}}.v'\rangle_E \subset M,$$

and set $r := a - b$. The proof of Proposition 10.6 will follow from the following lemmas.

Lemma 10.7:

(1) We have $t\mathcal{X}t^{-1} = \mathcal{X}^p$.
(2) We have

$$v = (*)\mathcal{X}^r t.v'$$
$$v' = (*)\mathcal{X}^{p-1-r} t.v,$$

where $(*)$ *is some element of* $E^\times$.

Proof: (1) Since E has characteristic p, we have

$$t\mathcal{X}t^{-1} = t\left(\begin{pmatrix}1&1\\0&1\end{pmatrix} - 1\right)t^{-1} = \begin{pmatrix}1&p\\0&1\end{pmatrix} - 1 = \left(\begin{pmatrix}1&1\\0&1\end{pmatrix} - 1\right)^p = \mathcal{X}^p.$$

(2) Note that, if $X^{r-i}Y^i$ is an element of $V = F(a,b) = E[X,Y]_{(r)}$ with $0 \le i \le r$, we have

$$\mathcal{X}.X^{r-i}Y^i = X^{r-i}(Y+X)^i - X^{r-i}Y^i = iX^{r-i+1}Y^{i-1} + Q(X,Y),$$

where $Q(X,Y)$ is a homogeneous polynomial of degree r, such that the degree of $Q(X,Y)$ as a polynomial in Y is strictly less than $i-1$. Applying this r times, we obtain

$$\mathcal{X}^r.X^{r-i}Y^i = \begin{cases} 0 & \text{if } i \neq r, \\ r!X^r & \text{if } i = r. \end{cases}$$

Hence, as $\det s = -1$, we have

$$\mathcal{X}^r t.v' = \mathcal{X}^r s\Pi.v' = \mathcal{X}^r s.v = (-1)^b r! v,$$

and by symmetry, we obtain

$$\mathcal{X}^{p-1-r} t.v = (-1)^a (p-1-r)! v'.$$

Note that the condition $0 \le r \le p-1$ implies $r!$ and $(p-1-r)!$ are nonzero in E. □

Lemma 10.8: *We have*

$$N^{I^+} = Ev, \qquad (N')^{I^+} = Ev'.$$

Remark 10.9: Note that N^{I^+} and $(N')^{I^+}$ are both nonzero, since I^+ is a pro-p group.

Proof: We prove the claim for N; the proof for N' is identical. By Lemma 10.7, we have

$$\begin{aligned} v &= (*)\mathcal{X}^r t\mathcal{X}^{p-1-r}t.v \\ &= (*)\mathcal{X}^{r+p(p-1-r)}t^2.v, \end{aligned}$$

where $(*)$ denotes an unspecified nonzero constant. Iterating this, we obtain

$$v = (*)\mathcal{X}^{e_n}t^{2n}.v,$$

where $e_n = (r+p(p-1-r))(1+p^2+\ldots+p^{2n-2})$. Since $v \in \pi^{I(1)}$, we have $\mathcal{X}.v = 0$; hence, $\mathcal{X}^{e_n}t^{2n}.v \neq 0$, but $\mathcal{X}^{e_n+1}t^{2n}.v = 0$.

Let $N_n := \langle I^+t^{2n}.v\rangle_E \subset N$. As $I^+ \cong \mathbb{Z}_p$, we have

$$I^+ = \begin{pmatrix} 1 & 1 \\ 0 & 1 \end{pmatrix}^{\mathbb{Z}_p} = (1+\mathcal{X})^{\mathbb{Z}_p}.$$

Since we may express the action of I^+ in terms of the operator $\mathcal{X}$, we have an isomorphism

$$\begin{array}{ccc} E[\mathcal{X}]/(\mathcal{X}^{e_n+1}) & \cong & N_n \\ 1 & \longmapsto & t^{2n}.v \end{array}$$

that is compatible with the action of $\mathcal{X}$. Note that $N_n^{I^+} = N_n[\mathcal{X}]$, the $\mathcal{X}$-torsion elements. The $\mathcal{X}$-torsion elements of $E[\mathcal{X}]/(\mathcal{X}^{e_n+1})$ are one-dimensional (spanned by $\mathcal{X}^{e_n}$), and correspond to the subspace $E\mathcal{X}^{e_n}t^{2n}.v = Ev$ of N_n.

Now, since $t^{2n}.v = (*)\mathcal{X}^{p^{2n}e_1}t^{2n+2}.v \in N_{n+1}$, we obtain a series of inclusions

$$N_0 \subset N_1 \subset \ldots \subset N = \bigcup_{n\geq 0} N_n.$$

This, combined with the fact that $N_n^{I^+} = Ev$ shown above, proves that $N^{I^+} = Ev$. □

Remark 10.10: The identity $I^+ = (1+\mathcal{X})^{\mathbb{Z}_p}$ relies crucially on the fact that $\mathbb{Z}_p$ is *pro-cyclic*. There is no analogous such statement for the ring of integers in a finite extension of $\mathbb{Q}_p$.

Lemma 10.11: *The subspaces M, N, and N' are I-stable.*

Proof: Recall that $M = \langle I^+t^{\mathbb{N}}.(Ev \oplus Ev')\rangle_E$. Therefore, by the Iwahori Decomposition (Lemma 10.5), we have

$$I.M = \langle I^+I^0I^-t^{\mathbb{N}}.(Ev \oplus Ev')\rangle_E.$$

By Lemma 10.5, we have $I^0 I^- t^n \subset t^n I^0 I^-$ for every $n \geq 0$, and moreover, the space $Ev \oplus Ev'$ is I-stable by definition. Hence

$$I.M \subset \langle I^+ t^{\mathbb{N}}.(Ev \oplus Ev')\rangle_E = M.$$

The same proof applies to N and N'. □

Lemma 10.12: *We have $t.v \in N'$ and $t.v' \in N$.*

Proof: By Lemma 10.7, we have $t.v = (*)\mathcal{X}^{pr} t^2.v' \in N'$. The other claim follows by symmetry. □

Lemma 10.13: *The K-subrepresentation generated by v is contained in M; that is,*

$$V = \langle K.v\rangle_E \subset M.$$

Proof: The Bruhat Decomposition for $G_p = \mathrm{GL}_2(\mathbb{F}_p)$ states

$$G_p = B_p \sqcup B_p s B_p,$$

which we inflate to a decomposition of $K = \mathrm{GL}_2(\mathbb{Z}_p)$:

$$K = I \sqcup IsI.$$

Since I acts on v by a character, we obtain

$$\begin{aligned}\langle K.v\rangle_E &= \langle I.v\rangle_E + \langle IsI.v\rangle_E \\ &= Ev + \langle Is.v\rangle_E \\ &= Ev + \langle It\Pi^{-1}.v\rangle_E \\ &= Ev + \langle It.v'\rangle_E.\end{aligned}$$

By Lemma 10.12, $t.v'$ is contained in $N \subset M$, and by Lemma 10.11, $It.v'$ is contained in M. □

Lemma 10.14: *The space M decomposes as*

$$M = N \oplus N'.$$

Proof: We clearly have $N + N' \subset M$; we prove the opposite inclusion. This follows easily from

$$I^+ t^{\mathbb{N}} = I^+ t^{2\mathbb{N}} \cup I^+ t^{2\mathbb{N}} t,$$

along with Lemma 10.12 and that N and N' are $I^+t^{2\mathbb{N}}$-stable.

Assume now that $N \cap N' \neq 0$. On the one hand, I^+ is a pro-p group, and therefore we obtain $(N \cap N')^{I^+} \neq 0$; on the other hand, Lemma 10.8 gives $(N \cap N')^{I^+} = N^{I^+} \cap (N')^{I^+} = Ev \cap Ev' = 0$, a contradiction. □

We are now ready to finish the proof of Proposition 10.6.

Proof: (Proof of Proposition 10.6) Part (2) of the proposition follows from Lemma 10.13. We prove part (1). It follows from Lemma 10.5 that M is $I^0I^+t^{\mathbb{N}}$-stable. Let M' be a nonzero submodule of M, stable by $I^0I^+t^{\mathbb{N}}$. By Lemmas 10.8 and 10.14, we have

$$M^{I^+} = Ev \oplus Ev'.$$

Note that $(M')^{I^+} \neq 0$. We consider two cases.

- If $0 < r < p-1$, then I^0 acts on v and v' by distinct characters. This implies that either v or v' is contained in $(M')^{I^+} \subset M'$, and the relation $v = (*)\mathcal{X}^r t.v'$ (resp. $v' = (*)\mathcal{X}^{p-1-r}t.v$) implies that both vectors must be contained in M'. Hence $M' = M$.
- If $r = 0$ or $r = p-1$, then $\lambda v + \mu v' \in (M')^{I^+} \subset M'$ for some $\lambda, \mu \in E$, not both zero. We assume $\lambda\mu \neq 0$, else we may proceed as above to conclude $M' = M$. Lemma 10.7 shows that applying $\mathcal{X}^{p-1}t$ to $\lambda v + \mu v'$ gives an element of M', equal to either $(*)v$ or $(*)v'$. Hence, we proceed as above and conclude $M' = M$. □

All that remains is to show π is admissible, which will occupy the remainder of these notes. To do this, we'll use the following lemmas.

Lemma 10.15: *Any quotient of an admissible I^+-representation is admissible.*

Proof: (Sketch of proof) This comes down to the fact that $E[\![I^+]\!] \cong E[\![\mathcal{X}]\!]$ is noetherian, where $E[\![I^+]\!]$ denotes the completed group ring of I^+. Suppose that M is an admissible I^+-representation. As M is smooth, M is naturally an $E[\![\mathcal{X}]\!]$-module, where $\mathcal{X} = \left(\begin{smallmatrix}1 & 1\\ 0 & 1\end{smallmatrix}\right) - 1$, as above. Hence $M^* := \mathrm{Hom}_E(M, E)$ inherits an $E[\![\mathcal{X}]\!]$-module structure as well. It is easy to show that

$$M^*/\mathcal{X}M^* \xrightarrow{\sim} (M[\mathcal{X}])^*.$$

Therefore M is admissible as an I^+-representation if and only if $M[\mathcal{X}] = M^{I^+}$ is finite dimensional (by Proposition 9.2), if and only if $M^*/\mathcal{X}M^*$ is finite-dimensional. As M^* is separated for the $\mathcal{X}$-adic topology (exercise), the latter condition is equivalent to M^* being finitely generated as an $E[\![\mathcal{X}]\!]$-module, by a variant of Nakayama's Lemma (see [20], Theorem 8.4). Thus, if $M \twoheadrightarrow N$ with M admissible, then $N^* \hookrightarrow M^*$ and N^* is finitely generated, as $E[\![\mathcal{X}]\!]$ is noetherian. □

Let $\sigma := N + \Pi.N'$. Then σ is G^+-stable by Proposition 4.12 of [21], where $G^+ := \{g \in G : \det g \in p^{2\mathbb{Z}} \cdot \mathbb{Z}_p^\times\}$.

Lemma 10.16: *The G^+-representation σ is admissible.*

Proof: Consider the exact sequence of I-representations

$$0 \longrightarrow \Pi.N' \longrightarrow \sigma \longrightarrow N/(N \cap \Pi.N') \longrightarrow 0. \tag{10.3}$$

We know that $(\Pi.N')^{I(1)} \cong (N')^{I(1)}$ is finite-dimensional by Lemma 10.8. On the other hand, N is I^+-admissible; hence $N/(N \cap \Pi.N')$ is I^+-admissible by Lemma 10.15, and thus also $I(1)$-admissible. It follows from (10.3) that $\sigma^{I(1)}$ is finite-dimensional, and we are done by Proposition 9.2. □

Proof: (Proof of Theorem 10.1 (admissibility)) Note that $(G : G^+) = 2$ and that $G = \langle G^+, \Pi \rangle$. It follows that $\sigma + \Pi.\sigma = \pi$, as this is a nonzero G-subrepresentation and we have already shown that π is irreducible. Similarly, $\sigma \cap \Pi.\sigma$ equals either 0 or π, hence $\pi = \sigma \oplus \Pi.\sigma$ or $\pi = \sigma$. In either case, we have $\dim_E \pi^{I(1)} < \infty$ by Lemma 10.16. (In fact, $\sigma \cap \Pi.\sigma = 0$ by Corollary 6.5 in [21].) □

We remark that this is *not* Breuil's original proof; his method relies on computing

$$(\mathrm{ind}_{KZ}^G(V)/(\mathrm{T}))^{I(1)}$$

using explicit calculations with the Bruhat-Tits tree $\mathfrak{X}$.

References

1. Abe, N., "On a classification of irreducible admissible modulo p representations of a p-adic split reductive group." *Compos. Math.*, 149, (2013), no. 12, 2139-2168.
2. Abe, N., Henniart, G., Herzig, F., and Vignéras, M.-F., "A classification of irreducible admissible mod p representations of p-adic reductive groups." Preprint (2014); Available at http://arxiv.org/abs/1412.0737

3. Barthel, L. and Livné, R., "Irreducible Modular Representations of GL_2 of a Local Field." *Duke Math J.*, 75, (1994), no. 2, 261-292.
4. Barthel, L. and Livné, R., "Modular Representations of GL_2 of a Local Field: The Ordinary, Unramified Case." *J. Number Theory*, 55, (1995), no. 1, 1-27.
5. Berger, L., "Central characters for smooth irreducible modular representations of $\mathrm{GL}_2(\mathbf{Q}_p)$." *Rend. Semin. Mat. Univ. Padova*, 128, (2012), 1-6.
6. Breuil, C., "Sur quelques représentations modulaires et p-adiques de $\mathrm{GL}_2(\mathbf{Q}_p)$ I." *Compos. Math.*, 138, (2003), 165-188.
7. Breuil, C., "Sur quelques représentations modulaires et p-adiques de $\mathrm{GL}_2(\mathbf{Q}_p)$ II." *J. Inst. Math. Jussieu*, 2, (2003), 1-36.
8. Breuil, C., "The emerging p-adic Langlands programme." *Proceedings of I.C.M. 2010*, Vol. II, (2010), 203-230.
9. Carter, R.W. and Lusztig, G., "Modular Representations of Finite Groups of Lie Type." *Proc. London Math. Soc.*, 32, (1976), 347-384.
10. Colmez, P., "Représentations de $\mathbf{GL}_2(\mathbf{Q}_p)$ et (φ, Γ)-modules." *Astérisque* 330, (2010), 281-509.
11. Emerton, M., "On a class of coherent rings, with applications to the smooth representation theory of $\mathrm{GL}_2(\mathbb{Q}_p)$ in characteristic p." Preprint (2008); Available at http://www.math.uchicago.edu/~emerton/pdffiles/frob.pdf
12. Emerton, M., "Local-global compatibility in the p-adic Langlands programme for $\mathrm{GL}_{2/\mathbb{Q}}$." Preprint (2010); Available at http://www.math.uchicago.edu/~emerton/pdffiles/lg.pdf
13. Harris, M. and Taylor, R., "The geometry and cohomology of some simple Shimura varieties." *Annals of Mathematics Studies, vol. 151.* Princeton University Press, Princeton, NJ, (2001).
14. Henniart, G., "Une preuve simple des conjectures de Langlands pour $\mathrm{GL}(n)$ sur un corps p-adique." *Invent. Math.*, 113, (2000), no. 2, 439-455.
15. Henniart, G., and Vignéras, M.-F., "A Satake isomorphism for representations modulo p of reductive groups over local fields." To appear in *J. Reine Angew. Math.*, (2013).
16. Herzig, F., "A Satake isomorphism in characteristic p." *Compos. Math.*, 147, (2011), no. 1, 263-283.
17. Herzig, F., "The classification of irreducible admissible mod p representations of a p-adic GL_n." *Invent. Math.*, 186, (2011), no. 2, 373-434.
18. Herzig, F., "The mod p representation theory of p-adic groups." Notes for a graduate course at The Fields Institute, typed by C. Johansson. Notes available at http://www.math.toronto.edu/~herzig/modpreptheory.pdf
19. Kisin, M., "Deformations of $G_{\mathbb{Q}_p}$ and $\mathrm{GL}_2(\mathbb{Q}_p)$ representations." *Astérisque*, 330, (2010), 511-528.
20. Matsumura, H., "Commutative Ring Theory." *Cambridge Studies in Advanced Mathematics 8*, Cambridge University Press, Cambridge, United Kingdom, (1989).
21. Paškūnas, V., "Extensions for supersingular representations of $\mathrm{GL}_2(\mathbb{Q}_p)$." *Astérisque*, 331, (2010), 317-353.
22. Paškūnas, V., "The image of Colmez's Montreal functor." *Publ. Math. Inst. Hautes Études Sci.*, 118 (2013), 1-191.

23. Scholze, P., "The Local Langlands Correspondence for GL_n over p-adic fields." *Invent. Math.*, 192 (2013), no. 3, 663-715.
24. Serre, J.-P., "Arbres, amalgames, SL_2." *Astérisque*, 46, (1977).
25. Tits, J., "Reductive Groups over Local Fields." *Automorphic Forms, Representations and L-function, Proc. Symp. Pure Math. XXXIII,* (1979), 29-69.

REPRESENTATION THEORY AND COHOMOLOGY OF KHOVANOV–LAUDA–ROUQUIER ALGEBRAS

Alexander S. Kleshchev
Department of Mathematics
University of Oregon
Eugene, OR 97403, USA
klesh@uoregon.edu

This expository chapter is based on the lectures given at the program 'Modular Representation Theory of Finite and p-Adic Groups' at the National University of Singapore. We are concerned with recent results on representation theory and cohomology of KLR algebras, with emphasis on standard module theory.

Contents

1. Set Up and Motivation

This expository chapter is based on the lectures given at the program 'Modular Representation Theory of Finite and p-Adic Groups' at the National University of Singapore. We are concerned with recent results on representation theory and cohomology of KLR algebra, with emphasis on standard module theory, as developed in [**21**], [**13**], [**27**], [17], [5]. Some proofs are given, but often we just review or illustrate the results. Other topics in the theory of KLR algebras are nicely reviewed in [**1**].

1.1. *KLR algebras*

In this chapter we will be mainly concerned with KLR algebras of finite Lie type. So let $\mathtt{C} = (\mathtt{c}_{ij})_{i,j\in I}$ be a *Cartan matrix* of *finite type*. As in [**10**, §1.1], let $(\mathfrak{h}, \Pi, \Pi^\vee)$ be a realization of the Cartan matrix $\mathtt{C}$, so we have simple roots $\{\alpha_i \mid i \in I\}$, simple coroots $\{\alpha_i^\vee \mid i \in I\}$, and a bilinear form $(\cdot,\cdot)$ on $\mathfrak{h}^*$ such that $\mathtt{c}_{ij} = 2(\alpha_i, \alpha_j)/(\alpha_i, \alpha_i)$ for all $i, j \in I$. We normalize $(\cdot,\cdot)$ so that $(\beta,\beta) = 2$ if β is a short root. The fundamental dominant weights $\{\Lambda_i \mid i \in I\}$ have the property that $\langle \Lambda_i, \alpha_j^\vee\rangle = \delta_{i,j}$, where $\langle\cdot,\cdot\rangle$ is the natural pairing between $\mathfrak{h}^*$ and $\mathfrak{h}$. We have the set of dominant weights $P_+ = \sum_{i\in I} \mathbb{Z}_{\geq 0} \cdot \Lambda_i$ and the positive part of the root lattice $Q_+ := \bigoplus_{i\in I} \mathbb{Z}_{\geq 0}\alpha_i$. For $\alpha \in Q_+$, we write $\mathrm{ht}(\alpha)$ for the sum of its coefficients when expanded in terms of the α_i's.

Denote $\mathscr{A} := \mathbb{Z}[q, q^{-1}]$ for an indeterminate q. For $n \in \mathbb{Z}_{\geq 0}$, define

$$[n]_q := \frac{q^n - q^{-n}}{q - q^{-1}}, \ [n]_q^! := \prod_{m=1}^{n} [m]_q.$$

Given in addition $\alpha \in Q_+$ and a simple root α_i, we let $d_\alpha := (\alpha, \alpha)/2$ and set

$$q_\alpha := q^{d_\alpha}, \ [n]_\alpha := [n]_{q_\alpha}, \ [n]_\alpha^! := [n]_{q_\alpha}^!, q_i := q_{\alpha_i} \ [n]_i := [n]_{\alpha_i}, \ [n]_i^! := [n]_{\alpha_i}^!.$$

Sequences of elements of I will be called *words*. The set of all words is denoted $\langle I \rangle$. If $\boldsymbol{i} = i_1 \dots i_d$ is a word, we denote $|\boldsymbol{i}| := \alpha_{i_1} + \dots + \alpha_{i_d} \in Q_+$. We refer to $|\boldsymbol{i}|$ as the *content* of the word $\boldsymbol{i}$. For any $\alpha \in Q_+$ we denote

$$\langle I \rangle_\alpha := \{\boldsymbol{i} \in \langle I \rangle \mid |\boldsymbol{i}| = \alpha\}.$$

If α is of height d, then the symmetric group $\mathfrak{S}_d$ with simple permutations $s_1, \dots, s_{d-1}$ acts transitively on $\langle I \rangle_\alpha$ from the left by place permutations.

Let F be an arbitrary field. Define the polynomials $\{Q_{ij}(u,v) \in F[u,v] \mid i,j \in I\}$ in the variables u, v as follows. Choose signs ε_{ij} for all $i, j \in I$ with $\mathtt{c}_{ij} < 0$ so that $\varepsilon_{ij}\varepsilon_{ji} = -1$. Then set:

$$Q_{ij}(u,v) := \begin{cases} 0 & \text{if } i = j; \\ 1 & \text{if } \mathtt{c}_{ij} = 0; \\ \varepsilon_{ij}(u^{-\mathtt{c}_{ij}} - v^{-\mathtt{c}_{ji}}) & \text{if } \mathtt{c}_{ij} < 0. \end{cases} \tag{1.1}$$

Fix $\alpha \in Q_+$ of height d. The *KLR-algebra* R_α is an associative graded unital F-algebra, given by the generators

$$\{1_{\boldsymbol{i}} \mid \boldsymbol{i} \in \langle I \rangle_\alpha\} \cup \{y_1, \dots, y_d\} \cup \{\psi_1, \dots, \psi_{d-1}\} \tag{1.2}$$

and the following relations for all $\boldsymbol{i}, \boldsymbol{j} \in \langle I \rangle_\alpha$ and all admissible r, t:

$$1_{\boldsymbol{i}} 1_{\boldsymbol{j}} = \delta_{\boldsymbol{i},\boldsymbol{j}} 1_{\boldsymbol{i}}, \quad \textstyle\sum_{\boldsymbol{i} \in \langle I \rangle_\alpha} 1_{\boldsymbol{i}} = 1; \tag{1.3}$$

$$y_r 1_{\boldsymbol{i}} = 1_{\boldsymbol{i}} y_r; \quad y_r y_t = y_t y_r; \tag{1.4}$$

$$\psi_r 1_{\boldsymbol{i}} = 1_{s_r \boldsymbol{i}} \psi_r; \tag{1.5}$$

$$(y_t \psi_r - \psi_r y_{s_r(t)}) 1_{\boldsymbol{i}} = \delta_{i_r, i_{r+1}} (\delta_{t,r+1} - \delta_{t,r}) 1_{\boldsymbol{i}}; \tag{1.6}$$

$$\psi_r^2 1_{\boldsymbol{i}} = Q_{i_r, i_{r+1}}(y_r, y_{r+1}) 1_{\boldsymbol{i}} \tag{1.7}$$

$$\psi_r \psi_t = \psi_t \psi_r \qquad (|r - t| > 1); \tag{1.8}$$

$$\begin{aligned}&(\psi_{r+1}\psi_r\psi_{r+1} - \psi_r\psi_{r+1}\psi_r)1_{\boldsymbol{i}}\\ &=\delta_{i_r,i_{r+2}}\frac{Q_{i_r,i_{r+1}}(y_{r+2},y_{r+1}) - Q_{i_r,i_{r+1}}(y_r,y_{r+1})}{y_{r+2}-y_r}1_{\boldsymbol{i}}.\end{aligned} \tag{1.9}$$

The *grading* on R_α is defined by setting:

$$\deg(1_{\boldsymbol{i}}) = 0, \quad \deg(y_r 1_{\boldsymbol{i}}) = (\alpha_{i_r}, \alpha_{i_r}), \quad \deg(\psi_r 1_{\boldsymbol{i}}) = -(\alpha_{i_r}, \alpha_{i_{r+1}}).$$

These algebras were defined in [**14, 15, 31**]. It is pointed out in [**15**] and [**31**, §3.2.4] that up to isomorphism the graded F-algebra R_α depends only on the Cartan matrix and α.

Fix in addition a dominant weight $\Lambda \in P_+$. The corresponding *cyclotomic KLR algebra* R_α^Λ is the quotient of R_α by the following ideal:

$$J_\alpha^\Lambda := (y_1^{\langle \Lambda, \alpha_{i_1}^\vee\rangle} 1_{\boldsymbol{i}} \mid \boldsymbol{i} = i_1 \dots i_d \in \langle I\rangle_\alpha). \tag{1.10}$$

For a graded algebra R, denote by R-Mod the abelian category of all graded left R-modules, denoting (degree-preserving) homomorphisms in this category by $\hom_R$. We write $\cong$ for the isomorphism in this category, and $\simeq$ for the isomorphism in the category of usual modules. Let R-mod denote the abelian subcategory of all finite dimensional graded left R-modules and R-proj denote the additive subcategory of all finitely generated projective graded left R-modules.

We also consider the Grothendieck groups $[R\text{-mod}]$ and $[R\text{-proj}]$. We view $[R\text{-mod}]$ and $[R\text{-proj}]$ as $\mathscr{A}$-modules via $q^m[M] := [q^m M]$, where $q^m M$ denotes the module obtained by shifting the grading up by m: $(q^m M)_n = M_{n-m}$. More generally, given a formal Laurent series $f(q) = \sum_{n\in\mathbb{Z}} f_n q^n$ with coefficients $f_n \in \mathbb{Z}_{\geq 0}$, $f(q)V$ denotes $\bigoplus_{n\in\mathbb{Z}} q^n V^{\oplus f_n}$. If V is a locally finite dimensional graded vector space (i.e. the dimension of each graded component V_n is finite), its *graded dimension* is $\dim_q V := \sum_{n\in\mathbb{Z}} (\dim V_n) q^n$. Given $M, L \in R$-mod with L irreducible, we write $[M : L]_q$ for the corresponding *graded composition multiplicity*, i.e. $[M : L]_q := \sum_{n\in\mathbb{Z}} a_n q^n$, where a_n is the multiplicity of $q^n L$ in a graded composition series of M.

Given $\alpha, \beta \in Q_+$, we set $R_{\alpha,\beta} := R_\alpha \otimes R_\beta$. There is an injective non-unital algebra homomorphism $R_{\alpha,\beta} \hookrightarrow R_{\alpha+\beta}$, $1_{\boldsymbol{i}} \otimes 1_{\boldsymbol{j}} \mapsto 1_{\boldsymbol{ij}}$, where $\boldsymbol{ij}$ is the concatenation of $\boldsymbol{i}$ and $\boldsymbol{j}$. The image of the identity element of $R_{\alpha,\beta}$ under this map is $1_{\alpha,\beta} := \sum_{\boldsymbol{i}\in\langle I\rangle_\alpha, \boldsymbol{j}\in\langle I\rangle_\beta} 1_{\boldsymbol{ij}}$. We consider the induction and restriction functors:

$$\begin{aligned}\mathrm{Ind}_{\alpha,\beta} &:= R_{\alpha+\beta} 1_{\alpha,\beta} \otimes_{R_{\alpha,\beta}} ? : R_{\alpha,\beta}\text{-Mod} \to R_{\alpha+\beta}\text{-Mod},\\ \mathrm{Res}_{\alpha,\beta} &:= 1_{\alpha,\beta} R_{\alpha+\beta} \otimes_{R_{\alpha+\beta}} ? : R_{\alpha+\beta}\text{-Mod} \to R_{\alpha,\beta}\text{-Mod},\end{aligned}$$

which preserve the categories of finite dimensional and finitely generated projective modules. For $M \in R_\alpha$-Mod, $N \in R_\beta$-Mod, denote $M \circ N := \mathrm{Ind}_{\alpha,\beta} M \boxtimes N$. Then $[R\text{-proj}] := \bigoplus_{\alpha \in Q_+} R_\alpha\text{-proj}$ and $[R\text{-mod}] := \bigoplus_{\alpha \in Q_+} R_\alpha\text{-mod}$ are Q_+-graded $\mathscr{A}$-algebras, with multiplication coming from the induction product $\circ$.

Example 1.11: For $m \geq 1$ and $i \in I$, the KLR algebra $R_{m\alpha_i}$ is the *nil-Hecke algebra* NH_m, which is given by generators $y_1, \dots, y_m$ and $\psi_1, \dots, \psi_{m-1}$ and relations: $y_i y_j = y_j y_i$, $\psi_i y_j = y_j \psi_i$ for $j \neq i, i+1$, $\psi_i y_{i+1} = y_i \psi_i + 1$, $y_{i+1}\psi_i = \psi_i y_i + 1$, $\psi_i^2 = 0$, together with the usual type $\mathtt{A_m}$ braid relations for $\psi_1, \dots, \psi_{m-1}$. It is well known that the nil-Hecke algebra is a matrix algebra over its center; see e.g. [**32**, §2] or [**19**, §4] for recent expositions. Moreover, writing w_0 for the longest element of $\mathfrak{S}_m$, the degree zero element

$$e_m := y_2 y_3^2 \cdots y_m^{m-1} \psi_{w_0} \tag{1.12}$$

is a primitive idempotent, hence $P(\alpha_i^m) := q_i^{m(m-1)/2} R_{m\alpha_i} e_m$ is an indecomposable projective $R_{m\alpha_i}$-module. The degree shift has been chosen so that irreducible head $L(\alpha_i^m)$ of $P(\alpha_i^m)$ has graded dimension $[m]_i^!$. Thus $R_{m\alpha_i} \cong [m]_i^! \, P(\alpha_i^m)$ as a left module.

1.2. *Some motivation*

The first reason why representation theory of KLR algebras is interesting is that it can be used to categorify quantum groups. One way to make this statement more precise is as follows. Let $\mathbf{f}$ be the quantized enveloping algebra over the field $\mathbb{Q}(q)$ associated to $\mathtt{C}$ with standard generators $\{\theta_i \mid i \in I\}$, cf. [**25**]. It is naturally Q_+-graded: $\mathbf{f} = \bigoplus_{\alpha \in Q_+} \mathbf{f}_\alpha$. Khovanov and Lauda showed that there is a unique Q_+-graded algebra isomorphism

$$\gamma : \mathbf{f} \xrightarrow{\sim} \mathbb{Q}(q) \otimes_{\mathscr{A}} [R\text{-proj}], \ \theta_i \mapsto [R_{\alpha_i}], \tag{1.13}$$

where R_{α_i} is the left regular module over the algebra R_{α_i}.

If $\mathtt{C}$ is symmetric and F has characteristic zero, Rouquier [**32**] and Varagnolo and Vasserot [**34**] have shown further that γ maps the canonical basis of $\mathbf{f}$ to the basis for $[R\text{-proj}]$ arising from the isomorphism classes of graded self-dual indecomposable projective modules. Taking a dual map to γ yields another algebra isomorphism

$$\gamma^* : \mathbb{Q}(q) \otimes_{\mathscr{A}} [R\text{-mod}] \xrightarrow{\sim} \mathbf{f}^*. \tag{1.14}$$

If $\mathtt{C}$ is symmetric and F has characteristic zero, this sends the basis for $[R\text{-mod}]$ arising from isomorphism classes of graded self-dual irreducible R_α-modules to the dual canonical basis for $\mathbf{f}$. For some further details concerning Khovanov–Lauda–Rouquier categorification see §2.4.

Another motivation for studying representation theory of KLR algebras is the following fact first proved in [**3**]: cyclotomic KLR algebras of finite and affine types $\mathtt{A}$ are explicitly isomorphic to blocks of cyclotomic Hecke algebras. The main reason this is interesting is that now we can transport the grading from KLR algebras to cyclotomic Hecke algebras, and the resulting grading on Hecke algebras turns out to be very important, see for example [**4, 6, 9**].

As yet another illustration, we now construct explicitly the irreducible modules for all semisimple cyclotomic Hecke algebras (both degenerate and non-degenerate). This is of course just a version of Young's orthogonal form, but the reader might appreciate how much simpler the construction via KLR algebras is.

We give the necessary definitions. Until the end of this subsection we assume that the Cartan matrix $\mathtt{C}$ is either of type $\mathtt{A}_\infty$ (this is equivalent to working with sufficiently large finite type $\mathtt{A}$) or of affine type $\mathtt{A}^{(1)}_{\mathtt{e}-1}$ (above we only defined the KLR algebras for finite Lie types, but the definition for $\mathtt{A}^{(1)}_{\mathtt{e}-1}$ is really the same). When $\mathtt{C} = \mathtt{A}_\infty$, we set $e = 0$ so that in both finite and affine types $\mathtt{A}$ we can identify the set I with $\mathbb{Z}/e\mathbb{Z}$.

Fix an ordered tuple $\kappa = (k_1, \dots, k_l) \in I^l$ such that $\Lambda = \Lambda_{k_1} + \dots + \Lambda_{k_l}$. An *l-multipartition* of d is an ordered l-tuple of partitions $\underline{\mu} = (\mu^{(1)}, \dots, \mu^{(l)})$ such that $\sum_{m=1}^{l} |\mu^{(m)}| = d$. We refer to $\mu^{(m)}$ as the mth component of $\underline{\mu}$. Let $\mathscr{P}^\kappa_d$ be the set of all l-multipartitions of d. Of course, $\mathscr{P}^\kappa_d$ only depends on l, and not on κ, but as soon as we consider contents of nodes of multipartitions, the dependence on κ becomes essential. The *Young diagram* of the multipartition $\underline{\mu} = (\mu^{(1)}, \dots, \mu^{(l)}) \in \mathscr{P}^\kappa$ is

$$\{(a, b, m) \in \mathbb{Z}_{>0} \times \mathbb{Z}_{>0} \times \{1, \dots, l\} \mid 1 \le b \le \mu^{(m)}_a\}.$$

The elements of this set are the *nodes* or *boxes of* $\underline{\mu}$. More generally, a *node* is any element of $\mathbb{Z}_{>0} \times \mathbb{Z}_{>0} \times \{1, \dots, l\}$. Usually, we identify the multipartition $\underline{\mu}$ with its Young diagram and visualize it as a column vector of Young diagrams. For example, $((3,1), \emptyset, (4,2))$ is the Young diagram

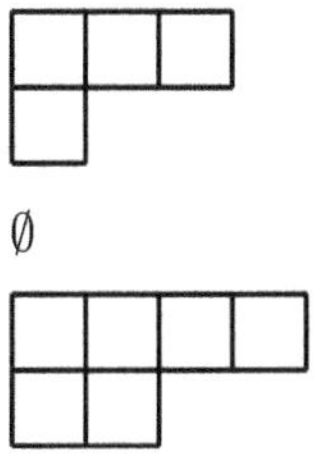

To each node $A = (a, b, m)$ we associate its *content*, which is an element of $I = \mathbb{Z}/e\mathbb{Z}$ defined as follows $\operatorname{cont} A := \operatorname{cont}^\kappa A = k_m + (b - a) \pmod{e}$. Define the *weight of* $\underline{\mu}$ to be $\operatorname{wt}(\underline{\mu}) := \sum_{A \in \underline{\mu}} \alpha_{\operatorname{cont} A} \in Q_+$. For $\alpha \in Q_+$, denote $\mathscr{P}^\kappa_\alpha := \{\underline{\mu} \in \mathscr{P}^\kappa \mid \operatorname{wt}(\underline{\mu}) = \alpha\}$. We call a partition *separating* if for any two nodes $A = (a_1, a_2, m)$, $B = (b_1, b_2, n)$ of $\underline{\mu}$, we have that $\operatorname{cont}^\kappa A = \operatorname{cont}^\kappa B$ implies that A and B are on the same diagonal of the same component, i.e. $m = n$ and $a_2 - a_1 = b_2 - b_1$.

Let $\underline{\mu} = (\mu^{(1)}, \dots, \mu^{(l)}) \in \mathscr{P}^\kappa_d$. A *$\underline{\mu}$-tableau* $\mathtt{T} = (\mathtt{T}^{(1)}, \dots, \mathtt{T}^{(l)})$ is obtained by inserting the integers $1, \dots, d$ into the boxes of $\underline{\mu}$, allowing no repeats. The group $\mathfrak{S}_d$ acts on the set of $\underline{\mu}$-tableaux from the left by acting on the entries of the tableaux. Let $\mathtt{T}^{\underline{\mu}}$ be the $\underline{\mu}$-tableau in which the numbers $1, 2, \dots, d$ appear in order from left to right along the successive rows, working from top row to bottom row. Set

$$\boldsymbol{i}^{\mathtt{T}} = \boldsymbol{i}^{\kappa,\mathtt{T}} = i_1^{\mathtt{T}} \dots i_d^{\mathtt{T}} \in I^d, \tag{1.15}$$

where $i_r^{\mathtt{T}}$ is the content of the node occupied by r in $\mathtt{T}$ for all $1 \le r \le d$. A $\underline{\mu}$-tableau $\mathtt{T}$ is called *standard* if its entries increase from left to right along the rows and from top to bottom along the columns within each component of $\mathtt{T}$. Let $\operatorname{St}(\underline{\mu})$ be the set of standard $\underline{\mu}$-tableaux.

Let $\alpha \in Q_+$ be of height d and fix a separating multipartition $\underline{\mu} \in \mathscr{P}^\kappa_\alpha$. Consider a formal vector space $S(\underline{\mu}) := \bigoplus_{\mathtt{T} \in \operatorname{St}(\underline{\mu})} F \cdot v_{\mathtt{T}}$ on basis $\{v_{\mathtt{T}} \mid \mathtt{T} \in \operatorname{St}(\underline{\mu})\}$ labeled by the standard $\underline{\mu}$-tableaux and concentrated in degree zero. Define the following action of the generators of R^Λ_α on $S(\underline{\mu})$:

$$1_{\boldsymbol{i}} v_{\mathtt{T}} = \delta_{\boldsymbol{i},\boldsymbol{i}^{\mathtt{T}}} v_{\mathtt{T}}, \quad y_s v_{\mathtt{T}} = 0, \quad \psi_r v_{\mathtt{T}} = \begin{cases} v_{s_r \mathtt{T}} & \text{if } s_r\mathtt{T} \text{ is standard,} \\ 0 & \text{otherwise.} \end{cases} \tag{1.16}$$

Theorem 1.17: *Suppose that $\underline{\mu}$ is separating. The formulas (1.16) define a (graded) action of R^Λ_α on $S(\underline{\mu})$. Moreover, $S(\underline{\mu})$ is an irreducible R_α-module, and $S(\underline{\mu}) \not\cong S(\underline{\nu})$ whenever $\underline{\mu} \neq \underline{\nu}$.*

Proof: To prove the first statement we need to observe that the defining relations of R_α hold for the linear operators defined by (1.16). The relations (1.3)–(1.5) are clear. To see that (1.6) holds it suffices to observe that in a standard tableau, r and $r+1$ can never occupy boxes on the same diagonal of the same component. As $\underline{\mu}$ is separating, it follows that $\boldsymbol{i}_r^{\mathtt{T}} \neq \boldsymbol{i}_{r+1}^{\mathtt{T}}$ for all $\mathtt{T} \in \mathrm{St}(\underline{\mu})$, which implies (1.6). To see (1.7), if r and $r+1$ occupy adjacent nodes in $\mathtt{T}$, then $s_r\mathtt{T}$ is not standard, and in this case we get $\psi_r^2 v_{\mathtt{T}} = 0 = Q_{i_r^{\mathtt{T}}, i_{r+1}^{\mathtt{T}}}(y_r, y_{r+1})v_{\mathtt{T}}$ as required. On the other hand, if r and $r+1$ do not occupy adjacent nodes in $\mathtt{T}$, then $s_r\mathtt{T}$ is standard, $\mathtt{c}_{i_r^{\mathtt{T}}, i_{r+1}^{\mathtt{T}}} = 0$, and $\psi_r^2 v_{\mathtt{T}} = v_{\mathtt{T}} = Q_{i_r^{\mathtt{T}}, i_{r+1}^{\mathtt{T}}}(y_r, y_{r+1})v_{\mathtt{T}}$, again as required. The relation (1.8) holds trivially. Finally, to check the relation (1.9), it is enough to notice that we never have $i_r^{\mathtt{T}} = i_{r+2}^{\mathtt{T}}$ for a standard $\underline{\mu}$-tableau $\mathtt{T}$ under the assumption that $\underline{\mu}$ is separating.

To see that $S(\underline{\mu})$ is irreducible, note first that $\mathtt{S} \neq \mathtt{T}$ for standard $\underline{\mu}$-tableaux $\mathtt{S}$ and $\mathtt{T}$ implies that $\boldsymbol{i}^{\mathtt{T}} \neq \boldsymbol{i}^{\mathtt{S}}$, so acting with the idempotents $1_{\boldsymbol{i}}$ yields projections to each 1-dimensional subspace $F \cdot v_{\mathtt{T}}$ spanned by the basis elements $v_{\mathtt{T}}$. So to prove the irreducibility of $S(\underline{\mu})$ it suffices to show that for any standard $\underline{\mu}$-tableaux $\mathtt{T}$ and $\mathtt{S}$, there exists a series of admissible transpositions which takes $\mathtt{T}$ to $\mathtt{S}$, which means that there exist $1 \leq k_1, \dots, k_l < d$ such that $s_{k_l}s_{k_{l-1}} \cdots s_{k_1}\mathtt{T} = \mathtt{S}$ and $s_{k_m}s_{k_{m-1}} \cdots s_{k_1}\mathtt{T}$ is standard for all $m = 1, \dots, l$. The existence of such a sequence follows from the following:

Claim. For any standard $\underline{\mu}$-tableau $\mathtt{T}$ there exists a series of admissible transpositions which takes $\mathtt{T}$ to $\mathtt{T}^{\underline{\mu}}$.

To prove the Claim, let A be the last box of the last row of $\underline{\mu}$. In $\mathtt{T}^{\underline{\mu}}$, the box A is occupied by d. In $\mathtt{T}$, the box A is occupied by some number $k \leq d$. Note that in $\mathtt{T}$, the numbers $k+1$ and k do not lie on adjacent diagonals. So we can apply an admissible transposition to swap k and $k+1$, then to swap $k+1$ and $k+2$, etc. As a result, we get a new standard $\underline{\mu}$-tableau in which A is occupied by d. Next, remove A together with d, and apply induction. □

The pair $(\Lambda, d) \in P_+ \times \mathbb{Z}_{\geq 0}$ is *separating* if all miltipartitions $\underline{\mu} \in \mathscr{P}_d^\kappa$ are separating. This notion is well-defined, since it does not depend on the choice of $\kappa = (k_1, \dots, k_l)$ such that $\Lambda = \Lambda_{k_1} + \cdots + \Lambda_{k_l}$. If (Λ, d) is separating, then all multipartitions $\underline{\mu} \in \mathscr{P}_d^\kappa$ have different contents, and the algebra $\bigoplus_{\alpha \in Q_+, \mathrm{ht}(\alpha) = d} R_\alpha^\Lambda$ is a semisimple algebra, with each R_α^Λ being zero or simple. We have mentioned above that by the main result of [**3**], this algebra is isomorphic to a cyclotomic Hecke algebra $\bigoplus_{\alpha\in Q_+,\mathrm{ht}(\alpha)=d} H_\alpha^\Lambda$. This cyclotomic Hecke algebra is semisimple if and only if (Λ, d) is separating. Thus in all cases where a cyclotomic Hecke algebra is semisimple, Theorem 1.17 yields an easy construction of all its irreducible representations via the isomorphism of [**3**].

2. Basic Representation Theory of KLR Algebras

2.1. *Semiperfect and Laurentian algebras*

We begin with some generalities on graded algebras. All gradings will be $\mathbb{Z}$-gradings. Let H be a graded algebra over a ground field F. All modules, ideals, etc. are assumed to be graded, unless otherwise stated. In particular, $\mathrm{rad}\, V$ (resp. $\mathrm{soc}\, V$) is the intersection of all maximal (graded) submodules (resp. the sum of all irreducible (graded) submodules) of V. All idempotents are assumed to be degree zero. We denote by $N(H)$ the (graded) Jacobson radical of H.

For modules U and V, we write $\mathrm{hom}_H(U,V)$ for homogeneous H-module homomorphisms, and set $\mathrm{HOM}_H(U,V) := \bigoplus_{n\in\mathbb{Z}} \mathrm{HOM}_H(U,V)_n$, where

$$\mathrm{HOM}_H(U,V)_n := \mathrm{hom}_H(q^n U, V) = \mathrm{hom}_H(U, q^{-n} V).$$

We define $\mathrm{ext}_H^d(U,V)$ and $\mathrm{EXT}_H^d(U,V)$ similarly. If U is finitely generated, then $\mathrm{HOM}_H(U,V) = \mathrm{Hom}_H(U,V)$, where $\mathrm{Hom}_H(U,V)$ denoted the homomorphisms in the ungraded category. We have a similar fact for Ext^d provided U has a resolution by finitely generated projective modules, in particular if U is finitely generated and H is Noetherian.

For an H-module V denote by $Z(V)$ the largest submodule of V with the trivial zero degree component, i.e. $Z(V)_0 = 0$. Define $\overline{V} := V/Z(V)$.

Lemma 2.1: *[28] Let V be an irreducible graded H-module, and W be an irreducible H_0-module.*

(i) If $V_n \neq 0$ for some $n \in \mathbb{Z}$, then V_n is irreducible as an H_0-module.
(ii) The graded H-module $X := \overline{H \otimes_{H_0} W}$ is irreducible, and $X_0 \cong W$ as H_0-modules.
(iii) If $V_0 \neq 0$, then we have $V \cong \overline{H \otimes_{H_0} V_0}$.

Proof: (i) is clear.

(ii) First of all note that

$$X_0 = (\overline{H \otimes_{H_0} W})_0 \cong (H \otimes_{H_0} W)_0 \cong H_0 \otimes_{H_0} W \cong W \neq 0.$$

Note that $H \otimes_{H_0} W$ is generated as an H-module by its degree zero part $1 \otimes W$, hence X is also generated by its degree zero part X_0. Moreover, X_0 is irreducible as an H_0-module, so X is generated by any non-zero vector in X_0. Now to prove the irreducibility of X it suffices to take any homogeneous vector v, say of degree n, and prove that $H_{-n}v \neq 0$. Well, otherwise Hv is a graded submodule of X which avoids X_0, a contradiction.

(iii) By (i) we have that V_0 is an irreducible H-module, and we have that the H-module $\overline{H \otimes_{H_0} W}$ is isomorphic to V, because it is irreducible by (ii) and surjects onto V. □

Now we assume that H is (graded) *semiperfect*, i.e. every finitely generated (graded) H-module has a (graded) projective cover. By [**7**], this is equivalent to H_0 being semiperfect, and is also equivalent to the fact that the following two properties hold: (1) $H/N(H)$ is (graded) semisimple Artinian; (2) idempotents lift from $H/N(H)$ to H. We fix a complete irredundant set of irreducible H-modules up to isomorphism and degree shift:

$$\{L(\pi) \mid \pi \in \Pi\},$$

and for each $\pi \in \Pi$, we fix a projective cover $P(\pi)$ of $L(\pi)$.

By the semiperfectness of H, we have $H/N(H)$ is (graded) left Artinian, so the set Π is finite. Moreover, if $\mathrm{End}_H(L(\pi))$ is finite dimensional over F then by the graded version of the Wedderburn-Artin Theorem [**28**, 2.10.10] the irreducible module $L(\pi)$ is finite dimensional. Finally, if $\mathrm{End}_H(L(\pi)) = F$ for all $\pi \in \Pi$, i.e. if H is *Schurian*, then $H/N(H)$ is a finite direct product of (graded) matrix algebras over F and we have

$$_HH = \bigoplus_{\pi \in \Pi} (\dim_q L(\pi)) P(\pi).$$

A graded algebra H is called *Laurentian* if each of its graded components H_n is finite dimensional and $H_n = 0$ for $n \ll 0$. In this case $\dim_q H$ as well as $\dim_q V$ for any finitely generated H-module are Laurent series.

Lemma 2.2: *Let H be a Laurentian algebra. Then:*

(i) H has only finitely many irreducible modules up to isomorphism and degree shift;
(ii) all irreducible H-modules are finite dimensional;
(iii) H is semiperfect.

Proof: (i) Since H_0 is finite dimensional, it has only finitely many irreducible modules. It now follows from Lemma 2.1 that up to a degree shift, H has only finitely many irreducible graded modules.

(ii) Let V be an irreducible H-module. Then each V_n is irreducible over H_0 by Lemma 2.1. So each V_n is finite dimensional. On the other hand, since V is cyclic and H is Laurentian, V has to be bounded below. Now, V also has to be bounded above, since it is irreducible and H is Laurentian.

(iii) follows from [**7**, Theorem 3.5] since H_0 is semiperfect being finite dimensional. □

2.2. *Formal characters*

Fix $\alpha \in Q_+$ with $\mathrm{ht}(\alpha) = d$. The results of the previous subsection apply to the KLR algebra R_α, since it is easily seen to be Schurian, see e.g. [**14**, Corollary 3.19], and is also Laurentian for example in view of the following Basis Theorem:

Theorem 2.3: *[**14**, Theorem 2.5], [**31**, Theorem 3.7] For each element $w \in \mathfrak{S}_d$ fix a reduced expression $w = s_{r_1} \dots s_{r_m}$ and set $\psi_w := \psi_{r_1} \dots \psi_{r_m}$. The elements*

$$\{\psi_w y_1^{m_1} \dots y_d^{m_d} 1_{\boldsymbol{i}} \mid w \in \mathfrak{S}_d,\ m_1, \dots, m_d \in \mathbb{Z}_{\geq 0},\ \boldsymbol{i} \in \langle I \rangle_\alpha\}$$

form an F-basis of R_α.

There exists a homogeneous algebra anti-involution

$$\tau : R_\alpha \longrightarrow R_\alpha, \quad 1_{\boldsymbol{i}} \mapsto 1_{\boldsymbol{i}}, \quad y_r \mapsto y_r, \quad \psi_s \mapsto \psi_s \tag{2.4}$$

for all $\boldsymbol{i} \in \langle I \rangle_\alpha$, $1 \leq r \leq d$, and $1 \leq s < d$. If $M = \bigoplus_{d \in \mathbb{Z}} M_d$ is a finite dimensional graded R_α-module, then the *graded dual* $M^{\circledast}$ is the graded R_α-module such that $(M^{\circledast})_n := \mathrm{Hom}_F(M_{-n}, F)$, for all $n \in \mathbb{Z}$, and the R_α-action is given by $(xf)(m) = f(\tau(x)m)$, for all $f \in M^{\circledast}, m \in M, x \in R_\alpha$.

For every irreducible module L, there is a unique choice of the grading shift so that we have $L^{\circledast} \cong L$ [**14**, §3.2]. When speaking of irreducible R_α-modules we often assume by fiat that the shift has been chosen in this way.

For $\boldsymbol{i} \in \langle I \rangle_\alpha$ and $M \in R_\alpha$-mod, the ***i**-word space* of M is $M_{\boldsymbol{i}} := 1_{\boldsymbol{i}} M$. We have the word space decomposition:

$$M = \bigoplus_{\boldsymbol{i} \in \langle I \rangle_\alpha} M_{\boldsymbol{i}}.$$

We say that $\boldsymbol{i}$ is a *word of* M if $M_{\boldsymbol{i}} \neq 0$. Note from the relations that $\psi_r M_{\boldsymbol{i}} \subset M_{s_r \boldsymbol{i}}$. Define the *(graded formal) character* of M as follows:

$$\mathrm{ch}_q\, M := \sum_{\boldsymbol{i} \in \langle I \rangle_\alpha} (\dim_q M_{\boldsymbol{i}}) \boldsymbol{i} \in \mathscr{A}\langle I \rangle_\alpha.$$

The character map $\mathrm{ch}_q\ : R_\alpha\text{-mod} \to \mathscr{A}\langle I \rangle_\alpha$ factors through to give an *injective* $\mathscr{A}$-linear map

$$\mathrm{ch}_q\ : [R_\alpha\text{-mod}] \to \mathscr{A}\langle I \rangle_\alpha,$$

see [**14**, Theorem 3.17].

Let $\boldsymbol{i} = i_1 \dots i_d$ and $\boldsymbol{j} = i_{d+1} \dots i_{d+f}$ be two elements of $\langle I \rangle$. Define the *quantum shuffle product*:

$$\boldsymbol{i} \circ \boldsymbol{j} := \sum q^{-e(\sigma)} i_{\sigma(1)} \dots i_{\sigma(d+f)} \in \mathscr{A}\langle I \rangle,$$

where the sum is over all $\sigma \in \mathfrak{S}_{d+f}$ such that $\sigma^{-1}(1) < \cdots < \sigma^{-1}(d)$ and $\sigma^{-1}(d+1) < \cdots < \sigma^{-1}(d+f)$, and

$$e(\sigma) := \sum_{k \leq d < m,\ \sigma^{-1}(k) > \sigma^{-1}(m)} \mathtt{c}_{i_{\sigma(k)}, i_{\sigma(m)}}.$$

This defines an $\mathscr{A}$-algebra structure on the $\mathscr{A}$-module $\mathscr{A}\langle I \rangle$, which consists of all finite formal $\mathscr{A}$-linear combinations of elements $\boldsymbol{i} \in \langle I \rangle$.

In view of [**14**, Lemma 2.20], we have

$$\mathrm{ch}_q\,(M_1 \circ \cdots \circ M_n) = \mathrm{ch}_q\,(M_1) \circ \cdots \circ \mathrm{ch}_q\,(M_n). \tag{2.5}$$

2.3. *Crystal operators and extremal words*

The theory of crystal operators has been developed in [**14**], [**22**] and [**11**] following ideas of Grojnowski [**8**], see also [**16**]. We review necessary facts for the reader's convenience.

Let $\alpha \in Q_+$ and $i \in I$. By Example 1.11, $R_{n\alpha_i}$ is a nil-Hecke algebra with unique irreducible module $L(\alpha_i^n)$ with $\dim_q L(\alpha_i^n) = [n]_i^!$. We have functors

$$e_i : R_\alpha\text{-mod} \to R_{\alpha-\alpha_i}\text{-mod},\ M \mapsto \mathrm{Res}^{R_{\alpha-\alpha_i,\alpha_i}}_{R_{\alpha-\alpha_i}} \circ \mathrm{Res}_{\alpha-\alpha_i,\alpha_i} M,$$

$$f_i : R_\alpha\text{-mod} \to R_{\alpha+\alpha_i}\text{-mod},\ M \mapsto \mathrm{Ind}_{\alpha,\alpha_i} M \boxtimes L(\alpha_i).$$

If $L \in R_\alpha$-mod is irreducible, we define

$$\tilde{f}_i L := \mathrm{head}(f_i L), \quad \tilde{e}_i L := \mathrm{soc}(e_i L).$$

A fundamental fact is that $\tilde{f}_i L$ is again irreducible and $\tilde{e}_i L$ is irreducible or zero. We refer to $\tilde{e}_i$ and $\tilde{f}_i$ as the crystal operators. These are operators on

$B \cup \{0\}$, where B is the set of isomorphism classes of irreducible R_α-modules for all $\alpha \in Q_+$. Define $\mathrm{wt} : B \to P$, $[L] \mapsto -\alpha$ if $L \in R_\alpha$-mod.

Theorem 2.6: *[22] The set B with the operators $\tilde{e}_i, \tilde{f}_i$ and the function* wt *is the crystal graph of the negative part $U_q(\mathfrak{n}_-)$ of the quantized enveloping algebra of $\mathfrak{g}$ of Lie type* C.

For any $M \in R_\alpha$-mod, we define

$$\varepsilon_i(M) := \max\{k \geq 0 \mid e_i^k(M) \neq 0\}.$$

Then $\varepsilon_i(M)$ is also the length of the longest 'i-tail' of words of M, i.e. the maximum of $k \geq 0$ such that $j_{d-k+1} = \cdots = j_d = i$ for some word $\boldsymbol{j} = j_1 \dots j_d$ of M.

Proposition 2.7: *[14], [22] Let L be an irreducible R_α-module, $i \in I$, and $\varepsilon = \varepsilon_i(L)$.*

(i) $\tilde{e}_i \tilde{f}_i L \cong L$ *and if* $\tilde{e}_i L \neq 0$ *then* $\tilde{f}_i \tilde{e}_i L \cong L$;
(ii) $\varepsilon = \max\{k \geq 0 \mid \tilde{e}_i^k(L) \neq 0\}$;
(iii) $\mathrm{Res}_{\alpha - \varepsilon\alpha_i, \varepsilon\alpha_i} L \cong \tilde{e}_i^\varepsilon L \boxtimes L(\alpha_i^\varepsilon)$.

Let $i \in I$. Consider the map $\theta_i^* : \langle I \rangle \to \langle I \rangle$ such that for $\boldsymbol{j} = j_1 \dots j_d \in \langle I \rangle$, we have

$$\theta_i^*(\boldsymbol{j}) = \begin{cases} j_1, \dots, j_{d-1} & \text{if } j_d = i; \\ 0 & \text{otherwise.} \end{cases} \tag{2.8}$$

We extend θ_i^* by linearity to a map $\theta_i^* : \mathscr{A}\langle I \rangle \to \mathscr{A}\langle I \rangle$.

Let x be an element of $\mathscr{A}\langle I \rangle$. Define

$$\varepsilon_i(x) := \max\{k \geq 0 \mid (\theta_i^*)^k(x) \neq 0\}.$$

A word $i_1^{a_1} \dots i_b^{a_b} \in \langle I \rangle$, with $a_1, \dots, a_b \in \mathbb{Z}_{\geq 0}$, is called *extremal* for x if

$$a_b = \varepsilon_{i_b}(x),\ a_{b-1} = \varepsilon_{i_{b-1}}((\theta_{i_b}^*)^{a_b}(x))\ ,\ \dots\ ,\ a_1 = \varepsilon_{i_1}\big((\theta_{i_2}^*)^{a_2} \dots (\theta_{i_b}^*)^{a_b}(x)\big).$$

A word $i_1^{a_1} \dots i_b^{a_b} \in \langle I \rangle_\alpha$ is called *extremal* for $M \in R_\alpha$-mod if it is an extremal word for $\mathrm{ch}_q M \in \mathscr{A}\langle I \rangle$, in other words, if

$$a_b = \varepsilon_{i_b}(M),\ a_{b-1} = \varepsilon_{i_{b-1}}(\tilde{e}_{i_b}^{a_b} M)\ ,\ \dots\ ,\ a_1 = \varepsilon_{i_1}(\tilde{e}_{i_2}^{a_2} \dots \tilde{e}_{i_b}^{a_b} M).$$

The following useful result, which is a version of [**2**, Corollary 2.17], describes the multiplicities of extremal word spaces in irreducible modules. We denote by 1_F the trivial module F over the trivial algebra $R_0 \cong F$.

Lemma 2.9: *Let L be an irreducible R_α-module, and $\boldsymbol{i} = i_1^{a_1} \dots i_b^{a_b} \in \langle I \rangle_\alpha$ be an extremal word for L. Then $\dim_q L_{\boldsymbol{i}} = [a_1]^!_{i_1} \cdots [a_b]^!_{i_b}$, and*

$$L \cong \tilde{f}_{i_b}^{a_b} \tilde{f}_{i_{b-1}}^{a_{b-1}} \dots \tilde{f}_{i_1}^{a_1} 1_F.$$

Moreover, $\boldsymbol{i}$ is not an extremal word for any irreducible module $L' \not\cong L$.

Proof: Follows easily from Proposition 2.7, cf. [**2**, Theorem 2.16]. $\square$

Corollary 2.10: *Let $M \in R_\alpha$-mod, and $\boldsymbol{i} = i_1^{a_1} \dots i_b^{a_b} \in \langle I \rangle_\alpha$ be an extremal word for M. Then we can write $\dim_q M_{\boldsymbol{i}} = m[a_1]^!_{i_1} \cdots [a_b]^!_{i_b}$ for some $m \in \mathscr{A}$. Moreover, if $L \cong \tilde{f}_{i_b}^{a_b} \tilde{f}_{i_{b-1}}^{a_{b-1}} \dots \tilde{f}_{i_1}^{a_1} 1_F$ and $L^\circledast \cong L$, then we have $[M : L]_q = m$.*

Proof: Apply Lemma 2.9, cf. [**2**, Corollary 2.17]. $\square$

Now we establish some useful 'multiplicity-one results'. The first one shows that in every irreducible module there is a word space with a one dimensional graded component:

Lemma 2.11: *Let L be an irreducible R_α-module, and $\boldsymbol{i} = i_1^{a_1} \dots i_b^{a_b} \in \langle I \rangle_\alpha$ be an extremal word for L. Set $N := \sum_{m=1}^{b} a_m(a_m - 1)(\alpha_{i_m}, \alpha_{i_m})/4$. Then $\dim 1_{\boldsymbol{i}} L_N = \dim 1_{\boldsymbol{i}} L_{-N} = 1$.*

Proof: This follows immediately from the equality $\dim_q 1_{\boldsymbol{i}} L = [a_1]^!_{i_1} \cdots [a_b]^!_{i_b}$, which comes from Lemma 2.9. $\square$

The following result shows that any induction product of irreducible modules always has a multiplicity one composition factor.

Proposition 2.12: *Suppose that $n \in \mathbb{Z}_{>0}$ and for $r = 1, \dots, n$, we have $\alpha^{(r)} \in Q_+$, an irreducible $R_{\alpha^{(r)}}$-module $L^{(r)}$, and $\boldsymbol{i}^{(r)} := i_1^{a_1^{(r)}} \dots i_k^{a_k^{(r)}} \in \langle I \rangle_{\alpha^{(r)}}$ is an extremal word for $L^{(r)}$. Denote $a_t := \sum_{r=1}^{n} a_t^{(r)}$ for all $1 \leq t \leq k$. Then $\boldsymbol{j} := i_1^{a_1} \dots i_k^{a_k}$ is an extremal word for $L^{(1)} \circ \cdots \circ L^{(n)}$, and the graded multiplicity of the $\circledast$-self-dual irreducible module*

$$N \simeq \tilde{f}_{i_k}^{a_k} \tilde{f}_{i_{k-1}}^{a_{k-1}} \dots \tilde{f}_{i_1}^{a_1} 1_F$$

in $L^{(1)} \circ \cdots \circ L^{(n)}$ is q^m, where

$$m := -\sum_{1 \leq t < u \leq n} \Big(\sum_{1 \leq r < s \leq k} a_r^{(u)} a_s^{(t)} (\alpha_{i_r}, \alpha_{i_s}) + \tfrac{1}{2} \sum_{r=1}^{k} a_r^{(t)} a_r^{(u)} (\alpha_{i_r}, \alpha_{i_r}) \Big).$$

In particular, the ungraded multiplicity of N in $L^{(1)} \circ \cdots \circ L^{(n)}$ is one.

Proof: By Lemma 2.9, the multiplicity of $\boldsymbol{i}^{(r)}$ in $\mathrm{ch}_q L^{(r)}$ is $[a_1^{(r)}]^!_{i_1} \cdots [a_k^{(r)}]^!_{i_k}$. By (2.5), we have

$$\mathrm{ch}_q (L^{(1)} \circ \cdots \circ L^{(n)}) = \mathrm{ch}_q (L^{(1)}) \circ \cdots \circ \mathrm{ch}_q (L^{(n)}).$$

It is easy to see that the word $\boldsymbol{j}$ is an extremal word for $L^{(1)} \circ \cdots \circ L^{(n)}$, and that $\boldsymbol{j}$ can be obtained only from the shuffle product $\boldsymbol{i}^{(1)} \circ \cdots \circ \boldsymbol{i}^{(n)}$. An elementary computation shows that $\boldsymbol{j}$ appears in $\boldsymbol{i}^{(1)} \circ \cdots \circ \boldsymbol{i}^{(n)}$ with multiplicity $q^m [a_1]^!_{i_1} \dots [a_k]^!_{i_k}$. Now apply Corollary 2.10. □

Corollary 2.13: *Let L be an irreducible R_α-module and $n \in \mathbb{Z}_{>0}$. Then there is an irreducible $R_{n\alpha}$-module N which appears in $L^{\circ n}$ with graded multiplicity $q_\alpha^{-n(n-1)/2}$. In particular, the ungraded multiplicity of N is one.*

Proof: Apply Proposition 2.12 with $L^{(1)} = \cdots = L^{(n)} = L$. □

2.4. *Khovanov–Lauda–Rouquier categorification*

We recall the Khovanov–Lauda–Rouquier categorification of the quantized enveloping algebra $\mathbf{f}$ obtained in [**14**,**15**,**31**], and briefly mentioned in §1.2. Let $\mathbf{f}_{\mathscr{A}} \subset \mathbf{f}$ be the $\mathscr{A}$-form of the Lusztig's quantum group $\mathbf{f}$ corresponding to the Cartan matrix $\mathtt{C}$. This $\mathscr{A}$-algebra is generated by the divided powers $\theta_i^{(n)} = \theta_i^n/[n]^!_i$ of the standard generators. The algebra $\mathbf{f}_{\mathscr{A}}$ has a Q_+-grading $\mathbf{f}_{\mathscr{A}} = \oplus_{\alpha \in Q_+} (\mathbf{f}_{\mathscr{A}})_\alpha$ so that each θ_i is in degree α_i.

There is a bilinear form $(\cdot,\cdot)$ on $\mathbf{f}$ defined in [**25**, §1.2.5, §33.1.2]. Let $\mathbf{f}^*_{\mathscr{A}} = \{y \in \mathbf{f} \mid (x,y) \in \mathscr{A} \text{ for all } x \in \mathbf{f}_{\mathscr{A}}\}$. Let $(\theta_i^*)^{(n)}$ be the map dual to the map $\mathbf{f}_{\mathscr{A}} \to \mathbf{f}_{\mathscr{A}},\ x \mapsto x\theta_i^{(n)}$. Finally, there is a coproduct r on $\mathbf{f}$ such that $\mathbf{f}$ is a twisted unital and counital bialgebra. Moreover, for all $x, y, z \in \mathbf{f}$ we have

$$(xy, z) = (x \otimes y, r(z)). \tag{2.14}$$

The field $\mathbb{Q}(q)$ possesses a unique automorphism called the *bar-involution* such that $\overline{q} = q^{-1}$. With respect to this involution, let $\mathtt{b} : \mathbf{f} \to \mathbf{f}$ be the anti-linear algebra automorphism such that $\mathtt{b}(\theta_i) = \theta_i$ for all $i \in I$. Also let $\mathtt{b}^* : \mathbf{f} \to \mathbf{f}$ be the adjoint anti-linear map to $\mathtt{b}$ with respect to Lusztig's form, so $(x, \mathtt{b}^*(y)) = \overline{(\mathtt{b}(x), y)}$ for all $x, y \in \mathbf{f}$. The maps $\mathtt{b}$ and $\mathtt{b}^*$ preserve $\mathbf{f}_{\mathscr{A}}$ and $\mathbf{f}^*_{\mathscr{A}}$, respectively.

Let $[R\text{-mod}] = \bigoplus_{\alpha \in Q_+} [R_\alpha\text{-mod}]$ denote the Grothendieck ring, which is an $\mathscr{A}$-algebra via induction product. Similarly the functors of restriction define a coproduct r on $[R\text{-mod}]$. This product and coproduct make $[R\text{-mod}]$ into a twisted unital and counital bialgebra [**14**, Proposition 3.2].

In [**14**,**15**] an explicit $\mathscr{A}$-bialgebra isomorphisms $\gamma^* : [R\text{-mod}] \xrightarrow{\sim} \mathbf{f}^*_{\mathscr{A}}$ is constructed (this has already been mentioned in (1.14)). In fact [**14**] establishes a dual isomorphism γ, see [**21**, Theorem 4.4] for all details on this.

Moreover, $\gamma^*([V^{\circledast}]) = \mathtt{b}^*(\gamma^*([V]))$, and we have a commutative triangle

$$\begin{array}{ccc} & \mathscr{A}\langle I\rangle & \\ {}^{\mathrm{ch}_q}\nearrow & & \nwarrow^{\iota} \\ [R\text{-mod}] & \xrightarrow{\ \ \gamma^*\ \ } & \mathbf{f}^*_{\mathscr{A}} \end{array}, \qquad (2.15)$$

where the map ι is defined as follows:

$$\iota(x) = \sum_{\boldsymbol{i}=i_1\dots i_d \in \langle I\rangle} (x, \theta_{i_1}\dots\theta_{i_d})\boldsymbol{i} \qquad (x \in \mathbf{f}^*_{\mathscr{A}}).$$

Lemma 2.16: *Let v^* be a dual canonical basis element of $\mathbf{f}$, and $\boldsymbol{i} = i_1^{a_1}\dots i_k^{a_k}$ be an extremal word of $\iota(v^*)$ in the sense of § 2.3. Then $\boldsymbol{i}$ appears in $\iota(v^*)$ with coefficient $[a_1]^!_{i_1}\cdots[a_k]^!_{i_k}$.*

Proof: Apply induction on $a_1+\dots+a_k$. The induction base is $a_1+\dots+a_k = 0$, in which case $v^* = 1 \in \mathbf{f}^*_{\mathscr{A}}$ and $\iota(1)$ is the empty word. Recall the map $\theta_i^* : \mathscr{A}\langle I\rangle \to \mathscr{A}\langle I\rangle$ from (2.8). For all $x \in \mathbf{f}^*_{\mathscr{A}}$ we have $\iota((\theta_i^*)^{(n)}(x)) = (\theta_i^*)^{(n)}(\iota(x))$, where in the right hand side $(\theta_i^*)^{(n)} = (\theta_i^*)^n/[n]^!_{\alpha_i}$. By [**12**, Proposition 5.3.1], $(\theta^*_{i_k})^{(a_{i_k})}(v^*)$ is again a dual canonical basis element, and by induction, the word $i_1^{a_1}\dots i_{k-1}^{a_{k-1}}$ appears in $\iota((\theta^*_{i_k})^{(a_{i_k})}(v^*))$ with coefficient $[a_1]^!_{i_1}\cdots[a_{k-1}]^!_{i_{k-1}}$. The result follows. □

3. Standard Module Theory

We want to classify the irreducible R_α-modules using a standard module theory. This was first done in [**21**] and then substantially developed and generalized in [**27**]. Here we mainly follow the approach of [**27**], with an occasional idea from [**17**].

3.1. *Convex orders and cuspidal systems*

The theory depends on a choice of a convex order on the set Φ_+ of *positive roots* (we always mean the system of positive roots corresponding to our fixed choice of the simple roots α_i). We also denote by W the Weyl group of the *root system* Φ. It is a Coxeter group with standard generators $\{r_i \mid i \in I\}$.

A *convex order* on Φ_+ is a total order $\prec$ such that

$$\beta, \gamma, \beta+\gamma \in \Phi_+,\ \beta \prec \gamma \quad \Rightarrow \quad \beta \prec \beta+\gamma \prec \gamma.$$

By [**29**], there is a bijection between convex orders on Φ_+ and reduced expressions for the longest element w_0 of W which works as follows: given a reduced expression $w_0 = r_{i_1} \cdots r_{i_N}$ the corresponding convex order on Φ_+ is given by

$$\alpha_{i_1} \prec r_{i_1}(\alpha_{i_2}) \prec r_{i_1} r_{i_2}(\alpha_{i_3}) \prec \cdots \prec r_{i_1} \cdots r_{i_{N-1}}(\alpha_{i_N}).$$

This allows one to prove the following easy lemma, see [**5**, Lemma 2.4]:

Lemma 3.1: *Suppose we are given positive roots $\alpha, \beta_1, \ldots, \beta_k, \gamma_1, \ldots, \gamma_l$ such that $\beta_i \preceq \alpha \preceq \gamma_j$ for all i and j. We have that $\beta_1 + \cdots + \beta_k = \gamma_1 + \cdots + \gamma_l$ if and only if $k = l$ and $\beta_1 = \cdots = \beta_k = \gamma_1 = \cdots = \gamma_l = \alpha$.*

Now we give a key definition:

Definition 3.2: A *cuspidal system* (for a fixed convex preorder) is the following data: an irreducible R_ρ-module L_ρ assigned to every positive $\rho \in \Phi_+^{\mathrm{re}}$, with the following property:

(Cus) if $\beta, \gamma \in Q_+$ are non-zero elements such that $\rho = \beta + \gamma$ and $\mathrm{Res}_{\beta,\gamma} L_\rho \neq 0$, then β is a sum of positive roots less than ρ and γ is a sum of positive roots greater than ρ.

It is not obvious that a cuspidal system exists or is unique (for a fixed convex order). This will be proved later.

Let us fix a convex order $\prec$ on Φ_+ and an element $\alpha \in Q_+$. A *root partition* of α is a weakly decreasing tuple

$$(\beta_1 \succeq \beta_2 \succeq \cdots \succeq \beta_n) \tag{3.3}$$

of positive roots such that $\beta_1 + \beta_2 + \cdots + \beta_n = \alpha$. The set of root partitions of α is denoted by $\Pi(\alpha)$. For example, if ρ is a positive root, there always is a trivial root partition $(\rho) \in \Pi(\rho)$.

Sometimes we use other notations for root partitions. Let

$$\rho_1 \succ \cdots \succ \rho_N$$

be all positive roots taken in decreasing order. Collecting together equal terms of the root partition (3.3), we can write it as

$$\pi = (\rho_1^{m_1}, \ldots, \rho_N^{m_N})$$

with $\sum_{n=1}^N m_n \rho_n$ or simply as a tuple

$$\pi = (m_1, \ldots, m_N)$$

of nonnegative integers such that $\alpha = \sum_{n=1}^N m_n \rho_n$.

The left lexicographic order on $\Pi(\alpha)$ is denoted $\leq_l$ and the right lexicographic order on $\Pi(\alpha)$ is denoted $\leq_r$. We will also use the following *bilexicographic* partial order on $\Pi(\alpha)$:

$$\pi \leq \sigma \qquad \text{if and only if} \qquad \pi \leq_l \sigma \text{ and } \pi \geq_r \sigma.$$

Let $\mathfrak{g}$ be the finite dimensional complex semisimple Lie algebra with Cartan matrix $\mathtt{C}$. The positive subalgebra $\mathfrak{n}_+ \subset \mathfrak{g}$ has a basis consisting of *root vectors* $\{E_\rho \mid \rho \in \Phi_+\}$. To a root partition $\pi = (m_1, \dots, m_N)$, we assign a PBW monomial

$$E_\pi := E_{\rho_1}^{m_1} \dots E_{\rho_N}^{m_N}.$$

Then $\{E_\pi \mid \pi \in \Pi(\alpha)\}$ is a basis of the weight space $U(\mathfrak{n}_+)_\alpha$. In particular, $|\Pi(\alpha)| = \dim U(\mathfrak{n}_+)_\alpha$. In view of the isomorphism γ^* from (1.14), we conclude:

Lemma 3.4: *The number of irreducible R_α-modules (up to isomorphism and degree shift) is $|\Pi(\alpha)|$.*

3.2. *Standard modules*

We continue to work with a fixed convex preorder $\prec$ on Φ_+. Let $\{L_\rho \mid \rho \in \Phi_+\}$ be a cuspidal system for $\prec$ (we are yet to prove that it exists!). Fix $\alpha \in Q_+$ and a root partition $\pi = (m_1, \dots, m_N) \in \Pi(\alpha)$. We define an integer

$$\mathbf{sh}(\pi) := \sum_{k=1}^{N} (\rho_k, \rho_k) m_k (m_k - 1)/4. \tag{3.5}$$

Set

$$|\pi| = (m_1\rho_1, \dots, m_N\rho_N) \in Q_+^N.$$

The corresponding parabolic subalgerba is

$$R_{|\pi|} = R_{m_1\rho_1, \dots, m_N\rho_N} \cong R_{m_1\rho_1} \otimes \dots \otimes R_{m_N\rho_N} \subseteq R_\alpha.$$

Next, we define the $R_{|\pi|}$-module

$$L_\pi := q^{\mathtt{sh}(\pi)} \, L_{\rho_1}^{\circ m_1} \boxtimes \dots \boxtimes L_{\rho_N}^{\circ m_N}, \tag{3.6}$$

and we define the *proper standard module*

$$\bar{\Delta}(\pi) := \mathrm{Ind}_{|\pi|} L_\pi \cong q^{\mathtt{sh}(\pi)} \, L_{\rho_1}^{\circ m_1} \circ \dots \circ L_{\rho_N}^{\circ m_N} \in R_\alpha\text{-mod}. \tag{3.7}$$

Also introduce the *proper costandard module*

$$\bar{\nabla}(\pi) := \bar{\Delta}(\pi)^{\circledast}. \tag{3.8}$$

It will become clear in Lemma 3.11 why we apply the shift by $\mathtt{sh}(\pi)$ in our definitions.

Lemma 3.9: *Let $\underline{\gamma} := (\gamma_1, \dots, \gamma_n) \in Q_+^n$, and $V_m \in R_{\gamma_m}$-mod for $m = 1, \dots, n$. Denote $d(\underline{\gamma}) = \sum_{1 \le m < k \le n} (\gamma_m, \gamma_k)$. Then $(V_1 \circ \cdots \circ V_n)^{\circledast} \cong q^{d(\underline{\gamma})}(V_n^{\circledast} \circ \cdots \circ V_1^{\circledast})$.*

Proof: Follows from [**22**, Theorem 2.2] by uniqueness of adjoint functors as in the proof of [**16**, Corollary 3.7.4]. □

Recall that for every irreducible module L, there is a unique choice of the grading shift so that we have $L^{\circledast} \cong L$, and unless otherwise stated, we assume that the shift has been chosen in this way. This in particular applies to the modules L_ρ of our cuspidal system.

Lemma 3.10: *Let $\rho \in \Phi_+^{\mathrm{re}}$, L_ρ be the corresponding cuspidal module, and $n \in \mathbb{Z}_{>0}$. Then*

$$(L_\rho^{\circ n})^{\circledast} \cong q_\rho^{n(n-1)} L_\rho^{\circ n}.$$

In particular, the module $q_\rho^{n(n-1)/2} L_\rho^{\circ n}$ is $\circledast$-self-dual.

Proof: Recall that our standard choice of shifts of irreducible modules is so that $L_\rho^{\circledast} \cong L_\rho$. Now the result follows from Lemma 3.9. □

Lemma 3.11: *We have $L_\pi^{\circledast} \cong L_\pi$.*

Proof: Recall that our standard choice of shifts of irreducible modules is so that $L_\rho^{\circledast} \cong L_\rho$. Let $\rho \in \Phi_+$ and $n \in \mathbb{Z}_{>0}$. Then by Lemma 3.10, we have that the module $q_\rho^{n(n-1)/2} L_\rho^{\circ n}$ is $\circledast$-self-dual. The result follows. □

3.3. *Restrictions of proper standard modules*

We recall the Mackey Theorem of Khovanov and Lauda [**14**, Proposition 2.18]. Given $x \in \mathfrak{S}_n$ and $\underline{\gamma} = (\gamma_1, \dots, \gamma_n) \in Q_+^n$, we denote

$$x\underline{\gamma} := (\gamma_{x^{-1}(1)}, \dots, \gamma_{x^{-1}(n)}) \in Q_+^n.$$

$$s(x, \underline{\gamma}) := - \sum_{1 \le m < k \le n,\ x(m) > x(k)} (\gamma_m, \gamma_k) \in \mathbb{Z}.$$

Writing $R_{\underline{\gamma}}$ for $R_{\gamma_1,\dots,\gamma_n}$, there is an obvious natural algebra isomorphism

$$\varphi^x : R_{x\underline{\gamma}} \to R_{\underline{\gamma}}$$

permuting the components. Composing with this isomorphism, we get a functor

$$R_{\underline{\gamma}}\text{-mod} \to R_{x\underline{\gamma}}\text{-mod},\ M \mapsto {}^{\varphi^x}M.$$

Making an additional shift, we get a functor

$$R_{\underline{\gamma}}\text{-mod} \to R_{x\underline{\gamma}}\text{-mod},\ M \mapsto {}^{x}M := q^{s(x,\underline{\gamma})}({}^{\varphi^x}M). \tag{3.12}$$

Theorem 3.13: *Let $\underline{\gamma} = (\gamma_1, \dots, \gamma_n) \in Q_+^n$ and $\underline{\beta} = (\beta_1, \dots, \beta_m) \in Q_+^m$ with $\gamma_1 + \cdots + \gamma_n = \beta_1 + \cdots + \beta_m =: \alpha$. Then for any $M \in R_{\underline{\gamma}}$-mod we have that $\mathrm{Res}_{\underline{\beta}}\,\mathrm{Ind}_{\underline{\gamma}} M$ has filtration with factors of the form*

$$\mathrm{Ind}^{\beta_1;\dots;\beta_m}_{\alpha_1^1,\dots,\alpha_1^n;\dots;\alpha_m^1,\dots,\alpha_m^n}\, {}^{x(\underline{\alpha})}\big(\mathrm{Res}^{\gamma_1;\dots;\gamma_n}_{\alpha_1^1,\dots,\alpha_m^1;\dots;\alpha_1^n,\dots,\alpha_m^n} M\big)$$

with $\underline{\alpha} = (\alpha_b^a)_{1\le a\le n,\ 1\le b\le m}$ running over all tuples of elements of Q_+ such that $\sum_{b=1}^m \alpha_b^a = \gamma_a$ for all $1 \le a \le n$ and $\sum_{a=1}^n \alpha_b^a = \beta_b$ for all $1 \le b \le m$, and $x(\underline{\alpha})$ is the permutation of mn which maps

$$(\alpha_1^1, \dots, \alpha_m^1; \alpha_1^2, \dots, \alpha_m^2; \dots; \alpha_1^n, \dots, \alpha_m^n)$$

to

$$(\alpha_1^1, \dots, \alpha_1^n; \alpha_2^1, \dots, \alpha_2^n; \dots; \alpha_m^1, \dots, \alpha_m^n).$$

We use the Mackey Theorem to study restrictions of proper standard modules:

Proposition 3.14: *Let $\pi, \sigma \in \Pi(\alpha)$. Then:*

(i) $\mathrm{Res}_{|\sigma|}\bar{\Delta}(\sigma) \cong L_\sigma$.
(ii) $\mathrm{Res}_{|\pi|}\bar{\Delta}(\sigma) \neq 0$ *implies* $\pi \le \sigma$.

Proof: Write $\pi = (m_1, \dots, m_N)$, $\sigma = (n_1, \dots, n_N)$. Let $\mathrm{Res}_{|\pi|}\bar{\Delta}(\sigma) \neq 0$. It suffices to prove that $\pi \ge_l \sigma$ or $\pi \le_r \sigma$ implies that $\pi = \sigma$ and $\mathrm{Res}_{|\pi|}\bar{\Delta}(\sigma) \cong L_\sigma$. We may assume that $\pi \ge_l \sigma$, the case $\pi \le_r \sigma$ being similar. We apply induction on $\mathrm{ht}(\alpha)$. Pick the minimal a with $m_a \neq 0$. Let

$$\pi' = (0, \dots, 0, m_{a+1}, \dots, m_N) \in \Pi(\alpha - m_a\rho_a)$$

and

$$\sigma' = (0, \dots, 0, n_{a+1}, \dots, n_N) \in \Pi(\alpha - n_a\rho_a).$$

By Theorem 3.13, $\mathrm{Res}_{|\pi|}\bar{\Delta}(\sigma)$ has filtration with factors of the form $\mathrm{Ind}^{m_a\rho_a;|\pi'|}_{\kappa_1,\dots,\kappa_c;\underline{\gamma}} V$, where $m_a\rho_a = \kappa_1 + \cdots + \kappa_c$, with $\kappa_1, \dots, \kappa_c \in Q_+ \setminus \{0\}$, and $\underline{\gamma}$ is a refinement of $|\pi'|$. Moreover, the module V is obtained by twisting and degree shifting as in (3.12) of a module obtained by restriction of $L_{\rho_1}^{\boxtimes n_1} \boxtimes \cdots \boxtimes L_{\rho_N}^{\boxtimes n_N}$ to a parabolic which has $\kappa_1, \dots, \kappa_c$ in the beginnings of the corresponding blocks. In particular, if $V \neq 0$, then for each $b = 1, \dots, c$ we have that $\mathrm{Res}_{\kappa_b, \rho_k - \kappa_b} L_{\rho_k} \neq 0$ for some $k = k(b)$ with $n_k \neq 0$.

Let $1 \leq b \leq c$. If $\mathrm{Res}_{\kappa_b, \rho_k - \kappa_b} L_{\rho_k} \neq 0$, then by the definition of cuspidal modules, κ_b is a sum of roots $\preceq \rho_k$. Moreover, since $\pi \geq_l \sigma$ and $n_k \neq 0$, we have that $\rho_k \preceq \rho_a$. Thus κ_b is a sum of roots $\preceq \rho_a$. Using Lemma 3.1, we conclude that $c = m_a$ and $\kappa_b = \rho_a = \rho_{k(b)}$ for all $b = 1, \dots, c$. Hence $n_a \geq m_a$. Since $\pi \geq_l \sigma$, we conclude that $n_a = m_a$, and

$$\mathrm{Res}_{|\pi|}\bar{\Delta}(\sigma) \simeq L_{\rho_a}^{\circ m_a} \boxtimes \mathrm{Res}_{|\pi'|}\bar{\Delta}(\sigma').$$

Since $\mathrm{ht}(\alpha - m_a\rho_a) < \mathrm{ht}(\alpha)$, we can apply the inductive hypothesis. □

3.4. *Classification of irreducible modules*

We continue to work with a fixed convex preorder $\preceq$ on Φ_+. In this subsection we prove the following theorem:

Theorem 3.15: *For a given convex preorder, there exists a unique cuspidal system $\{L_\rho \mid \rho \in \Phi_+\}$. Moreover:*

(i) *For every root partition π, the proper standard module $\bar{\Delta}(\pi)$ has an irreducible head; denote this irreducible module $L(\pi)$.*
(ii) *$\{L(\pi) \mid \pi \in \Pi(\alpha)\}$ is a complete and irredundant system of irreducible R_α-modules up to isomorphism.*
(iii) *$L(\pi)^{\circledast} \cong L(\pi)$.*
(iv) *$[\bar{\Delta}(\pi) : L(\pi)]_q = 1$, and $[\bar{\Delta}(\pi) : L(\sigma)]_q \neq 0$ implies $\sigma \leq \pi$.*
(v) *$\mathrm{Res}_{|\pi|} L(\pi) \cong L_\pi$ and $\mathrm{Res}_{|\sigma|} L(\pi) \neq 0$ implies $\sigma \leq \pi$.*
(vi) *$L_\rho^{\circ n}$ is irreducible for all $\rho \in \Phi_+$ and all $n \in \mathbb{Z}_{>0}$.*

The rest of §3.4 is devoted to the proof of Theorem 3.15, which goes by induction on $\mathrm{ht}(\alpha)$. To be more precise, we prove the following statements for all $\alpha \in Q_+$ by induction on $\mathrm{ht}(\alpha)$:

(1) For each $\rho \in \Phi_+$ with $\mathrm{ht}(\rho) \leq \mathrm{ht}(\alpha)$ there exists a unique up to isomorphism irreducible R_ρ-module L_ρ which satisfies the property (Cus) of Definition 3.2. Moreover, L_ρ also satisfies the property (vi) of Theorem 3.15 if $\mathrm{ht}(n\rho) \leq \mathrm{ht}(\alpha)$.

(2) The proper standard modules $\bar{\Delta}(\pi)$ for all $\pi \in \Pi(\alpha)$, defined as in (3.7) using the modules from (1), satisfy the properties (i)–(v) of Theorem 3.15.

The induction starts with $\mathrm{ht}(\alpha) = 0$, and for $\mathrm{ht}(\alpha) = 1$ the theorem is also clear since R_{α_i} is a polynomial algebra, which has only the trivial irreducible (graded) representation L_{α_i}. The inductive assumption will stay valid throughout §3.4.

3.4.1. *The case* $\pi \neq (\rho^n)$

In the following proposition, we exclude the case where the proper standard module is of the form $L_\rho^{\circ n}$. The excluded cases will be dealt with in §§3.4.2 and 3.4.3.

Proposition 3.16: *Let* $\pi = (m_1, \dots, m_N) \in \Pi(\alpha)$*, and suppose that there are* $1 \leq k \neq l \leq N$ *such that* $m_k \neq 0$ *and* $m_l \neq 0$.

(i) $\bar{\Delta}(\pi)$ *has an irreducible head; denote this irreducible module* $L(\pi)$.
(ii) If $\pi \neq \sigma$*, then* $L(\pi) \not\simeq L(\sigma)$.
(iii) $L(\pi)^{\circledast} \cong L(\pi)$.
(iv) $[\bar{\Delta}(\pi) : L(\pi)]_q = 1$*, and* $[\bar{\Delta}(\pi) : L(\sigma)]_q \neq 0$ *implies* $\sigma \leq \pi$.
(v) $\mathrm{Res}_{|\pi|} L(\pi) \simeq L_\pi$ *and* $\mathrm{Res}_{|\sigma|} L(\pi) \neq 0$ *implies* $\sigma \leq \pi$.

Proof: (i) and (v) If L is an irreducible quotient of $\bar{\Delta}(\pi) = \mathrm{Ind}_{|\pi|} L_\pi$, then by adjointness of $\mathrm{Ind}_{|\pi|}$ and $\mathrm{Res}_{|\pi|}$ and the irreducibility of the $R_{|\pi|}$-module L_π, which holds by the inductive assumption, we conclude that L_π is a submodule of $\mathrm{Res}_{|\pi|} L$. On the other hand, by Proposition 3.14(i) the multiplicity of L_π in $\mathrm{Res}_{|\pi|} \bar{\Delta}(\pi)$ is 1, so (i) follows. Note that we have also proved the first statement in (v), while the second statement in (v) follows from Proposition 3.14(ii) and the exactness of the functor $\mathrm{Res}_{|\pi|}$.

(iv) By (v), $\mathrm{Res}_{|\sigma|} L(\sigma) \cong L_\sigma \neq 0$. Therefore, if $L(\sigma)$ is a composition factor of $\bar{\Delta}(\pi)$, then $\mathrm{Res}_{|\sigma|} \bar{\Delta}(\pi) \neq 0$ by exactness of $\mathrm{Res}_{|\sigma|}$. By Proposition 3.14, we then have $\sigma \leq \pi$ and (iv).

(ii) If $L(\pi) \simeq L(\sigma)$, then we deduce from (iv) that $\pi \leq \sigma$ and $\sigma \leq \pi$, whence $\pi = \sigma$.

(iii) follows from (v) and Lemma 3.11. □

3.4.2. *The case* $\pi = (\rho)$

We now assume that $\alpha = \rho_k \in \Phi_+$. There is a *trivial* root partition $(\rho_k) \in \Pi(\alpha)$. Proposition 3.16 yields $|\Pi(\alpha)| - 1$ irreducible R_α-modules, namely the ones which correspond to the *non-trivial* root partitions $\pi \in \Pi(\alpha)$. We define the cuspidal module L_α to be the missing irreducible R_α-module, cf. Lemma 3.4. Then, of course, we have that $\{L(\pi) \mid \pi \in \Pi(\alpha)\}$ is a complete and irredundant system of irreducible R_α-modules up to isomorphism. We now prove that L_α satisfies the property (Cus) and is uniquely determined by it:

Lemma 3.17: *Let* $\alpha = \rho_k \in \Phi_+$. *If* $\beta, \gamma \in Q_+$ *are non-zero elements such that* $\alpha = \beta + \gamma$ *and* $\mathrm{Res}_{\beta,\gamma} L_\alpha \neq 0$, *then* β *is a sum of roots less than* α *and* γ *is a sum of roots greater than* α. *Moreover, this property characterizes* L_α *among the irreducible* R_α*-modules uniquely up to isomorphism and degree shift.*

Proof: We prove that β is a sum of roots less than α, the proof that γ is a sum of roots greater than α being similar. Let $L(\pi) \boxtimes L(\sigma)$ be an irreducible submodule of $\mathrm{Res}_{\beta,\gamma} L_\alpha$, so that $\pi = (m_1, \dots, m_N) \in \Pi(\beta)$ and $\sigma = (n_1, \dots, n_N) \in \Pi(\gamma)$. Let a be minimal with $m_a \neq 0$. Then $\mathrm{Res}_{\rho_a, \beta-\rho_a} L(\pi) \neq 0$, and hence $\mathrm{Res}_{\rho_a, \gamma+\beta-\rho_a} L_\alpha \neq 0$. If we can prove that ρ_a is a sum of roots less than α, then by convexity, ρ_a is a root less than α, whence, by the minimality of a, we have that β is a sum of roots less than α. So we may assume from the beginning that β is a root and $L(\pi) = L_\beta$. Moreover, we may assume that β is the maximal positive root for which $\mathrm{Res}_{\beta,\gamma} L_\alpha \neq 0$.

Now, let l be the minimal with $n_l \neq 0$. Then we have a non-zero map

$$L_\beta \boxtimes L_{\rho_l} \boxtimes V \to \mathrm{Res}_{\beta,\kappa,\gamma-\rho_l} L_\alpha,$$

for some $0 \neq V \in R_{\gamma-\rho_l}$-mod. By adjunction, this yields a non-zero map

$$f : (\mathrm{Ind}_{\beta,\rho_l} L_\beta \boxtimes L_{\rho_l}) \boxtimes V \to \mathrm{Res}_{\beta+\rho_l, \gamma-\rho_l} L_\alpha.$$

If $\rho_l = \gamma$, then we must have $\beta \prec \gamma$, for otherwise L_α is a quotient of the proper standard module $L_\beta \circ L_\gamma$, which contradicts the definition of the cuspidal module L_α. Now, since $\alpha = \beta + \rho_l$, we have by convexity that $\beta \prec \alpha \prec \gamma$, in particular $\beta \prec \alpha$ as desired.

Next, let $\rho_l \neq \gamma$, and pick a composition factor $L(\pi')$ of $\mathrm{Ind}_{\beta,\rho_l} L_\beta \boxtimes L_{\rho_l}$, which is not in the kernel of f. Write $\pi' = (m'_1, \dots, m'_N) \in \Pi(\beta + \rho_l)$. By the assumption on the maximality of β, we have $\rho_c \preceq \beta$ whenever $m'_c > 0$.

Thus $\beta + \rho_l$ is a sum of roots $\preceq \beta$. Lemma 3.1 implies that $\rho_l \preceq \beta$, and so by adjointness, L_α is a quotient of the proper standard module $L_\beta \circ \bar{\Delta}(\sigma)$, which is a contradiction.

The second statement of the lemma is clear since, in view of Proposition 3.16(v) and Lemma 3.1, the irreducible modules $L(\pi)$, corresponding to non-trivial root partitions $\pi \in \Pi(\alpha)$, do not satisfy the property (Cus). □

3.4.3. *The case* $\pi = (\rho^n)$

Assume now that $\alpha = n\rho_k$ for some $\rho_k \in \Phi_+$ and $n \in \mathbb{Z}_{>1}$.

Lemma 3.18: *The induced module* $L_{\rho_k}^{\circ n}$ *is irreducible.*

Proof: In view of Proposition 3.16, we have the irreducible modules $L(\pi)$ for all root partitions $\pi \in \Pi(\alpha)$, except for $\pi = \sigma := (\rho_k^n)$, for which $\bar{\Delta}(\sigma) = L_{\rho_k}^{\circ n}$. By Lemma 3.1, σ is the unique minimal element of $\Pi(\alpha)$. By Proposition 3.16(v), we conclude that $L_{\rho_k}^{\circ n}$ has only one composition factor L appearing with certain multiplicity $c(q) \in \mathscr{A}$, and such that $L \not\cong L(\pi)$ for all $\pi \in \Pi(\alpha) \setminus \{\sigma\}$. Finally, by Corollary 2.13, we conclude that $L_{\rho_k}^{\circ n} \simeq L$. □

The proof of Theorem 3.15 is now complete.

3.5. *Reduction modulo p*

In this subsection we work with two fields: F of characteristic $p > 0$ and K of characteristic 0. We use the corresponding indices to distinguish between the two situations. Given an irreducible $R_\alpha(K)$-module L_K for a root partition $\pi \in \Pi(\alpha)$ we can pick a (graded) $R_\alpha(\mathbb{Z})$-invariant lattice $L_\mathbb{Z}$ as follows: pick a homogeneous word vector $v \in L_K$ and set $L_\mathbb{Z} := R_\alpha(\mathbb{Z})v$. The lattice $L_\mathbb{Z}$ can be used to *reduce modulo p*:

$$\bar{L} := L_\mathbb{Z} \otimes_\mathbb{Z} F.$$

In general, the $R_\alpha(F)$-module $\bar{L}$ depends on the choice of the lattice $L_\mathbb{Z}$. However, we have $\mathrm{ch}_q \bar{L} = \mathrm{ch}_q L_K$, so by linear independence of characters of irreducible $R_\alpha(F)$-modules, composition multiplicities of irreducible $R_\alpha(F)$-modules in $\bar{L}$ are well-defined. In particular, we have well-defined *decomposition numbers*

$$d_{\pi,\sigma} := [\bar{L}(\pi) : L_F(\sigma)]_q \qquad (\pi, \sigma \in \Pi(\alpha)),$$

which depend only on the characteristic p of F, since prime fields are splitting fields for irreducible modules over KLR algebras.

Lemma 3.19: *Let L_K be an irreducible $R_\alpha(K)$-module and let $\boldsymbol{i} = i_1^{a_1} \dots i_b^{a_b}$ be an extremal word for L_K. Let N be the irreducible $\circledast$-selfdual $R_\alpha(F)$-module defined by $N := \tilde{f}_{i_k}^{a_k} \dots \tilde{f}_{i_1}^{a_1} 1_F$. Then $[\bar{L} : N]_q = 1$.*

Proof: Reduction modulo p preserves formal characters, so the result follows from Corollary 2.10. □

Proposition 3.20: *Let $\pi, \sigma \in \Pi(\alpha)$. Then $d_{\pi,\sigma} \neq 0$ implies $\sigma \leq \pi$. In particular, reduction modulo p of any cuspidal module is an irreducible cuspidal module again: $\bar{L}_\rho \simeq L_{\rho,F}$.*

Proof: By Theorem 3.15(v), which holds over any field, we conclude that any composition factor of $\bar{L}_\rho$ is isomorphic to $L_{\rho,F}$ up to a degree shift. Now use Lemma 3.19. □

3.6. *PBW bases and canonical bases*

We now return to the algebra $\mathbf{f}$ and recall some results on its PBW bases. For a fixed convex order on Φ_+, Lusztig used a certain braid group action to define *root vectors* $\{r_\rho \,|\, \rho \in \Phi_+\}$ in $\mathbf{f}$. The corresponding *dual root vectors*

$$r_\rho^* := (1 - q_\rho^2) r_\rho \qquad (\rho \in \Phi_+) \tag{3.21}$$

are invariant under $\mathtt{b}^*$.

For $\pi = (m_1, \dots, m_N) \in \Pi(\alpha)$, we set

$$r_\pi := \frac{r_{\rho_1}^{m_1}}{[m_1]_{\rho_1}^!} \cdots \frac{r_{\rho_N}^{m_N}}{[m_N]_{\rho_N}^!}, \qquad r_\pi^* := q^{\mathtt{sh}(\pi)} (r_{\rho_1}^*)^{m_1} \cdots (r_{\rho_N}^*)^{m_N}. \tag{3.22}$$

Theorem 3.23: *[**25**] Let $\alpha \in Q_+$. Then $\{r_\pi \mid \pi \in \Pi(\alpha)\}$ and $\{r_\pi^* \mid \pi \in \Pi(\alpha)\}$ are a pair of dual bases for the free $\mathscr{A}$-modules $(\mathbf{f}_{\mathscr{A}})_\alpha$ and $(\mathbf{f}_{\mathscr{A}}^*)_\alpha$ respectively.*

One can use the $\mathtt{b}^*$-invariance of the dual root vectors together with the Levendorskii-Soibelman formula [**23**, Proposition 5.5.2] or [**26**, Proposition 1.9] to deduce:

$$\mathtt{b}^*(r_\pi^*) = r_\pi^* + (\text{a } \mathbb{Z}[q, q^{-1}]\text{-linear combination of } r_\sigma^* \text{ for } \sigma < \pi). \tag{3.24}$$

$$\mathtt{b}(r_\pi) = r_\pi + (\text{a } \mathbb{Z}[q, q^{-1}]\text{-linear combination of } r_\sigma \text{ for } \sigma > \pi). \tag{3.25}$$

In view (3.24)–(3.25) and Lusztig's Lemma, there exist unique bases $\{b_\pi | \pi \in \Pi(\alpha)\}$ and $\{b^*_\pi \mid \pi \in \Pi(\alpha)\}$ for $(\mathbf{f}_{\mathscr{A}})_\alpha$ and $(\mathbf{f}^*_{\mathscr{A}})_\alpha$, respectively, such that

$$\mathsf{b}(b_\pi) = b_\pi, \quad b_\pi = r_\pi + (\text{a } q\mathbb{Z}[q]\text{-linear combination of } r_\sigma \text{ for } \sigma > \pi), \tag{3.26}$$

$$\mathsf{b}^*(b^*_\pi) = b^*_\pi, \quad b^*_\pi = r^*_\pi + (\text{a } q\mathbb{Z}[q]\text{-linear combination of } r^*_\sigma \text{ for } \sigma < \pi). \tag{3.27}$$

These are the canonical and dual canonical bases, respectively (cf. [**24**] in simply-laced types or [**33**] in non-simply-laced types).

3.7. *Cuspidal modules and dual PBW bases*

We continue to work with a fixed convex order $\prec$ on Φ_+. Suppose that we are given elements

$$\{E^*_\rho \in (\mathbf{f}^*_{\mathscr{A}})_\rho \mid \rho \in \Phi_+\}. \tag{3.28}$$

If $\pi = (m_1, \dots, m_N)$ is a root partition, define the corresponding *dual PBW monomial*

$$E^*_\pi := q^{\mathtt{sh}(\pi)} (E^*_{\rho_1})^{m_1} \dots (E^*_{\rho_N})^{m_N} \in \mathbf{f}^*_{\mathscr{A}}.$$

We say that (3.28) is a *dual PBW family* if the following properties are satisfied:

(i) ('convexity') if $\beta \succ \gamma$ are positive roots then $E^*_\gamma E^*_\beta - q^{-(\beta,\gamma)} E^*_\beta E^*_\gamma$ is an $\mathscr{A}$-linear combination of elements E^*_π with $\pi < (\beta, \gamma) \in \Pi(\beta + \gamma)$;
(ii) ('basis') $\{E^*_\pi \mid \pi \in \Pi(\alpha)\}$ is an $\mathscr{A}$-basis of $(\mathbf{f}^*_{\mathscr{A}})_\alpha$ for all $\alpha \in Q_+$;
(iii) ('orthogonality')

$$(E^*_\pi, E^*_\sigma) = \delta_{\pi,\sigma} \prod_{k=1}^{N} ((E^*_{\rho_k})^{m_k}, (E^*_{\rho_k})^{m_k});$$

(iv) ('bar-triangularity') $\mathsf{b}^*(E^*_\pi) = E^*_\pi +$ an $\mathscr{A}$-linear combination of dual PBW monomials E^*_σ for $\sigma < \pi$.

The following result shows in particular that the elements E^*_ρ of the dual PBW family are determined uniquely up to signs (for a fixed preorder $\preceq$):

Lemma 3.29: *Assume that (3.28) is a dual PBW family. Then:*

(i) The elements of (3.28) are b^**-invariant.*

(ii) *Suppose that we are given another family* $\{'E^*_\rho \in (\mathbf{f}^*_{\mathscr{A}})_\rho \mid \rho \in \Phi_+\}$ *of* $\mathtt{b}^*$*-invariant elements which satisfies the basis and orthogonality properties. Then* $E^*_\rho = \pm' E^*_\rho$ *for all* $\rho \in \Phi^{\mathrm{re}}_+$.

Proof: (i) The convexity of $\prec$ implies that for $\rho \in \Phi_+$ the root partition $(\rho) \in \Pi(\rho)$ is a minimal element of $\Pi(\rho)$. So the bar-triangularity property (iv) implies that the elements of a dual PBW family are $\mathtt{b}^*$-invariant.

(ii) We apply induction on $\mathrm{ht}(\rho)$, the induction base being clear. By the basis property of dual PBW families, we can write

$$'E^*_\rho = cE^*_\rho + \sum_{\pi \in \Pi(\rho)\setminus\{(\rho)\}} c_\pi E^*_\pi \qquad (c, c_\pi \in \mathscr{A}). \tag{3.30}$$

Fix for a moment a root partition $\pi \in \Pi(\rho)\setminus\{(\rho)\}$. By the orthogonality property of dual PBW families and non-degeneracy of the form $(\cdot,\cdot)$, the element

$$X_\pi := \frac{1}{(E^*_\pi, E^*_\pi)} E^*_\pi$$

satisfies $(E^*_\sigma, X_\pi) = \delta_{\sigma,\pi}$ for all $\sigma \in \Pi(\rho)$. So pairing the right hand side of (3.30) with X_π yields c_π. On the other hand, by the inductive assumption, $E^*_\pi = \pm' E^*_\pi$. So using the orthogonality property for the primed family in (ii), we must have $('E^*_\rho, X_\pi) = 0$ for all $\pi \in \Pi(\rho) \setminus \{(\rho)\}$. So $c_\pi = 0$. Thus $'E^*_\rho = cE^*_\rho$. Furthermore, the elements $'E^*_\rho$ and E^*_ρ belong to the algebra $\mathbf{f}^*_{\mathscr{A}}$ and are parts of its $\mathscr{A}$-bases, whence $'E^*_\rho = \pm q^n E^*_\rho$. Since both $'E^*_\rho$ and E^*_ρ are $\mathtt{b}^*$-invariant, we conclude that $n = 0$. □

Proposition 3.31: *The following set of elements in* $\mathbf{f}^*_{\mathscr{A}}$

$$\{E^*_\rho := \gamma^*([L_\rho]) \mid \rho \in \Phi_+\}$$

is a dual PBW family.

Proof: Under the categorification map γ^*, the graded duality $\circledast$ corresponds to $\mathtt{b}^*$, so $\gamma^*([L])$ is $\mathtt{b}^*$-invariant for any $\circledast$-self-dual R_α-module L. Moreover, under γ^*, the induction product corresponds to the product in $\mathbf{f}^*_{\mathscr{A}}$, so the convexity condition (i) follows from Theorem 3.15(iv) and Lemma 3.9. Now, note that $E^*_\pi = \gamma^*([\bar\Delta(\pi)])$, so the conditions (ii) and (iv) follow from Theorem 3.15(iv) again. It remains to establish the orthogonality property (iii). Let $\pi = (m_1, \dots, m_N)$. Under γ^*, the coproduct r corresponds to the map on the Grothendieck group induces by Res. So using (2.14), we get

$$(E^*_\pi, E^*_\sigma) = \big((E^*_{\rho_1})^{m_1} \otimes \cdots \otimes (E^*_{\rho_N})^{m_N}, \gamma^*([\mathrm{Res}_{|\pi|}\bar\Delta(\sigma)])\big).$$

By Proposition 3.14, $\mathrm{Res}_{|\pi|}\bar{\Delta}(\sigma) = 0$ unless $\pi \leq \sigma$, and for $\pi = \sigma$ we have

$$\mathrm{Res}_{|\pi|}\bar{\Delta}(\sigma) = L_{\rho_1}^{\circ m_1} \boxtimes \cdots \boxtimes L_{\rho_N}^{\circ m_N}.$$

Since the form $(\cdot,\cdot)$ is symmetric, the orthogonality follows from the preceding remarks. □

It is shown in Lusztig [**25**] and [**33**] that $\{r_\rho^* \mid \rho \in \Phi_+\}$ is a dual PBW family. Since the dual PBW families are unique up to a sign by Lemma 3.29, it follows that $\gamma^*([L_\rho]) = \pm r_\rho^*$ for all $\rho \in \Phi_+$. In fact:

Proposition 3.32: *For every $\rho \in \Phi_+$ we have that $\gamma^*([L_\rho]) = r_\rho^* = b_{(\rho)}^*$ is a dual canonical basis element.*

Proof: By (3.27), we have $r_\rho^* = b_{(\rho)}^*$ is a dual canonical basis element. Now, in view of the commutativity of the triangle (2.15), to show that $E_\rho^* = b_{(\rho)}^*$, it suffices to know that for an arbitrary element b^* of the dual canonical basis, there exists at least one word $\boldsymbol{i} \in \langle I \rangle$ such that the coefficient of $\boldsymbol{i}$ in $\iota(b^*)$ evaluated at $q = 1$ is positive. But this follows from Lemma 2.16. □

4. Homological Properties of KLR Algebras

We now review some 'standard homological properties' of KLR algebras of finite Lie type. We continue to work with a fixed convex order $\prec$ on Φ_+. We mainly follow [**5**], to where we refer the reader for detailed proofs.

4.1. *Finiteness of global dimension*

First of all, we record a key fundamental fact:

Theorem 4.1: *If the Cartan matrix $\mathtt{C}$ is of finite type, the KLR algebra $R_\alpha(\mathtt{C})$ has global dimension equal to* $\mathrm{ht}(\alpha)$ *(as a graded algebra).*

The finiteness of the global dimension of $R_\alpha(\mathtt{C})$ (as a graded algebra) was first proved by Kato [**13**] for the case $\mathrm{char}\, F = 0$ and $\mathtt{C}$ of finite $\mathtt{ADE}$ types. For an arbitrary F and $\mathtt{C}$ of finite $\mathtt{BCFG}$ types McNamara [**27**] computed the global dimension explicitly as $\mathrm{ht}(\alpha)$. In [**5**, Appendix], it was verified that the methods of [**27**] also lead to the same answer for finite $\mathtt{ADE}$ types over any field, not surprisingly the case E_8 being the most difficult.

Still for $\mathtt{C}$ of finite type, the algebras $R_\alpha(\mathcal{O})$ are affine quasi-hereditary. This is shown in [**19**] for finite type $\mathtt{A}$ and in [**18**] for other finite Lie types.

From this we have the following slight generalization: if $\mathcal{O}$ is a commutative ring of finite global dimension, then $R_\alpha(\mathcal{O}, \mathtt{C})$ also has finite global dimension, even as an ungraded algebra.

Finally, it can be checked that for a fixed $\mathtt{C}$, the algebras $R_\alpha(F, \mathtt{C})$ have finite global dimension for all $\alpha \in Q_+$ if and only if $\mathtt{C}$ is of finite type.

4.2. *Standard modules*

Throughout this subsection, $\rho \in \Phi_+$ is a fixed positive root. Recall the cuspidal module L_ρ. The proof of the following result relies on the finiteness of the global dimension of R_α.

Theorem 4.2: *[27, §4] Let $d \geq 1$. Then*

$$\dim_q \operatorname{EXT}^d_{R_\rho}(L_\rho, L_\rho) = \begin{cases} q_\rho^2 & \text{if } d = 1; \\ 0 & \text{if } d \geq 2. \end{cases}$$

This theorem allows one to extend L_ρ by $q_\rho^2 L_\rho$, then by $q_\rho^4 L_\rho$, etc. to get in the limit the modules $\Delta(\rho)$ with the following properties:

Theorem 4.3: *[5, Theorem 3.4, Corollary 3.4] There is a short exact sequence*

$$0 \longrightarrow q_\rho^2 \Delta(\rho) \longrightarrow \Delta(\rho) \longrightarrow L_\rho \longrightarrow 0. \tag{4.4}$$

Moreover:

(i) $\Delta(\rho)$ is a cyclic module, and in the Grothendieck group we have

$$[\Delta(\rho)] = \frac{1}{1 - q_\rho^2}[L_\rho]; \tag{4.5}$$

(ii) $\Delta(\rho)$ has irreducible head isomorphic to L_ρ;
(iii) we have that $\operatorname{EXT}^d_{R_\rho}(\Delta(\rho), V) = 0$ for $d \geq 1$ and any finitely generated R_ρ-module V with all composition factors $\simeq L_\rho$;
(iv) $\operatorname{END}_{R_\rho}(\Delta(\rho)) \cong F[x]$ for x in degree $2d_\rho$.
(v) The functor $\operatorname{HOM}_{R_\rho}(\Delta(\rho), -)$ defines an equivalence from the category of finitely generated graded R_ρ-modules with all composition factors $\simeq L_\rho$ to the category of finitely generated graded $F[x]$-modules (viewing $F[x]$ as a graded algebra with $\deg(x) = 2d_\rho$).

Remark 4.6: In $\mathtt{ADE}$ types, there is a more elementary construction of $\Delta(\rho)$. For any $\alpha \in Q_+$ of height d, let R'_α be the subalgebra of R_α generated by

$$\{1_{\boldsymbol{i}} \mid \boldsymbol{i} \in \langle I \rangle_\alpha\} \cup \{y_1 - y_2, \dots, y_{d-1} - y_d\} \cup \{\psi_1, \dots, \psi_{d-1}\}.$$

Denote by L'_ρ the restriction of L_ρ from R_ρ to R'_ρ. Then $\Delta(\rho) \cong R_\rho \otimes_{R'_\rho} L'_\rho$.

By (3.21) and (4.5), the module $\Delta(\rho)$ categorifies the root vector r_ρ. Compare this to Proposition 3.32, which shows that $\bar{\Delta}(\rho) = L_\rho$ categorifies the dual root vector r^*_ρ. Next, we explain how to category the divided powers $r^m_\rho/[m]^!_\rho$ for all $m \in \mathbb{Z}_{\geq 0}$. For this we need to compute the endomorphism algebra of $\Delta(\rho)^{\circ m}$.

Choose a non-zero homogeneous vector v_ρ of minimal degree in $\Delta(\rho)$. It generates $\Delta(\rho)$ as an R_ρ-module. The proof of the following lemma is based on the Mackey Theorem and splitting coming from Theorem 4.3(iii).

Lemma 4.7: *[**5**, Lemma 3.6] Let $w \in \mathfrak{S}_{2n}$ be the permutation mapping $(1, \dots, n, n+1, \dots, 2n)$ to $(n+1, \dots, 2n, 1, \dots, n)$. There is a unique $R_{2\rho}$-module homomorphism*

$$\tau : \Delta(\rho) \circ \Delta(\rho) \to \Delta(\rho) \circ \Delta(\rho)$$

of degree $-2d_\rho$ such that $\tau(1_{\rho,\rho} \otimes (v_\rho \otimes v_\rho)) = \psi_w 1_{\rho,\rho} \otimes (v_\rho \otimes v_\rho)$.

Now pick a non-zero endomorphism $x \in \mathrm{END}_{R_\rho}(\Delta(\rho))_{2d_\rho}$. By Theorem 4.3(iv) we have that $\mathrm{END}_{R_\rho}(\Delta(\rho)) = F[x]$, so x is unique up to a scalar. Now we have commuting endomorphisms $x_1, \dots, x_m \in \mathrm{END}_{R_{m\rho}}(\Delta(\rho)^{\circ m})_{2d_\rho}$ with

$$x_r := \mathrm{id}^{\circ(r-1)} \circ x \circ \mathrm{id}^{\circ(m-r)}.$$

Moreover, the endomorphism τ from the previous lemma yields $\tau_1, \dots, \tau_{m-1} \in \mathrm{END}_{R_{m\rho}}(\Delta(\rho)^{\circ m})_{-2d_\rho}$ with

$$\tau_r := \mathrm{id}^{\circ(r-1)} \circ \tau \circ \mathrm{id}^{\circ(m-r-1)}.$$

Now [**5**, Lemmas 3.7–3.9] yield:

Theorem 4.8: *For a unique choice of $x \in \mathrm{END}_{R_\rho}(\Delta(\rho))_{2d_\rho}$, there is an algebra isomorphism*

$$NH_m \xrightarrow{\sim} \mathrm{END}_{R_{m\rho}}(\Delta(\rho)^{\circ m})^{\mathrm{op}}, \qquad y_i \mapsto x_i,\ \psi_j \mapsto \tau_j.$$

By the theorem, we can view $\Delta(\rho)^{\circ m}$ as an $(R_{m\rho}, NH_m)$-bimodule. Finally define the *divided power module*

$$\Delta(\rho^m) := q_\rho^{m(m-1)/2} \Delta(\rho)^{\circ m} e_m \tag{4.9}$$

where $e_m \in NH_m$ is the idempotent (1.12).

Lemma 4.10: *[**5**, Lemmas 3.7–3.9] We have that $\Delta(\rho)^{\circ m} \cong [m]^!_\rho \Delta(\rho^m)$ as an $R_{m\rho}$-module. Moreover $\Delta(\rho^m)$ has irreducible head $L(\rho^m)$, and in the*

Grothendieck group we have

$$[\Delta(\rho^m)] = \frac{1}{(1-q_\rho^2)(1-q_\rho^4)\cdots(1-q_\rho^{2m})}\,[L(\rho^m)].$$

The lemma shows that in the Grothendieck group $[\Delta(\rho^m)]$ corresponds the to the divided power $r_\rho^m/[m]_\rho^!$ under the Khovanov–Lauda–Rouquier categorification. More generally, for a root partition $\pi = (m_1, \dots, m_N)$, define the *standard module*

$$\Delta(\pi) := \Delta(\rho_1^{m_1}) \circ \cdots \circ \Delta(\rho_N^{m_N}). \tag{4.11}$$

Theorem 4.12: *[5, Theorem 3.11] For a root partition $\pi = (m_1, \dots, m_N)$, the module $V_0 := \Delta(\pi)$ has an exhaustive filtration $V_0 \supset V_1 \supset V_2 \supset \cdots$ such that $V_0/V_1 \cong \bar{\Delta}(\pi)$ and all other sections of the form $q^{2m}\bar{\Delta}(\pi)$ for $m > 0$. Moreover, $\Delta(\pi)$ has irreducible head $\cong L(\pi)$, and in the Grothendieck group:*

$$[\Delta(\pi)] = [\bar{\Delta}(\pi)] \Big/ \prod_{k=1}^{N}\prod_{r=1}^{m_k}(1-q_{\rho_k}^{2r}).$$

4.3. *Homological properties of standard modules*

Now that we have constructed the standard modules, we list some of their homological properties. Throughout the subsection, $\alpha \in Q_+$ is fixed.

Theorem 4.13: *[13, Theorem 4.12], [5, Theorm 3.12] Let $\pi \in \Pi(\alpha)$.*

(i) If $\mathrm{EXT}^d_{R_\alpha}(\Delta(\pi), V) \neq 0$ for some $d \geq 1$ and a finitely generated R_α-module V, then V has a composition factor $\simeq L(\sigma)$ for $\sigma \succ \pi$.

(ii) We have for all $d \geq 0$ and $\sigma \in \Pi(\alpha)$:

$$\dim_q \mathrm{EXT}^d_{R_\alpha}(\Delta(\pi), \bar{\nabla}(\sigma)) = \delta_{d,0}\delta_{\pi,\sigma}.$$

We say that an R_α-module V has a *Δ-filtration*, written $V \in \mathtt{Fil}(\Delta)$, if there is a *finite* filtration $V = V_0 \supset V_1 \supset \cdots \supset V_n = 0$ such that $V_i/V_{i+1} \simeq \Delta(\pi^{(i)})$ for each $i = 1, \dots, n-1$ and some $\pi^{(i)} \in \Pi(\alpha)$. If $V \in \mathtt{Fil}(\Delta)$, then by Theorem 4.13(ii), the (graded) multiplicity of $\Delta(\pi)$ in a Δ-filtration of V is well-defined (i.e. independent of the Δ-filtration) and is equal to that

$$[V : \Delta(\pi)]_q = \overline{\dim_q \mathrm{HOM}_{R_\alpha}(V, \bar{\nabla}(\pi))} \qquad (\pi \in \Pi(\alpha)).$$

Theorem 4.14: *[5, Theorem 3.13] Let V be a finitely generated R_α-module. Then $V \in \mathtt{Fil}(\Delta)$ if and only if $\mathrm{EXT}^1_{R_\alpha}(V, \bar{\nabla}(\sigma)) = 0$ for all $\sigma \in \Pi(\alpha)$.*

An immediate corollary is the following version of the 'BGG reciprocity'. Note that the projective cover $P(\pi)$ of the irreducible module $L(\pi)$ exists in view of the general theory described in §2.1.

Corollary 4.15: *[**13**, Remark 4.17], [**5**, Corollary 4.17] Let $\pi, \sigma \in \Pi(\alpha)$. Then $P(\pi) \in \mathtt{Fil}(\Delta)$ and $[P(\pi) : \Delta(\sigma)]_q = [\bar{\Delta}(\sigma) : L(\pi)]_q$.*

This implies the following important dimension formula. A more elementary proof of this formula is given in [**18**].

Corollary 4.16: *[**5**, Corollary 3.15] We have that*

$$\dim_q R_\alpha = \sum_{\pi \in \Pi(\alpha)} (\dim_q \Delta(\pi))(\dim_q \bar{\Delta}(\pi))$$
$$= \sum_{\pi=(m_1,\dots,m_N)\in\Pi(\alpha)} \frac{(\dim_q \bar{\Delta}(\pi))^2}{\prod_{k=1}^{N}\prod_{r=1}^{m_k}(1-q_{\rho_k}^{2r})}.$$

The following corollary yields a description of the standard modules $\Delta(\pi)$ and $\bar{\Delta}(\pi)$ in spirit of standardly stratified algebras, cf. [**13**, Corollary 4.18].

Corollary 4.17: *[**5**, Corollary 3.15] Let $\pi \in \Pi(\alpha)$, and*

$$K(\pi) := \sum_{\sigma \not\leq \pi} \sum_{f \in \mathrm{HOM}_{R_\alpha}(P(\sigma),P(\pi))} \mathrm{im} f, \quad \bar{K}(\pi) := \sum_{\sigma \not< \pi} \sum_{f \in \mathrm{HOM}_{R_\alpha}(P(\sigma),P(\pi))} \mathrm{im} f.$$

Then $\Delta(\pi) \cong P(\pi)/K(\pi)$ and $\bar{\Delta}(\pi) \cong P(\pi)/\bar{K}(\pi)$.

5. Projective Resolutions of Standard Modules

We now explain how the standard modules $\Delta(\rho)$ fit into some short exact sequences, giving an alternative way to deduce their properties. This bounds the projective dimension of standard modules, and allows us to construct some projective resolutions of standard modules. As usual, we work with a fixed convex order $\prec$ on Φ_+ and denote by ρ an arbitrary positive root.

5.1. *Minimal pairs*

A two-term root partition $\pi = (\beta, \gamma) \in \Pi(\rho)$ is called a *minimal pair* for ρ, if it is a minimal element of $\Pi(\rho) \setminus \{(\rho)\}$. Equivalently, a minimal pair for ρ is a pair (β, γ) of positive roots with $\beta + \gamma = \rho$ and $\beta \succ \gamma$ such that there exists no other pair (β', γ') of positive roots with $\beta' + \gamma' = \rho$ and $\beta \succ \beta' \succ \rho \succ \gamma' \succ \gamma$. Let $\mathrm{MP}(\rho)$ denote the set of all minimal pairs for ρ.

For $\pi = (\beta,\gamma) \in \mathrm{MP}(\rho)$, it follows from Theorem 3.15 and the minimality of π that all composition factors of $\operatorname{rad}\bar{\Delta}(\pi)$ are $\simeq L_\rho$. Since $\bar{\Delta}(\pi) = L_\beta \circ L_\gamma$ and $(L_\beta \circ L_\gamma)^{\circledast} \cong q^{(\beta,\gamma)} L_\gamma \circ L_\beta$ by Lemma 3.9, there are short exact sequences

$$0 \longrightarrow q^{-(\beta,\gamma)} M^{\circledast} \longrightarrow L_\beta \circ L_\gamma \longrightarrow L(\pi) \longrightarrow 0, \tag{5.1}$$

$$0 \longrightarrow q^{-(\beta,\gamma)} L(\pi) \longrightarrow L_\gamma \circ L_\beta \longrightarrow M \longrightarrow 0, \tag{5.2}$$

where $M := q^{-(\beta,\gamma)}(\operatorname{rad}\bar{\Delta}(\pi))^{\circledast}$ is a finite dimensional module with all composition factors $\simeq L_\rho$. It turns out that one can be much more precise. Let $\Phi = \Phi_+ \sqcup -\Phi_+$ be the set of all roots. For any $\beta,\gamma \in \Phi$, define the number

$$p_{\beta,\gamma} := \max\{m \in \mathbb{Z} \mid \beta - m\gamma \in \Phi\}.$$

Theorem 5.3: *[**5**, Theorem 4.7, Corollary 4.3] Let $\pi = (\beta,\gamma) \in \mathrm{MP}(\rho)$.*

(i) There are short exact sequences

$$0 \longrightarrow q^{p_{\beta,\gamma}-(\beta,\gamma)} L_\rho \longrightarrow L_\beta \circ L_\gamma \longrightarrow L(\pi) \longrightarrow 0, \tag{5.4}$$

$$0 \longrightarrow q^{-(\beta,\gamma)} L(\pi) \longrightarrow L_\gamma \circ L_\beta \longrightarrow q^{-p_{\beta,\gamma}} L_\rho \longrightarrow 0. \tag{5.5}$$

(ii) In the Grothendieck group we have that

$$\left[\operatorname{Res}^{\rho}_{\gamma,\beta} L_\rho\right] = [p_{\beta,\gamma}+1]_q \left[L_\gamma \boxtimes L_\beta\right].$$

Moreover, $\operatorname{Res}_{\gamma,\beta} L_\rho$ is uniserial with socle $q^{p_{\beta,\gamma}} L_\gamma \boxtimes L_\beta$.

We now explain how minimal pairs allow one to get a generalization of high weight theory which was first developed in [**21**] for the so-called Lyndon convex orders. Fix an arbitrary minimal pair $\mathtt{mp}(\rho) \in \mathrm{MP}(\rho)$ for each non-simple positive root $\rho \in \Phi_+$. Dependent on this choice, we recursively define a word $\boldsymbol{i}_\rho \in \langle I \rangle_\rho$ and a bar-invariant Laurent polynomial $\kappa_\rho \in \mathscr{A}$ for all $\rho \in \Phi_+$ as follows. For $i \in I$ set $\boldsymbol{i}_{\alpha_i} := i$ and $\kappa_{\alpha_i} := 1$; then for non-simple $\rho \in \Phi_+$ suppose that $(\beta,\gamma) = \mathtt{mp}(\rho)$ and set

$$\boldsymbol{i}_\rho := \boldsymbol{i}_\gamma \boldsymbol{i}_\beta, \qquad \kappa_\rho := [p_{\beta,\gamma}+1]_q \, \kappa_\beta \kappa_\gamma. \tag{5.6}$$

For example, in simply-laced types we have that $\kappa_\rho = 1$ for all $\rho \in \Phi_+$; this is also the case in non-simply-laced types for *multiplicity-free* positive roots, i.e. roots $\rho = \sum_{i\in I} c_i\alpha_i$ with $c_i \in \{0,1\}$ for all i. Finally for a root partition $\pi = (\beta_1 \geq \cdots \geq \beta_l) = (\rho_1^{m_1}, \ldots, \rho_N^{m_N})$ let

$$\boldsymbol{i}_\pi := \boldsymbol{i}_{\beta_1} \cdots \boldsymbol{i}_{\beta_l}, \qquad \kappa_\pi := \prod_{k=1}^{N} [m_k]^!_{\rho_k} \kappa_{\rho_k}^{m_k}. \tag{5.7}$$

The following lemma shows that the words $\boldsymbol{i}_\pi$ distinguish irreducible modules, generalizing [**21**, Theorem 7.2(ii)].

Lemma 5.8: *[5, Lemma 4.5] Let $\alpha \in Q_+$ and $\pi, \sigma \in \Pi(\alpha)$. Then $\dim_q L(\pi)_{\boldsymbol{i}_\sigma} = 0$ if $\sigma \not\preceq \pi$, and $\dim_q L(\pi)_{\boldsymbol{i}_\pi} = \kappa_\pi$.*

5.2. *Projective resolutions*

Let $\rho \in \Phi_+$ be a non-simple positive root, (β, γ) be a minimal pair for ρ, and $m := \mathrm{ht}(\gamma)$. Let $w \in \mathfrak{S}_n$ be the permutation

$$(1, \dots, n) \mapsto (n - m + 1, \dots, n, 1, \dots, n - m),$$

so that $\psi_w 1_{\gamma,\beta} = 1_{\beta,\gamma}\psi_w$. It is proved in [**5**, Lemma 4.9] that there is a unique homogeneous homomorphism

$$\varphi : q^{-(\beta,\gamma)}\, \Delta(\beta) \circ \Delta(\gamma) \to \Delta(\gamma) \circ \Delta(\beta) \tag{5.9}$$

such that $\varphi(1_{\beta,\gamma} \otimes (v_1 \otimes v_2)) = \psi_w 1_{\gamma,\beta} \otimes (v_2 \otimes v_1)$ for all $v_1 \in \Delta(\beta), v_2 \in \Delta(\gamma)$.

Theorem 5.10: *[5, Theorem 4.10] For $(\beta, \gamma) \in \mathrm{MP}(\rho)$ there is a short exact sequence*

$$0 \longrightarrow q^{-(\beta,\gamma)}\, \Delta(\beta) \circ \Delta(\gamma) \xrightarrow{\varphi} \Delta(\gamma) \circ \Delta(\beta) \longrightarrow [p_{\beta,\gamma} + 1]_q\, \Delta(\rho) \longrightarrow 0.$$

Let us again fix a choice of minimal pairs $\mathtt{mp}(\rho) \in \mathrm{MP}(\rho)$ for each $\rho \in \Phi_+$ of height at least two, and recall κ_ρ and κ_π from (5.6). Let

$$\tilde{\Delta}(\rho) := \kappa_\rho \Delta(\rho), \qquad \tilde{\Delta}(\pi) := \kappa_\pi \Delta(\pi). \tag{5.11}$$

For simply laced $\mathtt{C}$ we have $\tilde{\Delta}(\rho) = \Delta(\rho)$. We want to construct a projective resolution $\mathbf{P}_*(\rho)$ of $\tilde{\Delta}(\rho)$ for each $\rho \in \Phi_+$. Then more generally, given $\alpha \in Q^+$ and $\pi = (\beta_1 \succeq \beta_2 \succeq \cdots \succeq \beta_l) \in \Pi(\alpha)$, the total complex of the '$\circ$'-product of the complexes $\mathbf{P}_*(\beta_1), \dots, \mathbf{P}_*(\beta_l)$ gives a projective resolution $\mathbf{P}_*(\pi)$ of $\tilde{\Delta}(\pi)$.

The resolution $\mathbf{P}_*(\rho)$ is going to be of the form

$$\mathbf{P}_*(\rho): \qquad 0 \to P_{n-1}(\rho) \longrightarrow \dots \longrightarrow P_1(\rho) \longrightarrow P_0(\rho) \longrightarrow \tilde{\Delta}(\rho) \to 0,$$

where $n = \mathrm{ht}(\rho)$. The construction of $\mathbf{P}_*(\rho)$ is recursive. For $i \in I$ we have $\tilde{\Delta}(\alpha_i) = R_{\alpha_i}$, which is projective already. So we just set $P_0(\alpha_i) := R_{\alpha_i}$ and $P_d(\alpha_i) := 0$ for $d \neq 0$ to obtain the required resolution. Now suppose that $\rho \in \Phi_+$ is of height at least two and let $(\beta, \gamma) := \mathtt{mp}(\rho)$, a fixed minimal pair for ρ. We may assume by induction that the projective resolutions

$\mathbf{P}_*(\beta)$ and $\mathbf{P}_*(\gamma)$ are already defined. Taking the total complex of their '$\circ$'-product using [**35**, Acyclic Assembly Lemma 2.7.3], we obtain a projective resolution $\mathbf{P}_*(\beta,\gamma)$ of $\tilde{\Delta}(\beta)\circ\tilde{\Delta}(\gamma)$ with

$$P_d(\beta,\gamma) := \bigoplus_{d_1+d_2=d} P_{d_1}(\beta)\circ P_{d_2}(\gamma),$$
$$\partial_d := \left(\mathrm{id}\circ\partial_{d_2} - (-1)^{d_2}\partial_{d_1}\circ\mathrm{id}\right)_{d_1+d_2=d} : P_d(\beta,\gamma)\to P_{d-1}(\beta,\gamma).$$

Similarly we obtain a projective resolution $\mathbf{P}_*(\gamma,\beta)$ of $\tilde{\Delta}(\gamma)\circ\tilde{\Delta}(\beta)$ with

$$P_d(\gamma,\beta) := \bigoplus_{d_1+d_2=d} P_{d_1}(\gamma)\circ P_{d_2}(\beta),$$
$$\partial_d := \left(\partial_{d_1}\circ\mathrm{id} + (-1)^{d_1}\,\mathrm{id}\circ\partial_{d_2}\right)_{d_1+d_2=d} : P_d(\gamma,\beta)\to P_{d-1}(\gamma,\beta).$$

There is an injective homomorphism

$$\tilde{\varphi} : q^{-(\beta,\gamma)}\tilde{\Delta}(\beta)\circ\tilde{\Delta}(\gamma)\hookrightarrow\tilde{\Delta}(\gamma)\circ\tilde{\Delta}(\beta)$$

defined in exactly the same way as the map φ in (5.9), indeed, it is just a direct sum of copies of the map φ from there. Applying [**35**, Comparision Theorem 2.2.6], $\tilde{\varphi}$ lifts to a chain map $\tilde{\varphi}_* : q^{-(\beta,\gamma)}\mathbf{P}_*(\beta,\gamma)\to\mathbf{P}_*(\gamma,\beta)$. Then we take the mapping cone of $\tilde{\varphi}_*$ to obtain a complex $\mathbf{P}_*(\rho)$ with

$$P_d(\rho) := P_d(\gamma,\beta)\oplus q^{-(\beta,\gamma)}P_{d-1}(\beta,\gamma),$$
$$\partial_d := (\partial_d, \partial_{d-1} + (-1)^{d-1}\tilde{\varphi}_{d-1}) : P_d(\rho)\to P_{d-1}(\rho).$$

In view of Theorem 5.10 and [**35**, Acyclic Assembly Lemma 2.7.3] once again, $\mathbf{P}_*(\rho)$ is a projective resolution of $\tilde{\Delta}(\rho)$.

Let us describe $\mathbf{P}_*(\rho)$ more explicitly. First, for $i\in I$ and the empty tuple $\boldsymbol{\sigma}$, set $\boldsymbol{i}_{\alpha_i,\boldsymbol{\sigma}} := i$. Now suppose that ρ is of height $n\geq 2$ and that $(\beta,\gamma)=\mathtt{mp}(\rho)$ with γ of height m. For $\boldsymbol{\sigma}=(\sigma_1,\dots,\sigma_{n-1})\in\{0,1\}^{n-1}$, let

$$|\boldsymbol{\sigma}| := \sigma_1+\dots+\sigma_{n-1},\ \boldsymbol{\sigma}_{<m} := (\sigma_1,\dots,\sigma_{m-1}),\ \boldsymbol{\sigma}_{>m} := (\sigma_{m+1},\dots,\sigma_{n-1}).$$

Define $\boldsymbol{i}_{\rho,\boldsymbol{\sigma}}\in\langle I\rangle_\rho$ and $d_{\rho,\boldsymbol{\sigma}}\in\mathbb{Z}_{\geq 0}$ recursively from

$$\boldsymbol{i}_{\rho,\boldsymbol{\sigma}} := \begin{cases}\boldsymbol{i}_{\gamma,\boldsymbol{\sigma}_{<m}}\boldsymbol{i}_{\beta,\boldsymbol{\sigma}_{>m}} & \text{if } \sigma_m=0,\\ \boldsymbol{i}_{\beta,\boldsymbol{\sigma}_{>m}}\boldsymbol{i}_{\gamma,\boldsymbol{\sigma}_{<m}} & \text{if } \sigma_m=1;\end{cases}$$

$$d_{\rho,\boldsymbol{\sigma}} := \begin{cases} d_{\beta,\boldsymbol{\sigma}_{>m}} + d_{\gamma,\boldsymbol{\sigma}_{<m}} & \text{if } \sigma_m=0,\\ d_{\beta,\boldsymbol{\sigma}_{>m}} + d_{\gamma,\boldsymbol{\sigma}_{<m}} - (\beta,\gamma) & \text{if } \sigma_m=1.\end{cases}$$

Note in particular that $d_{\rho,\boldsymbol{\sigma}}=|\boldsymbol{\sigma}|$ for simply-laced $\mathtt{C}$. Also if $\boldsymbol{\sigma}=(0,\dots,0)$ then $\boldsymbol{i}_{\rho,\boldsymbol{\sigma}}$ is the tuple $\boldsymbol{i}_\rho$ from (5.6) and $d_{\rho,\boldsymbol{\sigma}}=0$. Then we have that

$$P_d(\rho) = \bigoplus_{\substack{\boldsymbol{\sigma}\in\{0,1\}^{n-1}\\ |\boldsymbol{\sigma}|=d}} q^{d_{\rho,\boldsymbol{\sigma}}}R_\rho 1_{\boldsymbol{i}_{\rho,\boldsymbol{\sigma}}}. \tag{5.12}$$

For the differentials $\partial_d : P_d(\rho) \to P_{d-1}(\rho)$, there are elements $\psi_{\boldsymbol{\sigma},\boldsymbol{\tau}} \in 1_{\boldsymbol{i}_{\rho,\boldsymbol{\sigma}}} R_\rho 1_{\boldsymbol{i}_{\rho,\rho}}$ for each $\boldsymbol{\sigma},\boldsymbol{\tau} \in \{0,1\}^{n-1}$ with $|\boldsymbol{\sigma}| = d, |\boldsymbol{\tau}| = d-1$ such that, on viewing elements of (5.12) as row vectors, the differential ∂_d is defined by right multiplication by the matrix $(\psi_{\boldsymbol{\sigma},\boldsymbol{\tau}})_{|\boldsymbol{\sigma}|=d,|\boldsymbol{\tau}|=d-1}$. Moreover $\psi_{\boldsymbol{\sigma},\boldsymbol{\tau}} = 0$ unless the tuples $\boldsymbol{\sigma}$ and $\boldsymbol{\tau}$ differ in just one entry. We are able to give a very explicit description of the elements $\psi_{\boldsymbol{\sigma},\boldsymbol{\tau}}$ in the following special case.

Theorem 5.13: *Suppose that $\rho \in \Phi_+$ is multiplicity-free, so that $\kappa_\rho = 1$ and $\mathbf{P}_*(\rho)$ is a projective resolution of the root module $\Delta(\rho)$. Then the elements $\psi_{\boldsymbol{\sigma},\boldsymbol{\tau}}$ may be chosen so that*

$$\psi_{\boldsymbol{\sigma},\boldsymbol{\tau}} := (-1)^{\sigma_1+\cdots+\sigma_{r-1}}\psi_w$$

if $\boldsymbol{\sigma}$ and $\boldsymbol{\tau}$ differ just in the rth entry, where $w \in \mathfrak{S}_n$ is the unique *permutation with $1_{\boldsymbol{i}_{\rho,\boldsymbol{\sigma}}}\psi_w = \psi_w 1_{\boldsymbol{i}_{\rho,\boldsymbol{\tau}}}$.*

6. Type A

In this section, we sketch an elementary approach to the homological theory described above in type $\mathtt{C} = \mathtt{A}_\infty$, which is equivalent to working in an arbitrary finite type $\mathtt{A}$. The proofs we give are independent of the theory described in §§4 and 5.

6.1. *Set up*

Throughout the section $\mathcal{O}$ is an arbitrary commutative unital ring. Sometimes we need to assume that $\mathcal{O} = F$, i.e. that $\mathcal{O}$ is a field. We often work with a fixed arbitrary positive root of the form

$$\rho = \rho(k,l) := \alpha_k + \alpha_{k+1} + \cdots + \alpha_l \in \Phi_+ \qquad (k \le l).$$

We then have $d := \mathrm{ht}(\rho) = l - k + 1$. We refer to R_ρ as a *cuspidal block*. We also set $n := d - 1$, and define the word

$$\boldsymbol{i}_\rho = (k, k+1, \ldots, l).$$

We fix the convex order $\prec$ with $\alpha_k \prec \alpha_l$ if and only if $k < l$. Then $\rho(k,l) \prec \rho(r,s)$ if and only if $\boldsymbol{i}_{\rho(k,l)} < \boldsymbol{i}_{\rho(r,s)}$ in the usual lexicographic order.

For $\rho = \rho(k,l)$, the corresponding *cuspidal module* L_ρ is the rank one $\mathcal{O}$-module $\mathcal{O} \cdot v_\rho$ with the action of R_ρ on the basis vector v_ρ defined by

$$1_{\boldsymbol{j}} v_\rho = \delta_{\boldsymbol{j},\boldsymbol{i}_\rho} v_\rho, \quad y_r v_\rho = 0, \quad \psi_t v_\rho = 0$$

for all admissible $\boldsymbol{j}, r, t$.

For $\alpha \in Q_+$ recall the algebra R'_α defined in Remark 4.6. Denote

$$x_1 = y_1 - y_2,\ x_2 = y_2 - y_3, \dots, x_{d-1} = y_{d-1} - y_d,$$

where $d = \mathrm{ht}(\alpha)$. For $\boldsymbol{i} \in \langle I \rangle_\alpha$, define the projective modules

$$E(\boldsymbol{i}) := R_\alpha 1_{\boldsymbol{i}} \quad \text{and} \quad E'(\boldsymbol{i}) = R'_\alpha 1_{\boldsymbol{i}}$$

over R_α and R'_α respectively. It is easy to see that

$$E(\boldsymbol{i}) \circ E(\boldsymbol{j}) \cong E(\boldsymbol{ij}). \tag{6.1}$$

Pick any $\boldsymbol{i} = (i_1, \dots, i_d) \in \langle I \rangle_\alpha$, and consider the degree 2 element $z(\boldsymbol{i}) \in R_\alpha$ defined as the sum of *distinct* basis elements of the form $y_{u\cdot 1} 1_{u\cdot \boldsymbol{i}}$ with $u \in \mathfrak{S}_d$. In other words,

$$z(\boldsymbol{i}) = \sum_{\boldsymbol{j} \in \langle I \rangle_\alpha} \Big(\sum_{1 \le r \le d,\ j_r = i_1} y_r \Big) 1_{\boldsymbol{j}}. \tag{6.2}$$

For example:

$$z(112) = y_1 1_{112} + y_1 1_{121} + y_2 1_{112} + y_2 1_{211} + y_3 1_{121} + y_3 1_{211},$$
$$z(211) = y_1 1_{211} + y_2 1_{121} + y_3 1_{112}.$$

Note by [**14**, Theorem 2.9] that $z(\boldsymbol{i})$ is central in R_α.

Proposition 6.3: *Let $\alpha = \sum_{i \in I} m_i \alpha_i$, and $\mathcal{O}[X]$ be a polynomial ring in a variable X of degree 2. Suppose that $m_i \cdot 1_\mathcal{O}$ is a unit in $\mathcal{O}$ for some $i \in I$, and let $z_\alpha := z(\boldsymbol{i})$ for $\boldsymbol{i} \in \langle I \rangle_\alpha$ of the form $\boldsymbol{i} = (i^{m_i}, \dots)$, i.e. $\boldsymbol{i}$ begins with m_i lots of i's. Then there exists a homogeneous isomorphism of graded algebras*

$$R'_\alpha \otimes \mathcal{O}[X] \xrightarrow{\sim} R_\alpha,\ a \otimes X \mapsto a z_\alpha.$$

Proof: In view of the Basis Theorem, it suffices to prove that

$$\mathrm{span}_\mathcal{O}(x_1, \dots, x_{d-1}, z_\alpha) = \mathrm{span}_\mathcal{O}(y_1, \dots, y_d),$$

for which it is enough to see that $y_1 \in \mathrm{span}_\mathcal{O}(x_1, \dots, x_{d-1}, z_\alpha)$. By (6.2),

$$y_1 - (m_i \cdot 1_\mathcal{O})^{-1} z_\alpha = \sum_{\boldsymbol{j} \in \langle I \rangle_\alpha} \Big(y_1 - (m_i \cdot 1_\mathcal{O})^{-1} \sum_{1 \le r \le d,\ j_r = i} y_r \Big) 1_{\boldsymbol{j}}.$$

Note that $|\{1 \le r \le d,\ j_r = i\}| = m_i$ for all $\boldsymbol{j} \in \langle I \rangle_\alpha$, so the sum of the coefficients of y_k's in the expression above is zero, hence this expression is a linear combination of $x_1, \dots, x_{d-1}$. □

6.2. *Basic algebra B_n*

Let B be the unital $\mathcal{O}$-algebra generated by the elements $\{e(+), e(-), b\}$ subject only to the relations

$$e(\pm)e(\mp) = 0, \quad e(\pm)e(\pm) = e(\pm), \quad e(+) + e(-) = 1, \quad e(\pm)b = be(\mp).$$

In other words, B is the path algebra of the quiver $\overset{+}{\circ} \rightleftarrows \overset{-}{\circ}$. Setting $\deg e(\pm) := 0$, $\deg b := 1$ defines a grading on B. Note that $\{b^m e(\sigma) \mid m \in \mathbb{Z}_{\geq 0},\ \sigma \in \{+,-\}\}$ is a basis of B.

Let $P(\pm) := Be(\pm)$, and $L(\pm)$ be the rank one $\mathcal{O}$-module $\mathcal{O} \cdot v_\pm$ on the basis vector $v_\pm$, with the action of B defined by

$$e(\tau)v_\sigma = \delta_{\tau,\sigma} v_\sigma, \quad bv_\sigma = 0 \qquad (\sigma, \tau \in \{+,-\}).$$

If $\mathcal{O} = F$, then $L(+)$ and $L(-)$ are the irreducible B-modules with projective covers $P(+)$ and $P(-)$ respectively.

More generally, let $B_n = B^{\otimes n}$. For $1 \leq r \leq n$, set

$$b_r := 1 \otimes \cdots \otimes 1 \otimes b \otimes 1 \otimes \cdots \otimes 1,$$

with b in the rth position, and for $\boldsymbol{\sigma} = (\sigma_1, \dots, \sigma_n) \in \{\pm\}^n$ set

$$e(\boldsymbol{\sigma}) := e(\sigma_1) \otimes \cdots \otimes e(\sigma_n).$$

For $1 \leq r \leq n$, denote

$$\boldsymbol{\varepsilon}_r := (+ \cdots +, -, +, \dots, +),$$

with '$-$' in the rth position. For any $\boldsymbol{\sigma} = (\sigma_1, \dots, \sigma_n) \in \{\pm\}^n$, we allow ourselves to multiply

$$\boldsymbol{\varepsilon}_r \boldsymbol{\sigma} := (\sigma_1, \dots, \sigma_{r-1}, -\sigma_r, \sigma_{r+1}, \dots, \sigma_n).$$

Then B_n is generated by $\{e(\boldsymbol{\sigma}),\ b_r \mid \boldsymbol{\sigma} \in \{\pm\}^n,\ 1 \leq r \leq n\}$ subject only to the relations

$$e(\boldsymbol{\sigma})e(\boldsymbol{\tau}) = \delta_{\boldsymbol{\sigma},\boldsymbol{\tau}} e(\boldsymbol{\sigma}),\ \textstyle\sum_{\boldsymbol{\sigma} \in \{\pm\}^n} e(\boldsymbol{\sigma}) = 1, \tag{6.4}$$

$$b_r b_s = b_s b_r, \tag{6.5}$$

$$e(\boldsymbol{\sigma})b_r = b_r e(\boldsymbol{\varepsilon}_r \boldsymbol{\sigma}). \tag{6.6}$$

The algebra B_n is graded with $\deg e(\boldsymbol{\sigma}) = 0$, $\deg b_r = 1$, and has basis

$$\{b_1^{m_1} \dots b_n^{m_n} e(\boldsymbol{\sigma}) \mid m_1, \dots, m_n \in \mathbb{Z}_{\geq 0},\ \boldsymbol{\sigma} \in \{\pm\}^n\}. \tag{6.7}$$

For $\boldsymbol{\sigma} \in \{\pm\}^n$, let $P(\boldsymbol{\sigma}) := B_n e(\boldsymbol{\sigma})$, and $L(\boldsymbol{\sigma})$ be the rank one $\mathcal{O}$-module $\mathcal{O} \cdot v_{\boldsymbol{\sigma}}$ on the basis vector $b_{\boldsymbol{\sigma}}$, with the action of B_n defined by

$$e(\boldsymbol{\tau})v_{\boldsymbol{\sigma}} = \delta_{\boldsymbol{\tau},\boldsymbol{\sigma}} v_{\boldsymbol{\sigma}}, \quad b_r v_{\boldsymbol{\sigma}} = 0 \qquad (\boldsymbol{\tau} \in \{\pm\}^n,\ 1 \leq r \leq n).$$

If $\mathcal{O} = F$, then $\{L(\boldsymbol{\sigma}) \mid \boldsymbol{\sigma} \in \{\pm\}^n\}$ is a complete irredundant set of irreducible B_n-modules, and $P(\boldsymbol{\sigma})$ is a projective cover of $L(\boldsymbol{\sigma})$ for every $\boldsymbol{\sigma}$.

For $n = 1$, we have a linear minimal projective resolution $\mathbf{P}_*(\pm)$ of $L(\pm)$:

$$0 \longrightarrow qP(\mp) \xrightarrow{\partial} P(\pm) \longrightarrow L(\pm) \longrightarrow 0,$$

where the map ∂ is the right multiplication by b, i.e.

$$\partial(b^m e(\mp)) = b^{m+1} e(\pm) \qquad (m \in \mathbb{Z}_{\geq 0}).$$

For a general n and an arbitrary $\boldsymbol{\sigma} \in \{\pm\}^n$, we have a linear minimal projective resolution $\mathbf{P}_*(\boldsymbol{\sigma}) := \mathbf{P}_*(\sigma_1) \otimes \cdots \otimes \mathbf{P}_*(\sigma_n)$ of $L(\boldsymbol{\sigma})$:

$$0 \longrightarrow P_n(\boldsymbol{\sigma}) \xrightarrow{\partial_n} P_{n-1}(\boldsymbol{\sigma}) \xrightarrow{\partial_{n-1}} \cdots \xrightarrow{\partial_1} P_0(\boldsymbol{\sigma}) \longrightarrow L(\boldsymbol{\sigma}) \longrightarrow 0, \tag{6.8}$$

where

$$P_m(\boldsymbol{\sigma}) = \bigoplus_{1 \leq r_1 < \cdots < r_m \leq n} q^m Be(\boldsymbol{\varepsilon}_{r_1} \ldots \boldsymbol{\varepsilon}_{r_m} \boldsymbol{\sigma}) \qquad (0 \leq m \leq n),$$

and the map ∂_m is defined on the direct summand $Be(\boldsymbol{\varepsilon}_{r_1} \ldots \boldsymbol{\varepsilon}_{r_m} \boldsymbol{\sigma})$ of $P_m(\boldsymbol{\sigma})$ as the right multiplication by $\sum_{k=1}^{m} (\prod_{l=1}^{k-1} \sigma_{r_l}) b_{r_k}$ with $\prod_{l=1}^{k-1} \sigma_{r_l} \in \{\pm\}$ interpreted as $\pm 1_{\mathcal{O}}$. In other words,

$$\partial_m = \bigoplus_{1 \leq r_1 < \cdots < r_m \leq n} \partial_m^{r_1, \ldots, r_m} : P_m(\boldsymbol{\sigma}) \to P_{m-1}(\boldsymbol{\sigma})$$

where

$$\partial_m^{r_1, \ldots, r_m} = \sum_{k=1}^{m} \partial_{m, r_k} : Be(\boldsymbol{\varepsilon}_{r_1} \ldots \boldsymbol{\varepsilon}_{r_m} \boldsymbol{\sigma}) \to P_{m-1}(\boldsymbol{\sigma})$$

for the homomorphism

$$\partial_{m, r_k} : Be(\boldsymbol{\varepsilon}_{r_1} \ldots \boldsymbol{\varepsilon}_{r_m} \boldsymbol{\sigma}) \to Be(\boldsymbol{\varepsilon}_{r_1} \ldots \widehat{\boldsymbol{\varepsilon}_{r_k}} \ldots \boldsymbol{\varepsilon}_{r_m} \boldsymbol{\sigma})$$

which maps

$$b_1^{a_1} \ldots b_n^{a_n} e(\boldsymbol{\varepsilon}_{r_1} \ldots \boldsymbol{\varepsilon}_{r_m} \boldsymbol{\sigma}) \in Be(\boldsymbol{\varepsilon}_{r_1} \ldots \boldsymbol{\varepsilon}_{r_m} \boldsymbol{\sigma})$$

to

$$(\prod_{l=1}^{k-1} \sigma_{r_l}) b_1^{a_1} \ldots b_n^{a_n} b_{r_k} e(\boldsymbol{\varepsilon}_{r_1} \ldots \widehat{\boldsymbol{\varepsilon}_{r_k}} \ldots \boldsymbol{\varepsilon}_{r_m} \boldsymbol{\sigma}) \in Be(\boldsymbol{\varepsilon}_{r_1} \ldots \widehat{\boldsymbol{\varepsilon}_{r_k}} \ldots \boldsymbol{\varepsilon}_{r_m} \boldsymbol{\sigma}).$$

We compute extensions between irreducible B_n-modules. First, for $n = 1$ we have:

Lemma 6.9: *We have*

$$\mathrm{EXT}^m_{B_1}(L(\pm), L(\pm)) = \begin{cases} \mathcal{O} \text{ if } m = 0, \\ 0 \text{ otherwise,} \end{cases}$$

$$\mathrm{EXT}^m_{B_1}(L(\pm), L(\mp)) = \begin{cases} q^{-1}\mathcal{O} \text{ if } m = 1, \\ 0 \qquad \text{otherwise.} \end{cases}$$

Proof: This is obtained by applying an appropriate functor $\mathrm{HOM}_{B_1}(-, L(\tau))$ to the resolution $\mathbf{P}_*(\sigma)$ for $\sigma, \tau \in \{\pm\}$ and computing cohomology. For example, let us compute $\mathrm{EXT}^m_{B_1}(L(\pm), L(\mp))$. An application of the functor $\mathrm{HOM}_{B_1}(-, L(\mp))$ to the resolution $\mathbf{P}_*(\pm)$ yields:

$$0 \longrightarrow \mathrm{HOM}_{B_1}(qP(\mp), L(\mp))\rangle \xrightarrow{\partial^*} \mathrm{HOM}_{B_1}(P(\pm), L(\mp)) \longrightarrow 0.$$

But $\mathrm{HOM}_{B_1}(qP(\mp), L(\mp)) \cong q^{-1}\mathcal{O}$ and $\mathrm{HOM}_{B_1}(P(\pm), L(\mp)) = 0$, which immediately implies the required result. □

Now we can deal with general n.

Proposition 6.10: *Let $\boldsymbol{\sigma}, \boldsymbol{\tau} \in \{\pm\}^n$. Write $\boldsymbol{\sigma} = \varepsilon_{r_1} \dots \varepsilon_{r_m} \boldsymbol{\tau}$ for unique $1 \le r_1 < \dots < r_m \le n$. Then:*

$$\mathrm{EXT}^k_{B_n}(L(\boldsymbol{\sigma}), L(\boldsymbol{\tau})) = \begin{cases} q^{-k}\mathcal{O} \text{ if } k = m; \\ 0 \qquad \text{otherwise.} \end{cases}$$

Proof: This obtained using Künneth formula and Lemma 6.9, since $B_r \simeq B_1 \otimes \dots \otimes B_1$ and $L(\boldsymbol{\sigma}) \simeq L(\sigma_1) \boxtimes \dots \boxtimes L(\sigma_n)$. □

6.3. *Skew shapes*

Recall the standard notions concerning multipartitions from §1.2. We are now going consider the special case $l = 1$ and $\Lambda = \Lambda_0$, so l-multipartitions are just partitions and boxes on the main diagonal have content 0. If μ is a Young diagram contained in a Young diagram λ then $\lambda \setminus \mu$ is called a *skew shape*. Skew shapes are identified up to the shifts along diagonals. For example, the following picture, with boxes marked with their contents, illustrates why $(7, 7, 4, 1) \setminus (7, 3, 2, 1) = (6, 3) \setminus (2, 1)$:

0	1	2	3	4	5	6
-1	0	1	2	3	4	5
-2	-1	0	1			
-3						

$=$

0	1	2	3	4	5
-1	0	1			

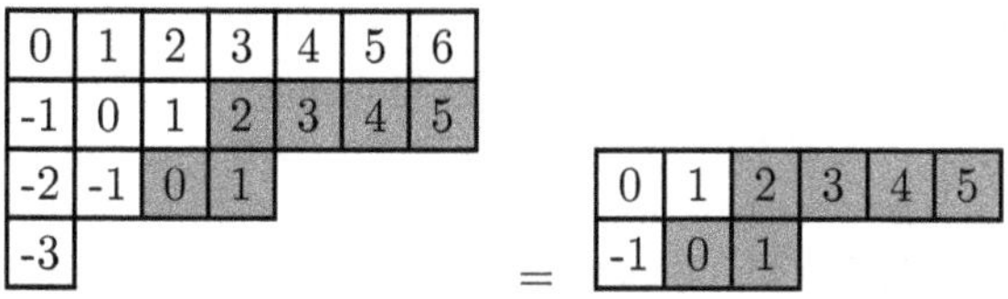

The given skew shape is determined by its Young diagram with box contents:

		2	3	4	5
0	1				

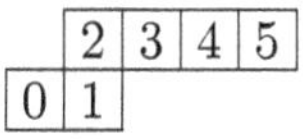

To the positive root $\rho = \rho(k, l)$, we associate the set $\mathscr{S}_\rho$ of all skew shapes containing exactly one box of each of the contents $k, k+1, \ldots, l$. For example, in the picture above the skew shape belongs to $\mathscr{S}_{\rho(0,5)}$.

As for partitions, for any $\lambda \in \mathscr{S}_\rho$, a λ-*tableau* T is an allocation of the numbers $1, 2, \ldots, d$ into the boxes of λ (recall that we have put $d := \mathrm{ht}(\rho)$). The symmetric group $\mathfrak{S}_d$ acts on the λ-tableaux by permutations of entries. For example $s_r \cdot$ T is T with the entries r and $r+1$ swapped. A λ-tableau is *standard* if its entries increase along the columns from top to bottom and along the rows from left to right. The set of the standard λ-tableaux is denoted $\mathrm{St}(\lambda)$.

The leading λ-tableaux T^λ is obtained by inserting the numbers $1, \ldots, d$ into the boxes of λ from left to right along the rows starting from the first row, then the second row, and so on. Given any $\mathrm{T} \in \mathrm{St}(\lambda)$, define

$$\boldsymbol{i}^{\mathrm{T}} := i_1 \ldots i_d \in \langle I \rangle_\rho,$$

where i_r is the content of the box occupied in T with r. Finally, set $\boldsymbol{i}^\lambda := \boldsymbol{i}^{\mathrm{T}^\lambda}$.

There is a one-to-one correspondence between the set $\mathscr{S}_\rho$ and the set $\{\pm\}^n$; in particular, $|\mathscr{S}_\rho| = 2^n$. To construct a bijection, number the boxes of a skew shape $\lambda \in \mathscr{S}_\rho$ by the numbers $1, \ldots, d$ from bottom-left to top-right, and for $r = 1, \ldots, n$, set $\sigma_r := +$ if the rth box of λ is not in the end of its row, and $\sigma_r := -$ otherwise. This will produce a sequence

$$\boldsymbol{\sigma}^\lambda = (\sigma_1, \ldots, \sigma_n) \in \{\pm\}^n.$$

The following is elementary:

Lemma 6.11: *The map* $\mathscr{S}_\rho \to \{\pm\}^n$, $\lambda \mapsto \boldsymbol{\sigma}^\lambda$ *is a bijection.*

For any $1 \le r \le n$, define the *row splitting operator* $\mathtt{sp}_r$ on the set $\mathscr{S}_\rho$ as follows. Recall that we number the boxes with the numbers $1, \dots, d$ from bottom-left to top-right. Let $\lambda \in \mathscr{S}_\rho$, and the rth box A of λ lie in the mth row; if A is not in the end of the mth row, then $\mathtt{sp}_r\lambda$ is the skew shape obtained from λ by splitting its mth row at A, so that A is now in the end of its row in $\mathtt{sp}_r\lambda$. On the other hand, if A is at the end of the mth row, then $\mathtt{sp}_r\lambda$ is the skew shape obtained from λ by attaching the $(m-1)$st row to the end of the mth row. For example:

$$\mathtt{sp}_4 \cdot \begin{array}{ccccc} & 3 & 4 & 5 & 6 \\ 1 & 2 & & & \end{array} = \begin{array}{ccccc} & & & 5 & 6 \\ & & 3 & 4 & \\ 1 & 2 & & & \end{array},$$

$$\mathtt{sp}_2 \cdot \begin{array}{ccccc} & 3 & 4 & 5 & 6 \\ 1 & 2 & & & \end{array} = \begin{array}{cccccc} 1 & 2 & 3 & 4 & 5 & 6 \end{array}.$$

The key properties of the row splitting are as follows:

Lemma 6.12: *Let $\lambda, \mu \in \mathscr{S}_\rho$ and $1, \le r, s \le n$. Then:*

(i) $\mathtt{sp}_r^2 = \mathrm{id}$;
(ii) $\mathtt{sp}_r\mathtt{sp}_s = \mathtt{sp}_s\mathtt{sp}_r$;
(iii) $\boldsymbol{\sigma}^{\mathtt{sp}_r\lambda} = \varepsilon_r\boldsymbol{\sigma}^\lambda$.
(iv) *There exist unique distinct numbers $r_1, \dots, r_l$ such that $1 \le r_1, \dots, r_l \le n$ and $\lambda = \mathtt{sp}_{r_1} \cdots \mathtt{sp}_{r_l}\mu$.*

Proof: Part (iii) is clear from the definitions. The rest follows from (iii) and Lemma 6.11. □

6.4. *The elements $\psi_{\lambda,\mu}$*

Since $\rho \in \Phi_+$ has coefficients at most 1 when decomposed as a linear combination of simple roots, the elements $\psi_u \in R_\rho$ are well-defined for all $u \in \mathfrak{S}_d$. Moreover, the action of the symmetric group $\mathfrak{S}_d$ on $\langle I\rangle_\rho$ is regular, and so for any $\lambda, \mu \in \mathscr{S}_\rho$, there is a unique element $w(\lambda, \mu) \in \mathfrak{S}_d$ such that $w(\lambda, \mu) \cdot \boldsymbol{i}^\mu = \boldsymbol{i}^\lambda$. This yields well-defined elements

$$\psi(\lambda, \mu) := \psi_{w(\lambda,\mu)} \qquad (\lambda, \mu \in \mathscr{S}_\rho).$$

Note that $\psi(\lambda, \mu)1_{\boldsymbol{i}^\mu} = 1_{\boldsymbol{i}^\lambda}\psi(\lambda, \mu)1_{\boldsymbol{i}^\mu} = 1_{\boldsymbol{i}^\lambda}\psi(\lambda, \mu)$.

Lemma 6.13: *Let $\mu \in \mathscr{S}_\rho$, $r_1, \dots, r_l$ be distinct numbers such that $1 \le r_1, \dots, r_l \le n$ and $\lambda = \mathtt{sp}_{r_1} \cdots \mathtt{sp}_{r_l}\mu$. For $k = 0, 1 \dots, l$, denote $\mu^{(k)} :=$*

$\prod_{m=1}^{k} \mathrm{sp}_{r_m}\mu$, *so that* $\mu = \mu^{(0)}$ *and* $\lambda = \mu^{(l)}$. *Then*

$$w(\lambda,\mu) = w(\mu^{(l)},\mu^{(l-1)})w(\mu^{(l-1)},\mu^{(l-2)})\dots w(\mu^{(1)},\mu^{(0)}).$$

Proof: The left hand side and the right hand side map $\boldsymbol{i}^{\mu}$ to $\boldsymbol{i}^{\lambda}$ and the symmetric group acts on $\langle I\rangle_{\rho}$ regularly. □

Proposition 6.14: *Let* $\rho \in \Phi_+$, $d = \mathrm{ht}(\rho)$, *and* $n = d-1$.

(i) *Let* $\nu \in \mathscr{S}_\rho$ *and* $1 \le r \le n$. *Denote the row number of the rth box in* ν *by* m. *Let* a *be the number of the leftmost box of the row* m *in* ν *and* b *be the number of the rightmost box of the row* m *in* ν.

(a) *If* $r < b$, *then*

$$\psi(\nu,\mathrm{sp}_r\nu)\psi(\mathrm{sp}_r\nu,\nu)1_{\boldsymbol{i}^\nu} = x_{d-b+r-a+1}1_{\boldsymbol{i}^\nu}.$$

(b) *If* $r = b$, *denote by* c *the number of the rightmost box in the row* $m-1$ *of* ν. *Then*

$$\psi(\nu,\mathrm{sp}_r\nu)\psi(\mathrm{sp}_r\nu,\nu)1_{\boldsymbol{i}^\nu} = -\sum_{k=d-c+1}^{d-a} x_k 1_{\boldsymbol{i}^\nu}.$$

(ii) *Let* $\mu \in \mathscr{S}_\rho$, $r_1,\dots,r_l$ *be distinct numbers such that* $1 \le r_1,\dots,r_l \le n$, *and* $\lambda = \mathrm{sp}_{r_1}\dots\mathrm{sp}_{r_l}\mu$. *For* $k = 0,1\dots,l$, *denote* $\mu^{(k)} := \prod_{m=1}^{k}\mathrm{sp}_{r_m}\mu$, *so that* $\mu = \mu^{(0)}$ *and* $\lambda = \mu^{(l)}$. *Then*

$$\psi(\lambda,\mu)1_{\boldsymbol{i}^\mu} = \psi(\lambda,\mu^{(l-1)})\psi(\mu^{(l-1)},\mu^{(l-2)})\dots\psi(\mu^{(1)},\mu)1_{\boldsymbol{i}^\mu}. \quad (6.15)$$

Proof: Assume without loss of generality that $\rho = \rho(1,d)$. Then the content of the tth box of any $\nu \in \mathscr{S}_\rho$ is t for all $t = 1,\dots,d$.

We first prove (i). The reader should keep in mind the following picture for r, a, b in ν:

$\nu =$ 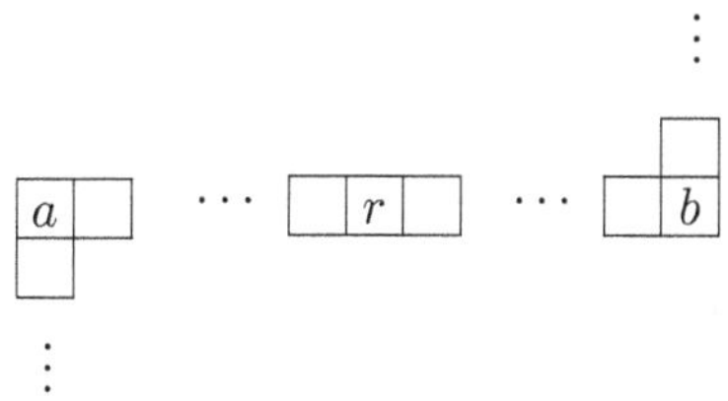

Also, if $r = b$, then:

$\nu =$ 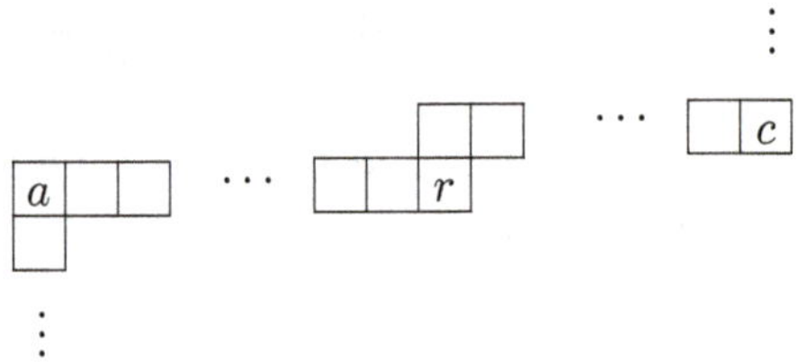

(a) If $r < b$, then, using the geometric presentation of the elements of R_ρ introduced in [**14**], we have

$\psi(\mathbf{sp}_r\nu, \nu)1_{\boldsymbol{i}^\nu} =$ 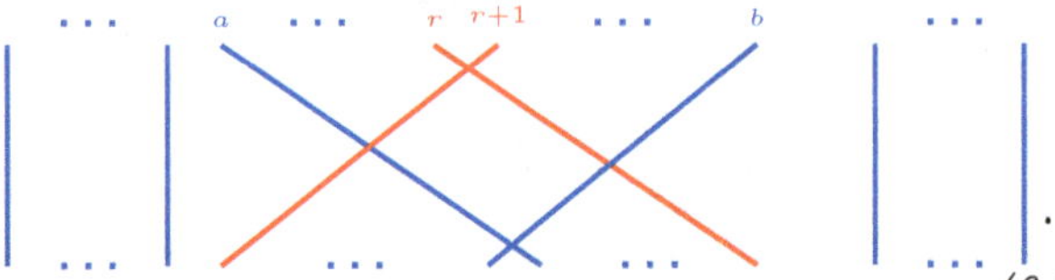

. (6.16)

Then $\psi(\nu, \mathbf{sp}_r\nu)\psi(\mathbf{sp}_r\nu, \nu)1_{\boldsymbol{i}^\nu}$ equals

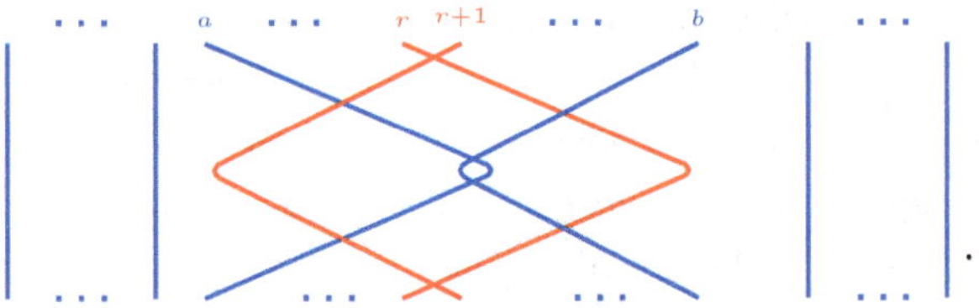

.

Using defining relations in R_ρ, we see that this element equals

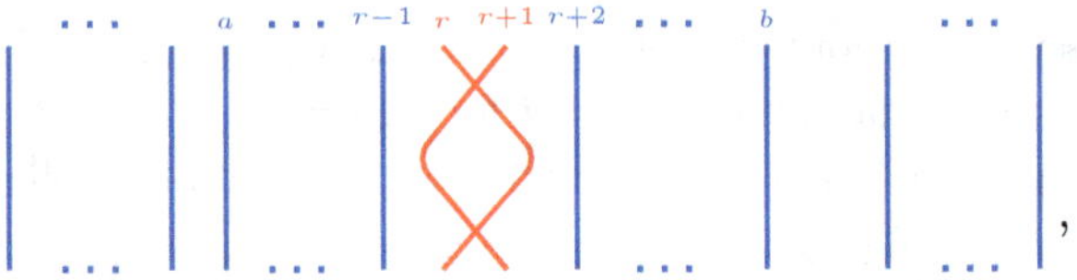

,

which equals

,

which is $(y_{d-b+r-a+1} - y_{d-b+r-a+2})1_{\boldsymbol{i}^\nu} = x_{d-b+r-a+1}1_{\boldsymbol{i}^\nu}$.

(b) If $r = b$, then

$$\psi(\mathtt{sp}_r\nu, \nu)1_{\boldsymbol{i}^\nu} = $$

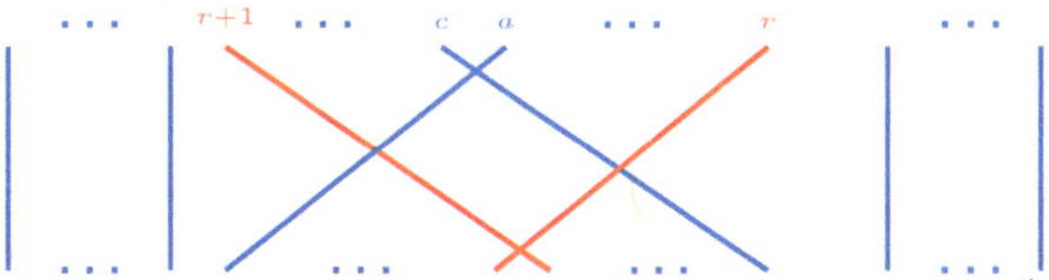

. (6.17)

Then

$$\psi(\nu, \mathtt{sp}_r\nu)\psi(\mathtt{sp}_r\nu, \nu)1_{\boldsymbol{i}^\nu} = $$

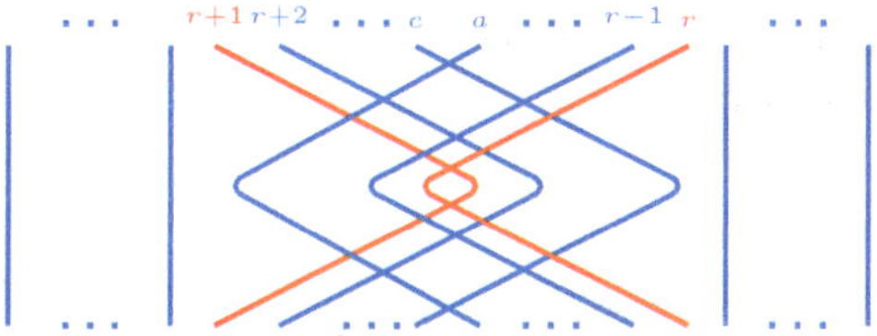

,

which equals

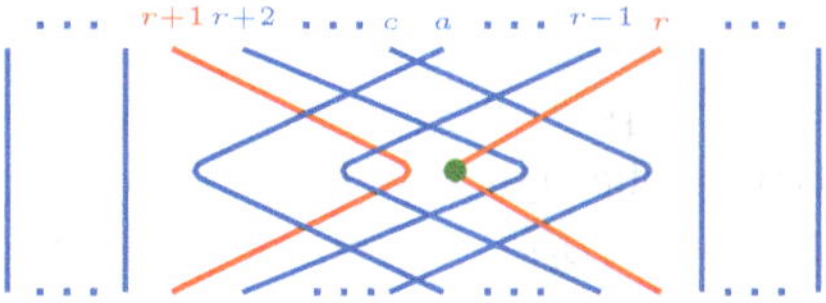

$-$

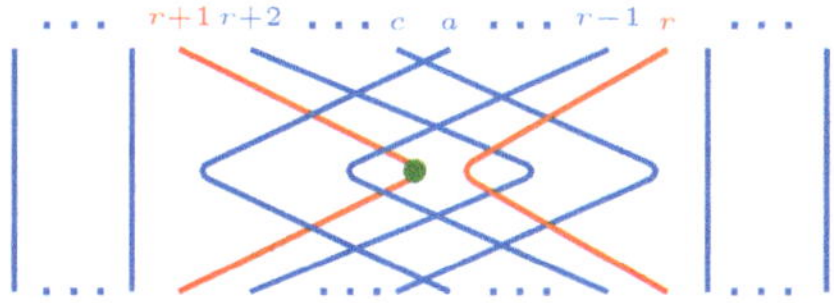

$=$

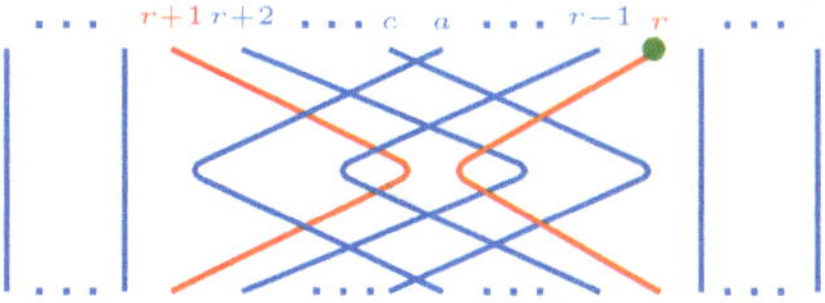

$-$

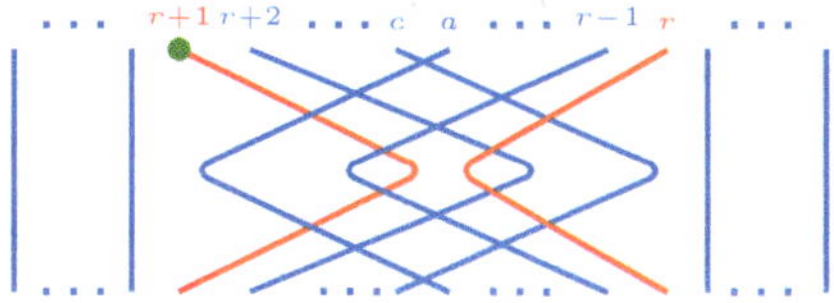

$=$

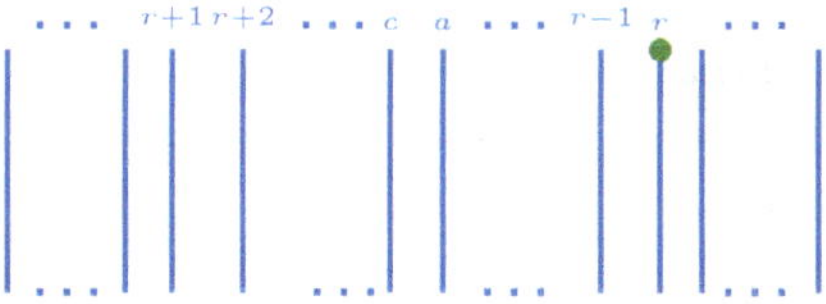

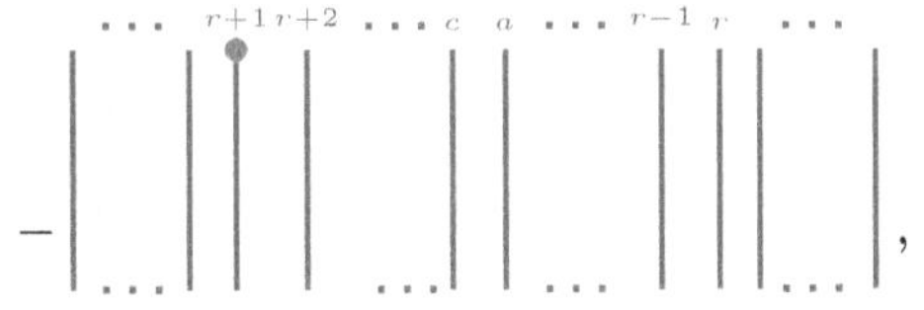

which is $(-y_{d-c+1}+y_{d-a+1})1_{\boldsymbol{i}^\nu} = -\sum_{k=d-c+1}^{d-a} x_k 1_{\boldsymbol{i}^\nu}$.

(ii) Note from (6.16) and (6.17) that r and $r+1$ is the only pair of neighboring entries in $\boldsymbol{i}^\nu$ that get permuted in the pictures above (the corresponding strings are colored red). This can be restated as the claim that we can write $\psi(\mathtt{sp}_r\nu,\nu)1_{\boldsymbol{i}^\nu}$ as a product of elements of the form $\psi_t 1_{\boldsymbol{j}}$ with $|j_t - j_{t+1}| > 1$ for all factors but one, and for that exceptional factor we have $\{j_t, t_{t+1}\} = \{r, r+1\}$.

Applying this observation repeatedly to the product in the right hand side of (6.15), we conclude that it can be written as a product of elements of the form $\psi_t 1_{\boldsymbol{j}}$ with $|j_t - j_{t+1}| > 1$ for all but l special factors, which are of the form, $\psi_{t_k} 1_{\boldsymbol{j}^{(k)}}$ for $k = 1, \dots, l$, and we have $\{j^{(k)}_{t_k}, j^{(k)}_{t_k+1}\} = \{r_k, r_k + 1\}$ for all $k = 1, \dots, l$. In particular, this means that we can get to the reduced decomposition of the right hand side using only braid relations of the form $\psi_t\psi_{t+1}\psi_t 1_{\boldsymbol{j}} = \psi_{t+1}\psi_t\psi_{t+1}1_{\boldsymbol{j}}$, and the quadratic relations of the form $\psi_t^2 1_{\boldsymbol{j}} = 1_{\boldsymbol{j}}$. Now part (ii) follows from Lemma 6.13. □

6.5. *Irreducibles and PIMs for cuspidal blocks*

As a special case of the main result of [**20**], we can describe all irreducible R_ρ-modules explicitly in spirit of Theorem 1.17. Given a skew shape $\lambda \in \mathscr{S}_\rho$, consider the free $\mathcal{O}$-module

$$L^\lambda = \bigoplus_{\mathtt{T}\in\mathtt{St}(\lambda)} \mathcal{O}\cdot v_{\mathtt{T}}$$

concentrated in degree 0 with $\mathcal{O}$-basis labelled by the standard λ-tableaux. The action of R_ρ on L^λ is defined by

$$1_{\boldsymbol{j}} v_{\mathtt{T}} = \delta_{\boldsymbol{j},\boldsymbol{i}^{\mathtt{T}}}, \quad y_r v_{\mathtt{T}} = 0, \quad \psi_t v_{\mathtt{T}} = \begin{cases} v_{s_r\cdot\mathtt{T}} & \text{if } s_r\cdot\mathtt{T}\in\mathrm{St}(\lambda), \\ 0 & \text{otherwise} \end{cases} \tag{6.18}$$

for all admissible $\boldsymbol{j}, r, t$.

Moreover, suppose that λ has m rows. For $k = 1, \dots, m$, define β_k to be the sum of the simple roots α_i where i runs over the contents of the boxes in the row k of λ. Then

$$\pi(\lambda) = (\beta_1, \dots, \beta_m) \in \Pi(\rho).$$

Theorem 6.19: *Let $\mathcal{O} = F$. Then:*

(i) *The formulas (6.18) define a structure of an irreducible R_ρ-module on L^λ, which is also irreducible on restriction to R'_ρ.*
(ii) *$\{L^\lambda \mid \lambda \in \mathscr{S}_\rho\}$ is a complete and irredundant set of irreducible R_ρ-modules up to degree zero isomorphism and degree shift.*
(iii) *For any $\lambda \in \mathscr{S}_\rho$, we have $L^\lambda \simeq L(\pi(\lambda))$.*
(iv) *For any $\lambda \in \mathscr{S}_\rho$, we have $\mathrm{ch}_q L^\lambda = \sum_{\mathtt{T} \in \mathrm{St}(\lambda)} \boldsymbol{i}^{\mathtt{T}}$.*
(v) *If $\mathtt{T} \in \mathrm{St}(\lambda)$ and $\mathtt{S} \in \mathrm{St}(\mu)$ for some $\lambda, \mu \in \mathscr{S}_\rho$ then $\boldsymbol{i}^{\mathtt{T}} = \boldsymbol{i}^{\mathtt{S}}$ if and only if $\lambda = \mu$ and $\mathtt{T} = \mathtt{S}$.*

Proof: In view of [**20**, Theorem 3.4], all irreducible R_ρ-modules are homogeneous. Now parts (i) and (ii) follow from [**20**, Theorem 3.6]. Part (iii) follows from part (ii) and Theorem 3.15. Finally, (iv) is clear from definition, while (v) follows again from [**20**, Theorem 3.4]. □

Lemma 6.20: *Let $\lambda \in \mathscr{S}_\rho$. Then $E(\boldsymbol{i}^\lambda)$ is a projective cover of L^λ as R_ρ-modules or R'_ρ-modules.*

Proof: Since $\boldsymbol{i}^\lambda$ is a weight of L^λ, and the corresponding weight space generates L^λ, the projective module $E(\boldsymbol{i}^\lambda)$ surjects onto L^λ. It remains to prove that $1_{\boldsymbol{i}^\lambda}$ is a primitive idempotent, for which, using change of scalars, we may assume that $\mathcal{O} = F$. In that case, it suffices to notice that the head of $E(\boldsymbol{i}^\lambda)$ is isomorphic to $L(\lambda)$. This fact follows from $\mathrm{HOM}_{R_\rho}(E(\boldsymbol{i}^\lambda), L^\lambda) \simeq 1_{\boldsymbol{i}^\lambda} L^\lambda \simeq F$ and $\mathrm{HOM}_{R_\rho}(E(\boldsymbol{i}^\lambda), L^\mu) \simeq 1_{\boldsymbol{i}^\lambda} L^\mu = 0$ for $\mu \neq \lambda$, using Theorem 6.19(v). The proof for R'_ρ is the same. □

Corollary 6.21: *$\{1_{\boldsymbol{i}^\lambda} \mid \lambda \in \mathscr{S}_\rho\}$ is a complete set of inequivalent primitive idempotents in R_ρ and R'_ρ.*

6.6. *Basic algebras of cuspidal blocks*

By Corollary 6.21, $\{1_{\boldsymbol{i}^\lambda} \mid \lambda \in \mathscr{S}_\rho\}$ is a complete set of inequivalent primitive idempotents in R_ρ and R'_ρ. So, setting,

$$e := \sum_{\lambda \in \mathscr{S}_\rho} 1_{\boldsymbol{i}^\lambda}, \tag{6.22}$$

we see that $eR_\rho e$ is a basic algebra Morita equivalent to R_ρ and $eR'_\rho e$ is a basic algebra Morita equivalent to R'_ρ.

Lemma 6.23: *$\{x_1^{m_1} \dots x_n^{m_n} \psi(\lambda, \mu) 1_{\boldsymbol{i}^\mu} \mid \lambda, \mu \in \mathscr{S}_\rho,\ m_1, \dots, m_n \in \mathbb{Z}_{\geq 0}\}$ is an $\mathcal{O}$-basis of $eR'_\rho e$.*

Proof: This follows from Theorem 2.3 and the defining relations for R_ρ. □

Recall the bijection $\mathscr{S}_\rho \xrightarrow{\sim} \{\pm\}^n$, $\lambda \mapsto \boldsymbol{\sigma}^\lambda$ and the algebra B_n from §6.2.

Theorem 6.24: *There is a homogeneous isomorphism of graded algebras*

$$\iota : B_n \to eR'_\rho e,\ e(\boldsymbol{\sigma}^\lambda) \mapsto 1_{\boldsymbol{i}^\lambda},\ b_r e(\boldsymbol{\sigma}^\lambda) \mapsto \psi(\mathtt{sp}_r\lambda, \lambda)1_{\boldsymbol{i}^\lambda}$$

for all $\lambda \in \mathscr{S}_\rho$ *and* $1 \leq r \leq n$.

Proof: To prove that there is a homomorphism ι as in the statement of the theorem, we check the defining relations of B_n for the images of its generators. The relations (6.4) are clear since $1_{\boldsymbol{i}^\lambda}$'s are orthogonal idempotents which sum to the identity e in the algebra $eR'_\rho e$.

Note that by linearity we must have $\iota(b_r) = \sum_{\mu\in\mathscr{S}_\rho} \psi(\mathtt{sp}_r\mu, \mu)1_{\boldsymbol{i}^\mu}$. Moreover, by Lemma 6.12(iii), $\iota(e(\boldsymbol{\varepsilon}_r\boldsymbol{\sigma}^\lambda)) = 1_{\boldsymbol{i}^{\mathtt{sp}_r\lambda}}$. So to check the relation (6.6), we have to prove that for all $\lambda \in \mathscr{S}_\rho$ and $1 \leq r \leq n$ we have that

$$1_{\boldsymbol{i}^\lambda} \sum_{\mu\in\mathscr{S}_\rho} \psi(\mathtt{sp}_r\mu, \mu)1_{\boldsymbol{i}^\mu} = \Big(\sum_{\mu\in\mathscr{S}_\rho} \psi(\mathtt{sp}_r\mu, \mu)1_{\boldsymbol{i}^\mu}\Big)1_{\boldsymbol{i}^{\mathtt{sp}_r\lambda}}.$$

But in view of the relation (1.5), both sides are equal to $\psi(\lambda, \mathtt{sp}_r\lambda)1_{\boldsymbol{i}^{\mathtt{sp}_r\lambda}}$.

Now, to check the relation (6.5), it suffices to verify that

$$\iota(b_r b_s e(\boldsymbol{\sigma}^\lambda)) = \iota(b_s b_r e(\boldsymbol{\sigma}^\lambda))$$

for all admissible r, s, λ, or

$$\psi(\mathtt{sp}_s\mathtt{sp}_r\lambda, \mathtt{sp}_r\lambda)\psi(\mathtt{sp}_r\lambda, \lambda)1_{\boldsymbol{i}^\lambda} = \psi(\mathtt{sp}_r\mathtt{sp}_s\lambda, \mathtt{sp}_s\lambda)\psi_{\mathtt{sp}_s\lambda,\lambda}1_{\boldsymbol{i}^\lambda}.$$

Of course, we may assume that $r \neq s$. Then, by Proposition 6.14(ii), the left hand side is equal to $\psi(\mathtt{sp}_s\mathtt{sp}_r\lambda, \lambda)1_{\boldsymbol{i}^\lambda}$ and the right hand side is equal to $\psi(\mathtt{sp}_r\mathtt{sp}_s\lambda, \lambda)1_{\boldsymbol{i}^\lambda}$. It remains to apply Lemma 6.12(ii).

Finally, to prove that ι is an isomorphism, we show that the basis (6.7) of B_n is mapped by ι to a basis of $eR'_\rho e$. Note that for $1 \leq r \leq n$, we have

$$\iota(b_r^2 e(\boldsymbol{\sigma}^\lambda)) = \psi(\lambda, \mathtt{sp}_r\lambda)\psi(\mathtt{sp}_r\lambda, \lambda)1_{\boldsymbol{i}^\lambda}.$$

The right hand side has been computed in Proposition 6.14(i) as follows. Denote the row number of the rth box in λ by m. Let $a(r)$ be the number of the leftmost box of the row m in λ and $b(r)$ be the number of the rightmost box of the row m in λ. If $r < b(r)$, then by Proposition 6.14(i)(a), we have

$$\iota(b_r^2 e(\boldsymbol{\sigma}^\lambda)) = x_{d-b(r)+r-a(r)+1}1_{\boldsymbol{i}^\lambda}.$$

If $r = b(r)$, denote by $c(r)$ the number of the rightmost box in the row $m-1$ of λ. Then by Proposition 6.14(i)(b), we have

$$\iota(b_r^2 e(\boldsymbol{\sigma}^\lambda)) = -\sum_{k=d-c(r)+1}^{d-a(r)} x_k 1_{\boldsymbol{i}^\lambda}.$$

It is now easy to see that

$$\mathrm{span}_{\mathcal{O}}\left(\iota(b_1^2 e(\boldsymbol{\sigma}^\lambda)), \dots, \iota(b_n^2 e(\boldsymbol{\sigma}^\lambda))\right) = \mathrm{span}_{\mathcal{O}}(x_1 1_{\boldsymbol{i}^\lambda}, \dots, x_n 1_{\boldsymbol{i}^\lambda}).$$

Therefore

$$\{\iota(b_1^{2m_1} \dots b_n^{2m_n} e(\boldsymbol{\sigma}^\lambda)) \mid m_1, \dots, m_n \in \mathbb{Z}_{\geq 0}\}$$

is an $\mathcal{O}$-basis of

$$\mathrm{span}_{\mathcal{O}}(x_1^{m_1} \dots x_n^{m_n} 1_{\boldsymbol{i}^\lambda} \mid m_1, \dots, m_n \in \mathbb{Z}_{\geq 0}).$$

Moreover, let $\mu \in \mathscr{S}_\rho$. Let $1 \leq r_1, \dots, r_l \leq n$ be distinct numbers such that

$$\lambda = \mathtt{sp}_{r_1} \dots \mathtt{sp}_{r_l} \mu,$$

see Lemma 6.12(iv). It follows from Proposition 6.14(ii) that

$$\psi(\lambda, \mu) 1_{\boldsymbol{i}^\mu} = \iota(b_{r_1} \cdots_{r_l} e(\boldsymbol{\sigma}^\mu)).$$

It remains to apply Lemma 6.23. □

Theorem 6.25: *For $\rho \in \Phi_+$ with $\mathrm{ht}(\rho) = n+1$, there is a Morita equivalence $\mathcal{F} : B_n\text{-Mod} \to R'_\rho\text{-Mod}$ such that*

$$\mathcal{F}(P(\boldsymbol{\sigma}^\lambda)) = E(\boldsymbol{i}^\lambda), \quad \mathcal{F}(L(\boldsymbol{\sigma}^\lambda)) = L^\lambda$$

for all $\lambda \in \mathscr{S}_\rho$.

Proof: Let e be as in (6.22). Since $\{1_{\boldsymbol{i}^\lambda} \mid \lambda \in \mathscr{S}_\rho\}$ is a complete system of orthogonal primitive idempotents by Corollary 6.21, we have Morita equivalence

$$\mathcal{F} : eR'_\rho e\text{-Mod} \to R'_\rho\text{-Mod}, \ V \mapsto R'_\rho e \otimes_{eR'_\rho e} V.$$

By Theorem 6.24, there is an isomorphism $\iota : B_n \to eR'_\rho e$, which maps $e(\boldsymbol{\sigma}^\lambda)$ to $1_{\boldsymbol{i}^\lambda}$ for all $\lambda \in \mathscr{S}_\rho$. The result follows. □

The next corollary should be compared to Theorem 4.2.

Corollary 6.26: *Let $\lambda, \mu \in \mathscr{S}_\rho$. Write $\lambda = \mathtt{sp}_{r_1} \dots \mathtt{sp}_{r_m} \mu$ for unique $1 \leq r_1 < \dots < r_m \leq n$. Then:*

$$\mathrm{EXT}^k_{R'_\rho}(L^\lambda, L^\mu) \cong \begin{cases} q^{-k}\mathcal{O} & \text{if } k = m; \\ 0 & \text{otherwise.} \end{cases}$$

$$\mathrm{EXT}^k_{R_\rho}(L^\lambda, L^\mu) \cong \begin{cases} q^{-k}\mathcal{O} & \text{if } k = m; \\ q^{-k+3}\mathcal{O} & \text{if } k = m+1; \\ 0 & \text{otherwise.} \end{cases}$$

Proof: The first equality follows from Theorem 6.25 and Proposition 6.10. The second equality follows from the first and the Künneth Formula, since $R_\rho \cong R'_\rho \otimes \mathcal{O}[X]$ with $\deg X = 2$, thanks to Proposition 6.3. □

6.7. *Resolutions for cuspidal blocks*

Let again $\rho \in \Phi_+$ be a positive root of height $d = n + 1$. Even though we have already computed extensions between the irreducible modules for the corresponding cuspidal block, we still need projective resolutions of them.

Recall that we have defined line-breaking operators $\mathtt{sp}_r$ for all $1 \leq r < d$. Now define also $\mathtt{sp}_d$ to be the trivial operator: $\mathtt{sp}_d\lambda := \lambda$ for all $\lambda \in \mathscr{S}_\rho$. For a subset $T = \{t_1 < \dots < t_m\} \subseteq [1, d]$ and $1 \leq r \leq d$ define

$$\mathtt{sp}_T := \mathtt{sp}_{t_1} \dots \mathtt{sp}_{t_m}, \tag{6.27}$$

$$s(T) := \begin{cases} |T| + 1 & \text{if } d \in T \\ |T| & \text{if } d \notin T \end{cases}, \tag{6.28}$$

$$\sigma(T, r) := (-1)^{|\{t \in T \mid t < r\}|}. \tag{6.29}$$

Now fix $\lambda \in \mathscr{S}_\rho$. The resolution $\mathbf{P}'_*(\lambda)$ is the projective resolution of the R'_ρ-module L^λ obtained by applying the Morita equivalence $\mathcal{F}$ of Theorem 6.25 to the resolution $\mathbf{P}_*(\boldsymbol{\sigma}^\lambda)$, i.e. $\mathbf{P}'_*(\lambda)$ is:

$$0 \longrightarrow P'_n \xrightarrow{\partial'_n} P'_{n-1} \xrightarrow{\partial'_{n-1}} \dots \xrightarrow{\partial'_1} P'_0 \longrightarrow L^\lambda \longrightarrow 0, \tag{6.30}$$

where

$$P'_m = \bigoplus_{T \subseteq [1,n],\ |T| = m} q^m E'(\boldsymbol{i}^{\mathtt{sp}_T\lambda}) \qquad (0 \leq m \leq n),$$

and the map ∂'_m is defined on the direct summand $q^m E'(\boldsymbol{i}^{\mathtt{sp}_T\lambda})$ of P'_m as the right multiplication by $\sum_{t \in T} \sigma(T, t)\psi(\mathtt{sp}_T\lambda, \mathtt{sp}_{T \setminus \{t\}}\lambda)$.

Finally, by Proposition 6.3, we have $R_\rho \simeq R'_\rho \otimes \mathcal{O}[X]$ where X is an indeterminate of degree 2, which under the isomorphism corresponds to the central element $z_\rho \in R_\rho$. Now, the irreducible R_ρ-module L^λ can be considered as the outer tensor product $L^\lambda \boxtimes \mathcal{O}$ where L^λ is the restriction from R_ρ to R'_ρ of L^λ, and $\mathcal{O}$ is the $\mathcal{O}[X]$ module of rank 1 with the trivial action of X. Tensoring $\mathbf{P}'_*(\lambda)$ with the following resolution $\mathbf{K}_*$ of $\mathcal{O}$:

$$0 \longrightarrow \mathcal{O}[X] \xrightarrow{X} \mathcal{O}[X] \longrightarrow \mathcal{O} \longrightarrow 0,$$

we obtain the resolution $\mathbf{P}_*(\lambda) := \mathbf{P}'_*(\lambda) \otimes \mathbf{K}_*$:

$$0 \longrightarrow P_d \xrightarrow{\partial_d} P_{d-1} \xrightarrow{\partial_{d-1}} \cdots \xrightarrow{\partial_1} P_0 \longrightarrow L^\lambda \longrightarrow 0, \tag{6.31}$$

where the modules P_m and the maps ∂_m are defined as follows. For a subset $T \subseteq [1, d]$ and $t \in T$ define

$$P_T := q^{s(T)} E(\boldsymbol{i}^{\mathtt{sp}_T\lambda}),$$

and the map

$$\partial_{T,t} : P_T \to P_{T\setminus\{t\}} \tag{6.32}$$

to be the right multiplication by

$$\sigma(T,t) \cdot \begin{cases} \psi(\mathtt{sp}_T\lambda, \mathtt{sp}_{T\setminus\{t\}}\lambda) & \text{if } t < d \\ z_\rho & \text{if } t = d. \end{cases}$$

In other words,

$$\begin{aligned} \partial_{T,t} : x1_{\boldsymbol{i}^{\mathtt{sp}_T\lambda}} &\mapsto \sigma(T,t)x1_{\boldsymbol{i}^{\mathtt{sp}_T\lambda}}\psi(\mathtt{sp}_T\lambda, \mathtt{sp}_{T\setminus\{t\}}\lambda) \\ &= \sigma(T,t)x\psi(\mathtt{sp}_T\lambda, \mathtt{sp}_{T\setminus\{t\}}\lambda)1_{\boldsymbol{i}^{\mathtt{sp}_{T\setminus\{t\}}\lambda}}, \end{aligned}$$

if $t < d$, and

$$\partial_{T,d} : x1_{\boldsymbol{i}^{\mathtt{sp}_T\lambda}} \mapsto \sigma(T,d)x1_{\boldsymbol{i}^{\mathtt{sp}_T\lambda}}z_\rho = \sigma(T,d)xz_\rho 1_{\boldsymbol{i}^{\mathtt{sp}_T\lambda}}.$$

Note that $\partial_{T,t}$ is a homogeneous map. Now we have

$$P_m := \bigoplus_{T\subseteq[1,d],\ |T|=m} P_T \qquad (0 \le m \le d),$$

and

$$\partial_m = \bigoplus_{T\subseteq[1,d],\ |T|=m} \partial_T$$

is a degree zero homomorphism where

$$\partial_T = \sum_{t\in T} \partial_{T,t} : P_T \to P_{m-1}.$$

Finally, the case of one row skew shape $\lambda = (d)$ is especially important, since in this case $L^{(d)}$ is the cuspidal module L_ρ, and we have the projective resolution $\mathbf{P}_*(L_\rho) := \mathbf{P}_*((d))$ of the cuspidal module.

6.8. *Resolving proper standard and costandard modules*

Now consider a root partition $\pi = (\beta_1 \succeq \cdots \succeq \beta_l) \in \Pi(\alpha)$. We have the proper standard module $\bar{\Delta}(\pi) := q^{\mathtt{sh}(\pi)} L_{\beta_1} \circ \cdots \circ L_{\beta_l}$. By taking (outer) tensor product of resolutions $\mathbf{P}_*(L_{\beta_1}), \dots, \mathbf{P}_*(L_{\beta_l})$, we get the projective resolution $\mathbf{P}_*(L_{\beta_1}) \boxtimes \cdots \boxtimes \mathbf{P}_*(L_{\beta_l})$ of the $R_{\beta_1,\dots,\beta_l}$-module $L_{\beta_1} \boxtimes \cdots \boxtimes L_{\beta_l}$. Since $\mathrm{Ind}^{\alpha}_{\beta_1,\dots,\beta_l}$ is exact and sends projectives to projectives, inducing to R_α and shifting grading by $\mathtt{sh}(\pi)$, yields a projective resolution

$$\mathbf{P}_*(\bar{\Delta}(\pi)) := q^{\mathtt{sh}(\pi)} \mathbf{P}_*(L_{\beta_1}) \circ \cdots \circ \mathbf{P}_*(L_{\beta_l})$$

of $\bar{\Delta}(\pi)$. If $d = \mathrm{ht}(\alpha)$, this resolution has length $\leq d$:

$$0 \longrightarrow P_d \xrightarrow{\partial_d} P_{d-1} \xrightarrow{\partial_{d-1}} \cdots \xrightarrow{\partial_1} P_0 \longrightarrow \bar{\Delta}(\pi) \longrightarrow 0. \qquad (6.33)$$

Recall also proper costandard modules from (3.8). By Lemma 3.9, we have

$$\bar{\nabla}(\pi) \cong q^{\mathtt{sh}'(\pi)} L(\beta_l) \circ \cdots \circ L(\beta_1)$$

for $\mathtt{sh}'(\pi) := -\mathtt{sh}(\pi) + \sum_{1 \leq r < s \leq l} (\beta_r, \beta_s)$. As for $\bar{\Delta}(\pi)$, we now get a projective resolution

$$\mathbf{P}_*(\bar{\nabla}(\pi)) := q^{\mathtt{sh}'(\pi)} \mathbf{P}_*(L_{\beta_l}) \circ \cdots \circ \mathbf{P}_*(L_{\beta_1})$$

of the proper costandard module $\bar{\nabla}(\pi)$ of length $\leq d$.

The resolutions $\mathbf{P}_*(\bar{\nabla}(\pi))$ and $\mathbf{P}_*(\bar{\Delta}(\pi))$ can be used to compute the global dimension of R_α. First of all, the presence of the polynomial subalgebra in d variables allows us to bound global dimension of R_α below as follows:

Lemma 6.34: *Let $\mathcal{O} = F$ and $\alpha \in Q_+$ with $\mathrm{ht}(\alpha) = d$. Then $\mathrm{gd}\, R_\alpha \geq d$.*

Proof: For the polynomial algebra with n variables, we have $\mathrm{gd}\, F[y_1, \dots, y_n] = n$ by the Hilbert Theorem on Syzygies, see for example [**30**, Theorem 8.37]. Note by the Basis Theorem 2.3 that R_α is a free module over its polynomial subalgebra. So the result comes from the McConnell-Roos Theorem [**30**, Theorem 8.40]. □

Theorem 6.35: *Let $\alpha \in Q_+$ with $\mathrm{ht}(\alpha) = d$, $\mathcal{O} = F$ and L be an irreducible R_α-module. Then $\mathrm{pd}\, L \leq d$.*

Proof: We have $L = L(\pi)$ for some $\pi \in \Pi(\alpha)$. We prove the theorem by the upward induction on the convex order. To start the induction, note that if π is minimal, then $L(\pi) = \bar{\Delta}(\pi)$ by Theorem 3.15(iv), and so it has a projective resolution $\mathbf{P}_*(\bar{\Delta}(\pi))$ of length d.

For the inductive step, let $\pi \in \Pi(\alpha)$ and assume that the theorem has been proved for all $\pi' < \pi$. We also have a projective resolution $\mathbf{P}_*(\bar{\nabla}(\pi))$ of length d. It follows that $\mathrm{EXT}^k_{R_\alpha}(\bar{\nabla}(\pi), M) = 0$ for all $k > d$ and all $M \in R_\alpha$-Mod. By the inductive assumption, we also have $\mathrm{EXT}^k_{R_\alpha}(L(\pi'), M) = 0$ for all $k > d$, $M \in R_\alpha$-Mod, and $\pi' < \pi$.

By Theorem 3.15(iv), all composition factors of $\bar{\nabla}(\pi)/L(\pi)$ are of the form $L(\pi')$ for some $\pi' < \pi$. So, using long exact sequences in cohomology, we conclude that $\mathrm{EXT}^k(\bar{\nabla}(\pi)/L(\pi), M) = 0$ for all $k > d$ and $M \in R_\alpha$-Mod. Now the long exact sequence corresponding to the short exact sequence

$$0 \longrightarrow L(\pi) \longrightarrow \bar{\nabla}(\pi) \longrightarrow \bar{\nabla}(\pi)/L(\pi) \longrightarrow 0$$

looks like

$$\begin{aligned}\dots \longrightarrow \mathrm{EXT}^k_{R_\alpha}(\bar{\nabla}(\pi)/L(\pi), M) &\longrightarrow \mathrm{EXT}^k_{R_\alpha}(\bar{\nabla}(\pi), M)\\ \longrightarrow \mathrm{EXT}^k_{R_\alpha}(L(\pi), M) &\longrightarrow \mathrm{EXT}^{k+1}_{R_\alpha}(\bar{\nabla}(\pi)/L(\pi), M) \longrightarrow \dots.\end{aligned}$$

We can now conclude that $\mathrm{EXT}^k_{R_\alpha}(L(\pi), M) = 0$ for all $k > d$ and $M \in R_\alpha$-Mod. Theorefore $\mathrm{pd}\, L(\pi) \le d$ by [**30**, Proposition 8.6]. □

A general argument as explained for example in [**27**] now yields:

Corollary 6.36: *Let* $\alpha \in Q_+$ *with* $\mathrm{ht}(\alpha) = d$. *Then* $\mathrm{gd}\, R_\alpha = d$.

A more explicit analysis of the terms P_k of the resolution (6.33) implies:

Proposition 6.37: *If* $\mathrm{EXT}^k(\bar{\Delta}(\pi), V) \neq 0$ *for some finitely generated* R_α-*module* V *and* $k > 0$, *then* V *has a composition factor* $\simeq L(\sigma)$ *for* $\pi \le \sigma$.

Corollary 6.38: *Let* $\pi, \sigma \in \Pi(\alpha)$. *If* $\pi \neq \sigma$, *then* $\mathrm{EXT}^k_{R_\alpha}(\bar{\Delta}(\pi), \bar{\nabla}(\sigma)) = 0$ *for all* $k \ge 0$.

Proof: If $\pi \not\le \sigma$, then, the Theorem holds by Proposition 6.37. If $\pi < \sigma$, then

$$\begin{aligned}\mathrm{EXT}^k_{R_\alpha}\left(\bar{\Delta}(\pi), \bar{\nabla}(\sigma)\right) &\cong \mathrm{EXT}^k_{R_\alpha}\left(\bar{\nabla}(\sigma)^{\circledast}, \bar{\Delta}(\pi)^{\circledast}\right)\\ &\cong \mathrm{EXT}^k_{R_\alpha}\left(\bar{\Delta}(\sigma), \bar{\nabla}(\pi)\right) = 0\end{aligned}$$

by Proposition 6.37 again. □

Acknowledgements

This research was supported by the NSF grant no. DMS-1161094 and the Humboldt Foundation.

References

1. J. Brundan, Quiver Hecke algebras and categorification, in: *Advances in Representation Theory of Algebras*, pp. 103–133, EMS Ser. Congr. Rep., Eur. Math. Soc., Zürich, 2013.
2. J. Brundan and A. Kleshchev, Representation theory of symmetric groups and their double covers, in: *Groups, Combinatorics and Geometry (Durham, 2001)*, pp. 31–53, World Scientific Publishing, River Edge, NJ, 2003.
3. J. Brundan and A. Kleshchev, Blocks of cyclotomic Hecke algebras and Khovanov-Lauda algebras, *Invent. Math.* **178** (2009), 451–484.
4. J. Brundan and A. Kleshchev, Graded decomposition numbers for cyclotomic Hecke algebras, *Adv. Math.* **222** (2009), 1883–1942.
5. J. Brundan, A, Kleshchev and P.J. McNamara, Homological properties of finite type Khovanov-Lauda-Rouquier algebras, *Duke Math. J.* **163** (2014), 1353–1404.
6. J. Brundan, A. Kleshchev, and W. Wang, Graded Specht modules, *J. Reine Angew. Math.*, **655** (2011), 61–87.
7. S. Dăscălescu, Graded semiperfect rings *Bull. Math. de le Soc. Math. Roum.* **36** (1992), 247–255.
8. I. Grojnowski, Affine $\mathfrak{sl}_p$ controls the representation theory of the symmetric group and related Hecke algebras, `arXiv:math.RT/9907129`.
9. J. Hu and A. Mathas, Graded cellular bases for the cyclotomic Khovanov-Lauda-Rouquier algebras of type A, *Adv. Math.* **225** (2010), 598–642.
10. V. G. Kac, *Infinite Dimensional Lie Algebras*, Cambridge University Press, Cambridge, 1990.
11. S.-J. Kang and M. Kashiwara, Categorification of highest weight modules via Khovanov-Lauda-Rouquier algebras, *Invent. Math.* **190** (2012), 699–742.
12. M. Kashiwara, Global crystal bases of quantum groups, *Duke Math. J.* **69** (1993), 455–485.
13. S. Kato, Poincaré-Birkhoff-Witt bases and Khovanov-Lauda-Rouquier algebras, *Duke Math. J.* **163** (2014), 619–663.
14. M. Khovanov and A. Lauda, A diagrammatic approach to categorification of quantum groups I, *Represent. Theory* **13** (2009), 309–347.
15. M. Khovanov and A. Lauda, A diagrammatic approach to categorification of quantum groups II, *Trans. Amer. Math. Soc.* **363** (2011), 2685–2700.
16. A. Kleshchev, *Linear and Projective Representations of Symmetric Groups*, Cambridge University Press, Cambridge, 2005.
17. A. Kleshchev, Cuspidal systems for affine Khovanov-Lauda-Rouquier algebras, *Math. Z.* **276** (2014), 691–726.
18. A. Kleshchev and J. Loubert, Affine Cellularity of Khovanov-Lauda-Rouquier algebras of finite type, IMRN, to appear, `arXiv:1310.4467`.

19. A. Kleshchev, J. Loubert, and V. Miemietz, Affine Cellularity of Khovanov-Lauda-Rouquier algebras in type A, *J. Lond. Math. Soc. (2)* **88** (2013), 338–358.
20. A. Kleshchev and A. Ram, Homogeneous representations of Khovanov-Lauda algebras, *J. Eur. Math. Soc.* **12** (2010), 1293–1306.
21. A. Kleshchev and A. Ram, Representations of Khovanov-Lauda-Rouquier algebras and combinatorics of Lyndon words, *Math. Ann.* **349** (2011), 943–975.
22. A. Lauda and M. Vazirani, Crystals from categorified quantum groups, *Advances Math.* **228** (2011), 803–861.
23. S. Levendorskii and S. Soibelman, Algebras of functions on compact quantum groups, Schubert cells and quantum tori, *Comm. Math. Phys.* **139** (1991), 141–170.
24. G. Lusztig, Canonical bases arising from quantized enveloping algebras, *J. Amer. Math. Soc.* **3** (1990), 447–498.
25. G. Lusztig, *Introduction to Quantum Groups*, Birkhäuser, 1993.
26. G. Lusztig, Braid group action and canonical bases, *Advances Math.* **122** (1996), 237–261.
27. P. McNamara, Finite dimensional representations of Khovanov-Lauda-Rouquier algebras I: finite type, *J. Reine. Angew. Math.*, to appear; `arxiv:1207.5860`.
28. C. Năstăsescu and F. Van Oystaeyen, Methods of Graded Rings, Springer, 2004.
29. P. Papi, A characterization of a special ordering in a root system, *Proc. Amer. Math. Soc.* **120** (1994), 661–665.
30. J.J Rotman, *An Introduction to Homological Algebra.* Second Edition, Springer, 2009.
31. R. Rouquier, 2-Kac-Moody algebras; `arXiv:0812.5023`.
32. R. Rouquier, Quiver Hecke algebras and 2-Lie algebras, *Algebra Colloq.* **19** (2012), 359–410.
33. Y. Saito, PBW basis of quantized universal enveloping algebras, *Publ. RIMS. Kyoto Univ.* **30** (1994), 209–232.
34. M. Varagnolo and E. Vasserot, Canonical bases and KLR-algebras, *J. Reine Angew. Math.* **659** (2011), 67–100.
35. C. Weibel, *An Introduction to Homological Algebra*, CUP, 1994.

CYCLOTOMIC QUIVER HECKE ALGEBRAS OF TYPE A

Andrew Mathas

School of Mathematics and Statistics
The University of Sydney
New South Wales 2006
Australia
andrew.mathas@sydney.edu.au

Introduction

The *cyclotomic Hecke algebras* of type A are a much studied class of algebras that include, as special cases, the group algebras of the symmetric groups and the Iwahori-Hecke algebras of types A and B. They have a rich representation theory that can be approached using algebraic combinatorics, standard tools of representation theory, or via the theory of Lie and algebraic groups, which brings deep methods from geometry into play.

In 2008 Khovanov and Lauda [74, 75] and Rouquier [121] introduced the *quiver Hecke algebras*, or *KLR algebras*. These are a remarkable family $\{\mathscr{R}_n(\Gamma) \mid n \geq 0\}$ of $\mathbb{Z}$-graded algebras defined by generators and relations depending on a quiver Γ. The motivation for defining and studying these algebras came, at least in part, from questions in geometry and 2-representation theory. The algebras $\mathscr{R}_n(\Gamma)$ categorify the negative part of the associated quantum group $U_q(\mathfrak{g}_\Gamma)$ [122, 132]. That is, there are natural isomorphisms

$$U_q^-(\mathfrak{g}_\Gamma) \cong \bigoplus_{n\geq 0}[\mathrm{Proj}(\mathscr{R}_n(\Gamma))],$$

where $[\mathrm{Proj}(\mathscr{R}_n(\Gamma))]$ is the Grothendieck group of the category of finitely generated graded projective $\mathscr{R}_n(\Gamma)$-modules. For each dominant weight Λ the quiver Hecke algebra $\mathscr{R}_n(\Gamma)$ has a cyclotomic quotient $\mathscr{R}_n^\Lambda(\Gamma)$ that categorifies the highest weight module $L(\Lambda)$ [67, 122, 134]. These results can be thought of as far reaching generalizations of Ariki's Categorification Theorem in type A [3].

Spectacularly, Brundan and Kleshchev [21, 121] proved that each cyclotomic Hecke algebra of type A is isomorphic to a cyclotomic quotient of a quiver Hecke algebra of type A. Thus, the KLR algebras give a new window for understanding the cyclotomic Hecke algebras of type A. This chapter is an attempt to open this window and show how the "classical" ungraded representation theory and the emerging graded representation theory of the cyclotomic quiver Hecke algebras interact.

With the advent of the KLR algebras the cyclotomic Hecke algebras can now be studied from many different perspectives including:

a) As ungraded cyclotomic Hecke algebras.
b) As graded cyclotomic quiver Hecke algebras or KLR algebras.
c) Geometrically as the ext-algebras of Lusztig sheaves [98, 123, 132].
d) Through the lens of 2-representation theory using Rouquier's theory of 2-Kac Moody algebras [67, 121, 134].

Here we focus on (a) and (b) taking an unashamedly combinatorial approach, although we will see shadows of geometry and 2-representation theory.

For every quiver there is a corresponding family of KLR algebras, however, the quiver Hecke algebras attached to the quivers of type A are special because these are the only quiver Hecke algebras that existed in the literature prior to [74, 121] — all of the other quiver Hecke algebras are "new" algebras. In type A, when we are working over a field, the quiver Hecke algebras are isomorphic to affine Hecke algebras of type A [122] and the cyclotomic quiver Hecke algebras are isomorphic to the cyclotomic Hecke algebras of type A [21]. The cyclotomic Hecke algebras of type A have a uniform description but, historically, they have been studied either as *Ariki-Koike* algebras ($v \neq 1$), or as degenerate Ariki-Koike algebras ($v = 1$). The existence of gradings on Hecke algebras, at least in the "abelian defect case", was predicted by Rouquier [120, Remark 3.11] and Turner [130].

The cyclotomic quiver Hecke algebras of type A are better understood than other types because we already know a lot about the isomorphic, but ungraded, cyclotomic Hecke algebras [107]. For example, by piggybacking on the ungraded theory, homogeneous bases have been constructed for the cyclotomic quiver Hecke algebras of type A [54] but such bases are not yet known in other types. Many of the major results for general quiver Hecke algebras were first proved in type A and then generalized to other types. In fact, the type A algebras, through Ariki's theorem and Chuang and Rouquier's seminal work on $\mathfrak{sl}_2$-categorifications [28], has motivated many of these developments.

The first section starts by giving a uniform description of the degenerate and non-degenerate cyclotomic Hecke algebras, recalling some structural results from the ungraded representation theory of these algebras. Everything mentioned in this section is applied later in the graded setting.

The second section introduces the cyclotomic KLR algebras as abstract algebras given by generators and relations. We use the relations in a series of extended examples to give the reader a feel for these algebras. In particular, using just the relations we show that the semisimple cyclotomic quiver Hecke algebras of type A are direct sums of matrix rings. From this we deduce Brundan and Kleshchev's Graded Isomorphism Theorem in the semisimple case.

The third section starts with Brundan and Kleshchev's Graded Isomorphism Theorem [21]. We develop the representation theory of the cyclotomic quiver Hecke algebras as graded cellular algebras, focusing on the graded Specht modules. The highlight of this section is a self-contained proof of Brundan and Kleshchev's Graded Categorification Theorem [22], starting from the graded branching rules for the graded Specht modules and then using Ariki's Categorification Theorem [3] to make the link with canonical bases. We also give a new treatment of graded adjustment matrices using a cellular algebra approach.

In the final section we sketch one way of proving Brundan and Kleshchev's Graded Isomorphism Theorem using the classical theory of seminormal forms. As an application we describe how to construct a new graded cellular basis for the cyclotomic quiver Hecke algebras that appears to have remarkable properties. We end with a conjecture for the q-characters of the graded simple modules.

Although the experts will find some new results, most of the novelty is in our approach and the arguments that we use. We include many examples and a comprehensive survey of the literature. For a different perspective we recommend Kleshchev's survey article [80] on the applications of quiver Hecke algebras to symmetric groups.

Acknowledgements

This chapter grew out of a series of lectures at the IMS at the National University of Singapore. I thank the organizers for inviting me to give these lectures and to write this chapter. The direction taken in these notes, and the conjecture formulated in §4.4, is partly motivated by the author's joint work with Jun Hu and I thank him for his implicit contributions. I thank Susumu Ariki, Anton Evseev, Matthew Fayers, Jun Hu, Kai Meng Tan and the referee for their comments on earlier versions of this manuscript and Jon Brundan for some helpful discussions about crystal bases. This chapter was written while visiting Universität Stuttgart and Charles University, Prague.

1 Cyclotomic Hecke Algebras of Type A

This section surveys the representation theory of the cyclotomic Hecke algebras of type A and, at the same time, introduces the results and the combinatorics that we need later.

1.1 *Cyclotomic Hecke algebras and Ariki-Koike algebras*

Hecke algebras of the complex reflections groups $G_{\ell,n} = \mathbb{Z}/\ell\mathbb{Z} \wr \mathfrak{S}_n$ of type $G(\ell, 1, n)$ were introduced by Ariki-Koike [10], motivated by the Iwahori-Hecke algebras of Coxeter groups [58]. Soon afterwards, Broué and Malle [16] defined Hecke algebras for arbitrary complex reflection groups. The following refinement of the definition of these algebras unifies the treatment of the degenerate and non-degenerate cyclotomic Hecke algebras of type $G(\ell, 1, n)$.

Let $\mathcal{Z}$ be a commutative domain with one.

Definition 1.1.1 (Hu-Mathas [57, Definition 2.2]**).** *Fix integers $n \geq 0$ and $\ell \geq 1$. The* ***cyclotomic Hecke algebra of type*** A*, with* ***Hecke parameter*** $v \in \mathcal{Z}^\times$ *and* ***cyclotomic parameters*** $Q_1, \dots, Q_\ell \in \mathcal{Z}$*, is the unital associative $\mathcal{Z}$-algebra $\mathscr{H}_n = \mathscr{H}_n(\mathcal{Z}, v, Q_1, \dots, Q_\ell)$ with generators $L_1, \dots, L_n$, $T_1, \dots, T_{n-1}$ and relations*

$$\prod_{l=1}^{\ell}(L_1 - Q_l) = 0, \quad (T_r + v^{-1})(T_r - v) = 0, \quad L_{r+1} = T_r L_r T_r + T_r,$$
$$L_r L_t = L_t L_r, \qquad T_r T_s = T_s T_r \text{ if } |r - s| > 1,$$
$$T_s T_{s+1} T_s = T_{s+1} T_s T_{s+1}, \qquad T_r L_t = L_t T_r \text{ if } t \neq r, r+1,$$

where $1 \leq r < n$, $1 \leq s < n-1$ and $1 \leq t \leq n$.

By definition, $\mathscr{H}_n$ is generated by $L_1, T_1, \dots, T_{n-1}$ but we prefer including $L_2, \dots, L_n$ in the generating set.

Let $\mathfrak{S}_n$ be the **symmetric group** on n letters. For $1 \leq r < n$ let $s_r = (r, r+1)$ be the corresponding simple transposition. Then $\{s_1, \dots, s_{n-1}\}$ is the standard set of Coxeter generators for $\mathfrak{S}_n$. A **reduced expression** for $w \in \mathfrak{S}_n$ is a word $w = s_{r_1} \dots s_{r_k}$ with k minimal and $1 \leq r_j < n$ for $1 \leq j \leq k$. If $w = s_{r_1} \dots s_{r_k}$ is reduced then set $T_w = T_{r_1} \dots T_{r_k}$. Then T_w

is independent of the choice of reduced expression by Matsumoto's Monoid Lemma [110] since the braid relations hold in $\mathscr{H}_n$; see, for example, [104, Theorem 1.8]. Arguing as in [10, Theorem 3.3], it follows that $\mathscr{H}_n$ is free as a $\mathcal{Z}$-module with basis

(1.1.2) $$\{\, L_1^{a_1} \dots L_n^{a_n} T_w \mid 0 \le a_1, \dots, a_n < \ell \text{ and } w \in \mathfrak{S}_n \,\}.$$

Consequently, $\mathscr{H}_n$ is free as a $\mathcal{Z}$-module of rank $\ell^n n!$, which is the order of the complex reflection group $G_{\ell,n} = \mathbb{Z}/\ell\mathbb{Z} \wr \mathfrak{S}_n$ of type $G(\ell, 1, n)$.

Definition 1.1.1 is different from Ariki and Koike's [10] definition of the cyclotomic Hecke algebras of type $G(\ell, 1, n)$ because we have changed the commutation relation for T_r and L_r. Ariki and Koike [10] defined their algebra to be the unital associative algebra generated by $T_0, T_1, \dots, T_{n-1}$ subject to the relations

$$\prod_{l=1}^{\ell}(T_0 - Q_l') = 0, \qquad (T_r + v^{-1})(T_r - v) = 0,$$
$$T_0T_1T_0T_1 = T_1T_0T_1T_0 \qquad T_sT_{s+1}T_s = T_{s+1}T_sT_{s+1},$$
$$T_rT_s = T_sT_r \text{ if } |r - s| > 1.$$

We have renormalised the quadratic relation for the T_r, for $1 \le r < n$, so that $q = v^2$ in the notation of [10]. Ariki and Koike then defined $L_1' = T_0$ and set $L_{r+1}' = T_rL_r'T_r$ for $1 \le r < n$. In fact, if $v - v^{-1}$ is invertible in $\mathcal{Z}$ then $\mathscr{H}_n$ is (isomorphic to) the Ariki-Koike algebra with parameters $Q_l' = 1 + (v - v^{-1})Q_l$ for $1 \le l \le \ell$. To see this set $L_r' = 1 + (v - v^{-1})L_r$ in $\mathscr{H}_n$, for $1 \le r \le n$. Then $T_rL_r'T_r = (v - v^{-1})T_rL_rT_r + T_r^2 = L_{r+1}'$, which implies our claim. Therefore, over a field, $\mathscr{H}_n$ is an Ariki-Koike algebra whenever $v^2 \ne 1$. On the other hand, if $v^2 = 1$ then $\mathscr{H}_n$ is a *degenerate* cyclotomic Hecke algebra [13, 79].

We note that the Ariki-Koike algebras with $v^2 = 1$ include as a special the group algebras $\mathcal{Z}G_{\ell,n}$ of the complex reflection groups $G_{\ell,n}$, for $n \ge 0$. One consequence of the last paragraph is that $\mathcal{Z}G_{\ell,n}$ is *not* a specialization of $\mathscr{H}_n$. This said, if F is a field such that $\mathscr{H}_n$ and $FG_{\ell,n}$ are both split semisimple then $\mathscr{H}_n \cong FG_{\ell,n}$. On the other hand, the algebras $\mathscr{H}_n$ always fit into the *spetses* framework of Broué, Malle and Michel [17].

The algebras $\mathscr{H}_n$ with $v^2 = 1$ are the *degenerate* cyclotomic Hecke algebras of type $G(\ell, 1, n)$ whereas if $v^2 \ne 1$ then $\mathscr{H}_n$ is an Ariki-Koike algebra in the sense of [10]. Our definition of $\mathscr{H}_n$ is more natural in the sense that many features of the algebras $\mathscr{H}_n$ have a uniform description in both the degenerate and non-degenerate cases:

- The centre of $\mathscr{H}_n$ is the set of symmetric polynomials in $L_1, \dots, L_n$ (Brundan [19] in the degenerate case when $v^2 = 1$ and announced when $v^2 \ne 1$ by Graham and Francis building on [42]).

- The blocks of $\mathscr{H}_n$ are indexed by the same combinatorial data (Lyle and Mathas [96] when $v^2 \neq 1$ and Brundan [19] when $v^2 = 1$).
- The irreducible $\mathscr{H}_n$-modules are indexed by the crystal graph of the integral highest weight module $L(\Lambda)$ for $U_q(\widehat{\mathfrak{sl}}_e)$ (Ariki [3] when $v^2 \neq 1$ and Brundan and Kleshchev [23] when $v^2 = 1$).
- The algebras $\mathscr{H}_n$ categorify $L(\Lambda)$. Moreover, in characteristic zero the projective indecomposable $\mathscr{H}_n$-modules correspond to the canonical basis of $L(\Lambda)$. (Ariki [3] when $v^2 \neq 1$ and Brundan and Kleshchev [23] when $v^2 = 1$.)
- The algebra $\mathscr{H}_n$ is isomorphic to a cyclotomic quiver Hecke algebra of type A (Brundan and Kleshchev [21]).

In contrast, the Ariki-Koike algebras with $v^2 = 1$ do not share any of these properties: their center can be larger than the set of symmetric polynomials in $L_1, \dots, L_n$ (Ariki [3]); if $\ell > 1$ then they have only one block (Lyle and Mathas [96]); their irreducible modules are indexed by a different set (Mathas [103]); they do not categorify $L(\Lambda)$ and no non-trivial grading on these algebras is known. In this sense, the definition of the Ariki-Koike algebras from [10] gives the wrong algebras when $v^2 = 1$. Definition 1.1.1 corrects for this.

Historically, many results for the cyclotomic Hecke algebras $\mathscr{H}_n$ were proved separately in the degenerate ($v^2 = 1$) and non-degenerate cases ($v^2 \neq 1$). Using Definition 1.1.1 it should now be possible to give uniform proofs of all of these results. In fact, in the cases that we have checked uniforms arguments can now be given for the degenerate and non-degenerate cases.

1.2 *Quivers of type A and integral parameters*

Rather than work with arbitrary cyclotomic parameters $Q_1, \dots, Q_\ell$, as in Definition 1.1.1, we now specialize to the *integral case* using the Morita equivalence results of Dipper and the author [32] (when $v^2 \neq 1$) and Brundan and Kleshchev [20] (when $v^2 = 1$). First, however, we need to introduce quivers and quantum integers.

Fix $e \in \{1, 2, 3, 4, \dots\} \cup \{\infty\}$ and let Γ_e be the quiver with vertex set $I_e = \mathbb{Z}/e\mathbb{Z}$ and edges $i \longrightarrow i+1$, for $i \in I_e$, where we adopt the convention that $e\mathbb{Z} = \{0\}$ when $e = \infty$. If $i, j \in I_e$ and i and j are not connected by an edge in Γ_e then we write $i \not\!\!- j$. When e is fixed we write $\Gamma = \Gamma_e$ and $I = I_e$. Hence, we are considering either the linear quiver $\mathbb{Z}$ ($e = \infty$) or a cyclic quiver ($e < \infty$):

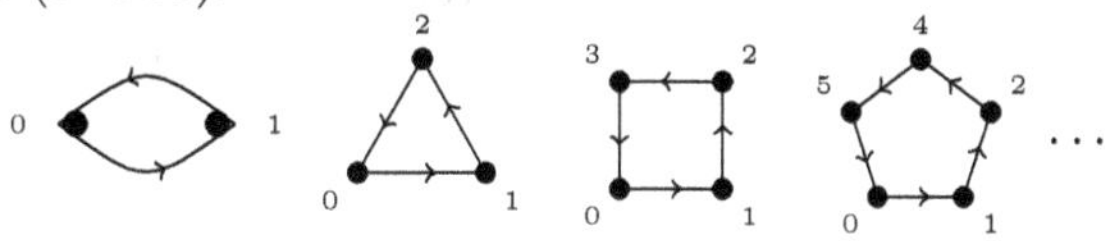

$e = 2$ $e = 3$ $e = 4$ $e = 5$ $\cdots$

In the literature the case $e = \infty$ is often written as $e = 0$, however, we prefer $e = \infty$ because then $e = |I_e|$. There are also several results that hold when $e > n$ — using the "$e = 0$ convention" this condition must be written as $e > n$ or $e = 0$. We write $e \geq n$ to mean $e \in \{n, n+1, n+2, \dots\} \cup \{\infty\}$.

To the quiver Γ_e we attach the symmetric Cartan matrix $(c_{ij})_{i,j \in I}$, where

$$c_{ij} = \begin{cases} 2, & \text{if } i = j, \\ -1, & \text{if } i \to j \text{ or } i \leftarrow j, \\ -2, & \text{if } i \leftrightarrows j, \\ 0, & \text{otherwise,} \end{cases}$$

Following [66, Chapter 1], let $\widehat{\mathfrak{sl}}_e$ be the Kac-Moody algebra of Γ_e [66] with simple roots $\{\, \alpha_i \mid i \in I \,\}$, fundamental weights $\{\, \Lambda_i \mid i \in I \,\}$, positive weight lattice $P^+ = \bigoplus_{i \in I} \mathbb{N}\Lambda_i$ and positive root lattice $Q^+ = \bigoplus_{i \in I} \mathbb{N}\alpha_i$. Let $(\cdot, \cdot)$ be the usual invariant form associated with this data, normalised so that $(\alpha_i, \alpha_j) = c_{ij}$ and $(\Lambda_i, \alpha_j) = \delta_{ij}$, for $i, j \in I$.

Fix a sequence $\boldsymbol{\kappa} = (\kappa_1, \dots, \kappa_\ell) \in \mathbb{Z}^\ell$, the **multicharge**, and define $\Lambda = \Lambda(\boldsymbol{\kappa}) = \Lambda_{\overline{\kappa}_1} + \cdots + \Lambda_{\overline{\kappa}_\ell}$, where $\overline{a} = a + e\mathbb{Z} \in I$ for $a \in \mathbb{Z}$. Then $\Lambda \in P^+$ is dominant weight of **level** ℓ. The integral cyclotomic Hecke algebras defined below depend only on Λ, however, the bases and our combinatorics often depends upon the choice of multicharge $\boldsymbol{\kappa}$.

Recall that $\mathcal{Z}$ is an integral domain. For $t \in \mathcal{Z}^\times$ and $k \in \mathbb{Z}$ define the **t-quantum integer** $[k]_t$ by

$$[k]_t = \begin{cases} t + t^3 + \cdots + t^{2k-1}, & \text{if } k \geq 0, \\ -(t^{-1} + t^{-3} + \cdots + t^{2k+1}), & \text{if } k < 0. \end{cases}$$

When t is understood we simply write $[k] = [k]_t$. Unpacking the definition, if $t^2 \neq 1$ then $[k] = (t^{2k} - 1)/(t - t^{-1})$ whereas $[k] = \pm k$ if $t = \pm 1$.

The **quantum characteristic** of v is the smallest element of $e \in \{2, 3, 4, 5, \dots\} \cup \{\infty\}$ such that $[e]_v = 0$, where we set $e = \infty$ if $[k]_v \neq 0$ for all $k > 0$.

Definition 1.2.1. Suppose that $\Lambda = \Lambda(\boldsymbol{\kappa}) \in P^+$, for $\boldsymbol{\kappa} \in \mathbb{Z}^\ell$, and that $v \in \mathcal{Z}$ has quantum characteristic e. The **integral** cyclotomic Hecke algebra of type A of weight Λ is the cyclotomic Hecke algebra $\mathscr{H}_n^\Lambda = \mathscr{H}_n(\mathcal{Z}, v, Q_1, \dots, Q_r)$ with Hecke parameter v and cyclotomic parameters $Q_r = [\kappa_r]_v$, for $1 \le r \le \ell$.

When $v^2 \neq 1$ the parameter Q_r corresponds to the Ariki-Koike parameters $Q'_r = v^{2\kappa_r}$, for $1 \le r \le \ell$, where we use the notation of §1.1.

As observed in [57, §2.2], translating the Morita equivalence theorems of [32, Theorem 1.1] and [20, Theorem 5.19] into the current setting explains the significance of the integral cyclotomic Hecke algebras.

Theorem 1.2.2 (Dipper-Mathas [32], Brundan-Kleshchev [20]). *Every cyclotomic quiver Hecke algebra $\mathscr{H}_n$ is Morita equivalent to a direct sum of tensor products of integral cyclotomic Hecke algebras.*

Brundan and Kleshchev treated the degenerate case when $v^2 = 1$ using very different arguments to those in [32]. With the benefit of Definition 1.1.1 the argument of [32] now applies uniformly to both the degenerate and non-degenerate cases. The Morita equivalences in [20, 32] are described explicitly, with the equivalence being determined by orbits of the cyclotomic parameters. See [20, 32] for more details.

In view of Theorem 1.2.2, it is enough to consider the integral cyclotomic Hecke algebras $\mathscr{H}_n^\Lambda$ where $v \in \mathcal{Z}^\times$ has quantum characteristic e and $\Lambda \in P^+$. This said, for most of Section 1 we consider the general case of a not necessarily integral cyclotomic Hecke algebra because we will need this generality in §4.2.

1.3 *Cellular algebras*

We recall Graham and Lehrer's cellular algebra framework [48]. This will allow us to define Specht modules for $\mathscr{H}_n$ as cell modules. Significantly, the cellular algebra machinery endows the Specht modules with an associative bilinear form. Here is the definition.

Definition 1.3.1 (Graham and Lehrer [48]). *Suppose that A is a $\mathcal{Z}$-algebra that is $\mathcal{Z}$-free and of finite rank as a $\mathcal{Z}$-module. A **cell datum** for A is an ordered triple $(\mathcal{P}, T, C)$, where $(\mathcal{P}, \rhd)$ is the **weight poset**, $T(\lambda)$ is a finite set for $\lambda \in \mathcal{P}$, and*

$$C\colon \coprod_{\lambda\in\mathcal{P}} T(\lambda) \times T(\lambda) \longrightarrow A; (\mathsf{s}, \mathsf{t}) \mapsto c_{\mathsf{st}},$$

is an injective map of sets such that:

(GC_1) $\{\, c_{\mathsf{st}} \mid \mathsf{s},\mathsf{t} \in T(\lambda)$ for $\lambda \in \mathcal{P} \,\}$ is a $\mathcal{Z}$-basis of A.

(GC_2) If $\mathsf{s},\mathsf{t} \in T(\lambda)$, for some $\lambda \in \mathcal{P}$, and $a \in A$ then there exist scalars $r_{\mathsf{tv}}(a)$, which do not depend on s, such that

$$c_{\mathsf{st}}a = \sum_{\mathsf{v}\in T(\lambda)} r_{\mathsf{tv}}(a)c_{\mathsf{sv}} \pmod{A^{\rhd\lambda}},$$

where $A^{\rhd\lambda} = \langle\, c_{\mathsf{ab}} \mid \mu \rhd \lambda$ and $\mathsf{a},\mathsf{b} \in T(\mu) \,\rangle_{\mathcal{Z}}$.

(GC_3) The $\mathcal{Z}$-linear map $* : A \longrightarrow A$ determined by $c_{\mathsf{st}}^* = c_{\mathsf{ts}}$, for all $\lambda \in \mathcal{P}$ and all $\mathsf{s},\mathsf{t} \in T(\lambda)$, is an anti-isomorphism of A.

*A **cellular algebra** is an algebra that has a cell datum. If A is a cellular algebra with cell datum $(\mathcal{P}, T, C)$ then the basis $\{\, c_{\mathsf{st}} \mid \lambda \in \mathcal{P}$ and $\mathsf{s},\mathsf{t} \in T(\lambda) \,\}$ is a **cellular basis** of A with cellular algebra anti-isomorphism $*$.*

König and Xi [86] have given an equivalent definition of cellular algebras that does not depend upon a choice of basis. Goodman and Graber [45] have shown that (GC_3) can be relaxed to the requirement that $(c_{\mathsf{st}})^* \equiv c_{\mathsf{ts}} \pmod{A^{\rhd\lambda}}$ for some anti-isomorphism $*$ of A.

The prototypical example of a cellular algebra is a matrix algebra with its basis of matrix units, which we call a *Wedderburn basis*. As any split semisimple algebra is isomorphic to a direct sum of matrix algebras it follows that every split semisimple algebra is cellular. The cellular algebra framework is, however, most useful in studying non-semisimple algebras that are not isomorphic to a direct sum of matrix rings. In general, a cellular basis can be thought of as an approximation, or weakening, of a basis of matrix units. (This idea is made more explicit in [108].)

The cellular basis axioms determine a filtration of the cellular algebra, via the ideals $A^{\rhd\lambda}$. As we will see, this leads to a quick construction of the irreducible representations.

For $\lambda \in \mathcal{P}$, let $A^{\unrhd\lambda} = \langle\, c_{\mathsf{ab}} \mid \mu \unrhd \lambda$ and $\mathsf{a},\mathsf{b} \in T(\mu) \,\rangle_{\mathcal{Z}}$. Then it follows from Definition 1.3.1 that $A^{\unrhd\lambda}$ is a two-sided ideal of A.

Fix $\lambda \in \mathcal{P}$. The **cell module** $\underline{C}^\lambda$ is the (right) A-module with basis $\{\, c_{\mathsf{t}} \mid \mathsf{t} \in T(\lambda) \,\}$ and where $a \in A$ acts on $\underline{C}^\lambda$ by:

$$c_{\mathsf{t}}a = \sum_{\mathsf{v}\in T(\lambda)} r_{\mathsf{tv}}(a)c_{\mathsf{v}}, \qquad \text{for } \mathsf{t} \in T(\lambda),$$

where the scalars $r_{\mathsf{tv}}(a) \in \mathcal{Z}$ are those appearing in (GC_2). It follows immediately from Definition 1.3.1 that $\underline{C}^\lambda$ is an A-module. Indeed, if $\mathsf{s} \in T(\lambda)$ then $\underline{C}^\lambda$ is isomorphic to the submodule $(c_{\mathsf{st}} + A^{\rhd\lambda})A$ of A/A^λ via the map $c_{\mathsf{t}} \mapsto c_{\mathsf{st}} + A^\lambda$, for $\mathsf{t} \in T(\lambda)$. The cell module $\underline{C}^\lambda$ comes with a symmetric bilinear form $\langle\ ,\ \rangle_\lambda$ that is uniquely determined by

$$\langle c_{\mathsf{t}}, c_{\mathsf{v}}\rangle_\lambda c_{\mathsf{ab}} \equiv c_{\mathsf{at}}c_{\mathsf{vb}} \pmod{A^{\rhd\lambda}}, \tag{1.3.2}$$

for $\mathsf{a}, \mathsf{b}, \mathsf{t}, \mathsf{v} \in T(\lambda)$. By (GC$_2$) of Definition 1.3.1, the inner product $\langle c_{\mathsf{t}}, c_{\mathsf{v}} \rangle_\lambda$ depends only on t and v, and not on the choices of a and b. In addition, $\langle xa, y \rangle_\lambda = \langle x, ya^* \rangle_\lambda$, for all $x, y \in \underline{C}^\lambda$ and $a \in A$. Therefore,

(1.3.3) $$\operatorname{rad} \underline{C}^\lambda = \{\, x \in \underline{C}^\lambda \mid \langle x, y \rangle_\lambda = 0 \text{ for all } y \in \underline{C}^\lambda \,\}$$

is an A-submodule of $\underline{C}^\lambda$. Set $\underline{D}^\lambda = \underline{C}^\lambda / \operatorname{rad} \underline{C}^\lambda$. Then $\underline{D}^\lambda$ is an A-module.

The following theorem summarizes some of the main properties of a cellular algebra. The proof is surprisingly easy given the strength of the result. In applications the main difficulty is in showing that a given algebra is cellular.

If M is an A-module and D is an irreducible A-module, let $[M : D]$ be the decomposition multiplicity of D in M.

Theorem 1.3.4 (Graham and Lehrer [48]). *Suppose that $\mathcal{Z} = F$ is a field. Then:*

a) *If $\mu \in \mathcal{P}$ then $\underline{D}^\mu$ is either zero or absolutely irreducible.*
b) *Let $\mathcal{K} = \{\, \mu \in \mathcal{P} \mid \underline{D}^\mu \neq 0 \,\}$. Then $\{\, \underline{D}^\mu \mid \mu \in \mathcal{K} \,\}$ is a complete set of pairwise non-isomorphic irreducible A-modules.*
c) *If $\lambda \in \mathcal{P}$ and $\mu \in \mathcal{K}$ then $[\underline{C}^\lambda : \underline{D}^\mu] \neq 0$ only if $\lambda \trianglerighteq \mu$. Moreover, $[\underline{C}^\mu : \underline{D}^\mu] = 1$.*

If $\mu \in \mathcal{K}$ let $\underline{P}^\mu$ be the projective cover of $\underline{D}^\mu$. It follows from Definition 1.3.1 that $\underline{P}^\mu$ has a filtration in which the quotients are cell modules such that $\underline{C}^\lambda$ appears with multiplicity $[\underline{C}^\lambda : \underline{D}^\mu]$. Consequently, an analogue of Brauer-Humphreys reciprocity holds for A. In particular, the Cartan matrix of A is symmetric.

1.4 *Multipartitions and tableaux*

A **partition** of m is a weakly decreasing sequence $\lambda = (\lambda_1, \lambda_2, \dots)$ of non-negative integers such that $|\lambda| = \lambda_1 + \lambda_2 + \cdots = m$. An ($\ell$-)**multipartition** of n is an ℓ-tuple $\boldsymbol{\lambda} = (\lambda^{(1)}, \dots, \lambda^{(\ell)})$ of partitions such that $|\lambda^{(1)}| + \cdots + |\lambda^{(\ell)}| = n$. We identify the multipartition $\boldsymbol{\lambda}$ with its **diagram**, which is the set of **nodes** $[\![\boldsymbol{\lambda}]\!] = \{\, (l, r, c) \mid 1 \le c \le \lambda^{(l)}_r \text{ for } 1 \le l \le \ell \,\}$. In this way, we think of $\boldsymbol{\lambda}$ as an ordered ℓ-tuple of arrays of boxes in the plane and we talk of the **components** of $\boldsymbol{\lambda}$. Similarly, by the **rows** and **columns** of $\boldsymbol{\lambda}$ we will mean the rows and columns in each component. For example, if $\boldsymbol{\lambda} = (3, 1^2 | 2, 1 | 3, 2)$ then

$\boldsymbol{\lambda} = [\![\boldsymbol{\lambda}]\!] =$ 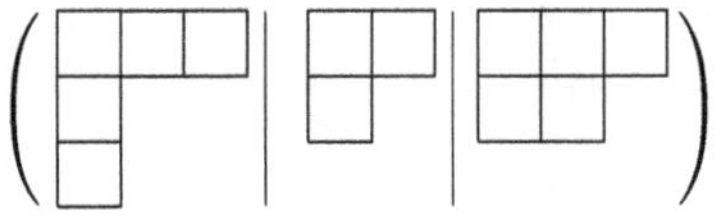 .

A node A is an **addable node** of $\boldsymbol{\lambda}$ if $A \notin \boldsymbol{\lambda}$ and $\boldsymbol{\lambda} \cup \{A\}$ is the (diagram of) a multipartition of $n+1$. Similarly, a node B is a **removable node** of $\boldsymbol{\lambda}$ if $B \in \boldsymbol{\lambda}$ and $\boldsymbol{\lambda} \setminus \{B\}$ is a multipartition of $n-1$. If A is an addable node of $\boldsymbol{\lambda}$ let $\boldsymbol{\lambda}+A$ be the multipartition $\boldsymbol{\lambda} \cup \{A\}$ and, similarly, if B is a removable node let $\boldsymbol{\lambda}-B = \boldsymbol{\lambda} \setminus \{B\}$. Order the nodes lexicographically by $\leq$.

The set of multipartitions of n becomes a poset under dominance where $\boldsymbol{\lambda}$ **dominates** $\boldsymbol{\mu}$, written as $\boldsymbol{\lambda} \trianglerighteq \boldsymbol{\mu}$, if

$$\sum_{k=1}^{l-1} |\lambda^{(k)}| + \sum_{j=1}^{i} \lambda_j^{(l)} \geq \sum_{k=1}^{l-1} |\mu^{(k)}| + \sum_{j=1}^{i} \mu_j^{(l)},$$

for $1 \leq l \leq \ell$ and $i \geq 1$. If $\boldsymbol{\lambda} \trianglerighteq \boldsymbol{\mu}$ and $\boldsymbol{\lambda} \neq \boldsymbol{\mu}$ then write $\boldsymbol{\lambda} \triangleright \boldsymbol{\mu}$. Let $\mathcal{P}_n^\Lambda = \mathcal{P}_{\ell,n}^\Lambda$ be the set of multipartitions of n. We consider $\mathcal{P}_n^\Lambda$ as a poset ordered by dominance.

Fix $\boldsymbol{\lambda} \in \mathcal{P}_n^\Lambda$. A **$\boldsymbol{\lambda}$-tableau** is a bijective map $\mathsf{t}\colon [\![\boldsymbol{\lambda}]\!] \longrightarrow \{1, 2, \dots, n\}$, which we identify with a labelling of (the diagram of) $\boldsymbol{\lambda}$ by $\{1, 2, \dots, n\}$. For example,

$$\left(\begin{array}{|c|c|c|}\hline 1&2&3\\\hline 4\\\cline{1-1} 5\\\cline{1-1}\end{array}\;\middle|\;\begin{array}{|c|c|}\hline 6&7\\\hline 8\\\cline{1-1}\end{array}\;\middle|\;\begin{array}{|c|c|c|}\hline 9&10&11\\\hline 12&13\\\cline{1-2}\end{array}\right) \quad \text{and} \quad \left(\begin{array}{|c|c|c|}\hline 9&12&13\\\hline 10\\\cline{1-1} 11\\\cline{1-1}\end{array}\;\middle|\;\begin{array}{|c|c|}\hline 6&8\\\hline 7\\\cline{1-1}\end{array}\;\middle|\;\begin{array}{|c|c|c|}\hline 1&3&5\\\hline 2&4\\\cline{1-2}\end{array}\right)$$

are both $\boldsymbol{\lambda}$-tableaux when $\boldsymbol{\lambda} = (3, 1^2|2, 1|3, 2)$.

A $\boldsymbol{\lambda}$-tableau is **standard** if its entries increase along rows and down columns in each component. For example, the two tableaux above are standard. Let $\operatorname{Std}(\boldsymbol{\lambda})$ be the set of standard $\boldsymbol{\lambda}$-tableaux. If $\mathcal{P}$ is any set of multipartitions let $\operatorname{Std}(\mathcal{P}) = \bigcup_{\boldsymbol{\lambda} \in \mathcal{P}} \operatorname{Std}(\boldsymbol{\lambda})$. Similarly set $\operatorname{Std}^2(\mathcal{P}) = \{\, (\mathsf{s}, \mathsf{t}) \mid \mathsf{s}, \mathsf{t} \in \operatorname{Std}(\boldsymbol{\lambda}) \text{ for } \boldsymbol{\lambda} \in \mathcal{P} \,\}$.

If t is a $\boldsymbol{\lambda}$-tableau set $\operatorname{Shape}(\mathsf{t}) = \boldsymbol{\lambda}$ and let $\mathsf{t}_{\downarrow m}$ be the subtableau of t that contains the numbers $\{1, 2, \dots, m\}$. If t is a standard $\boldsymbol{\lambda}$-tableau then $\operatorname{Shape}(\mathsf{t}_{\downarrow m})$ is a multipartition for all $m \geq 0$. We extend the dominance ordering to $\operatorname{Std}(\mathcal{P}_n^\Lambda)$, the set of all standard tableaux, by defining $\mathsf{s} \trianglerighteq \mathsf{t}$ if $\operatorname{Shape}(\mathsf{s}_{\downarrow m}) \trianglerighteq \operatorname{Shape}(\mathsf{t}_{\downarrow m})$, for $1 \leq m \leq n$. As before, write $\mathsf{s} \triangleright \mathsf{t}$ if $\mathsf{s} \trianglerighteq \mathsf{t}$ and $\mathsf{s} \neq \mathsf{t}$. Finally, define the **strong dominance ordering** on $\operatorname{Std}^2(\mathcal{P}_n^\Lambda)$ by $(\mathsf{s}, \mathsf{t}) \underline{\blacktriangleright} (\mathsf{u}, \mathsf{v})$ if $\mathsf{s} \trianglerighteq \mathsf{u}$ and $\mathsf{t} \trianglerighteq \mathsf{v}$. Similarly, $(\mathsf{s}, \mathsf{t}) \blacktriangleright (\mathsf{u}, \mathsf{v})$ if $(\mathsf{s}, \mathsf{t}) \underline{\blacktriangleright} (\mathsf{u}, \mathsf{v})$ and $(\mathsf{s}, \mathsf{t}) \neq (\mathsf{u}, \mathsf{v})$.

It is easy to see that there are unique standard $\boldsymbol{\lambda}$-tableaux $\mathsf{t}^{\boldsymbol{\lambda}}$ and $\mathsf{t}_{\boldsymbol{\lambda}}$ such that $\mathsf{t}^{\boldsymbol{\lambda}} \trianglerighteq \mathsf{t} \trianglerighteq \mathsf{t}_{\boldsymbol{\lambda}}$, for all $\mathsf{t} \in \operatorname{Std}(\boldsymbol{\lambda})$. The tableau $\mathsf{t}^{\boldsymbol{\lambda}}$ has the numbers $1, 2, \dots, n$ entered in order from left to right along the rows of $\mathsf{t}^{\lambda^{(1)}}$, and then $\mathsf{t}^{\lambda^{(2)}}, \dots, \mathsf{t}^{\lambda^{(\ell)}}$. Similarly, $\mathsf{t}_{\boldsymbol{\lambda}}$ is the tableau with the numbers $1, \dots, n$ entered in order down the columns of $\mathfrak{t}^{\lambda^{(\ell)}}, \dots, \mathfrak{t}^{\lambda^{(2)}}, \mathfrak{t}^{\lambda^{(1)}}$. If $\boldsymbol{\lambda} = (3, 1^2|2, 1|3, 2)$ then the two $\boldsymbol{\lambda}$-tableaux displayed above are $\mathfrak{t}^{\boldsymbol{\lambda}}$ and $\mathfrak{t}_{\boldsymbol{\lambda}}$, respectively.

Given a standard $\boldsymbol{\lambda}$-tableau $\mathfrak{t}$ define permutations $d(\mathfrak{t}), d'(\mathfrak{t}) \in \mathfrak{S}_n$ by $\mathfrak{t}^{\boldsymbol{\lambda}} d(\mathfrak{t}) = \mathfrak{t} = \mathfrak{t}_{\boldsymbol{\lambda}} d'(\mathfrak{t})$. Then $d(\mathfrak{t}) d'(\mathfrak{t})^{-1} = d(\mathfrak{t}_{\boldsymbol{\lambda}})$ with $\ell(d(\mathfrak{t})) + \ell(d'(\mathfrak{t})) = \ell(d(\mathfrak{t}_{\boldsymbol{\lambda}}))$, for all $\mathfrak{t} \in \operatorname{Std}(\boldsymbol{\lambda})$. Let $\leq$ be the Bruhat order on $\mathfrak{S}_n$ with the convention that $1 \leq w$ for all $w \in \mathfrak{S}_n$. Independently, Ehresmann and James [59] showed that if $\mathfrak{s}, \mathfrak{t} \in \operatorname{Std}(\boldsymbol{\lambda})$ then $\mathfrak{s} \trianglerighteq \mathfrak{t}$ if and only if $d(\mathfrak{s}) \leq d(\mathfrak{t})$ and if and only if $d'(\mathfrak{t}) \leq d'(\mathfrak{s})$. A proof can be found, for example, in [104, Theorem 3.8].

Finally, we will need to know how to conjugate multipartitions and tableaux. The conjugate of a partition λ is the partition $\lambda' = (\lambda'_1, \lambda'_2, \dots)$ where $\lambda'_r = \#\{\, s \geq 1 \mid \lambda_s \geq r \,\}$. That is, we swap the rows and columns of λ. The **conjugate** of a multipartition $\boldsymbol{\lambda} = (\lambda^{(1)}|\dots|\lambda^{(\ell)})$ is the multipartition $\boldsymbol{\lambda}' = (\lambda^{(\ell)\prime}|\dots|\lambda^{(1)\prime})$. Similarly, the conjugate of a $\boldsymbol{\lambda}$-tableau $\mathfrak{t} = (\mathfrak{t}^{(1)}|\dots|\mathfrak{t}^{(\ell)})$ is the $\boldsymbol{\lambda}'$-tableau $\mathfrak{t}' = (\mathfrak{t}^{(\ell)\prime}|\dots|\mathfrak{t}^{(1)\prime})$ where $\mathfrak{t}^{(k)\prime}$ is the tableau obtained by swapping the rows and columns of $\mathfrak{t}^{(k)}$, for $1 \leq k \leq \ell$. Then $\boldsymbol{\lambda} \trianglerighteq \boldsymbol{\mu}$ if and only if $\boldsymbol{\mu}' \trianglerighteq \boldsymbol{\lambda}'$, and $\mathfrak{s} \trianglerighteq \mathfrak{t}$ if and only if $\mathfrak{t}' \trianglerighteq \mathfrak{s}'$.

1.5 *The Murphy basis of $\mathscr{H}_n^\Lambda$*

Graham and Lehrer [48] showed that the cyclotomic Hecke algebras (when $v^2 \neq 1$) are cellular algebras. In this section we recall another cellular basis for these algebras that was constructed in [31] when $v^2 \neq 1$ and in [13] when $v^2 = 1$. When $\ell = 1$ these results are due to Murphy [113].

First observe that Definition 1.1.1 implies that there is a unique anti-isomorphism $*$ on $\mathscr{H}_n$ that fixes each of the generators $T_1, \dots, T_{n-1}, L_1, \dots, L_n$ of $\mathscr{H}_n$. It is easy to see that $T_w^* = T_{w^{-1}}$, for $w \in \mathfrak{S}_n$.

Fix a multipartition $\boldsymbol{\lambda} \in \mathcal{P}_n^\Lambda$. Following [31, Definition 3.14] and [13, §6], if $\mathfrak{s}, \mathfrak{t} \in \operatorname{Std}(\boldsymbol{\lambda})$ define $m_{\mathfrak{s}\mathfrak{t}} = T_{d(\mathfrak{s})^{-1}} m_{\boldsymbol{\lambda}} T_{d(\mathfrak{t})}$, where $m_{\boldsymbol{\lambda}} = u_{\boldsymbol{\lambda}} x_{\boldsymbol{\lambda}}$,

$$u_{\boldsymbol{\lambda}} = \prod_{1 \leq l < \ell} \prod_{r=1}^{|\lambda^{(1)}| + \dots + |\lambda^{(l)}|} \frac{1}{Q'_{l+1}} (L_r - [\kappa_{l+1}]) \quad \text{and} \quad x_{\boldsymbol{\lambda}} = \sum_{w \in \mathfrak{S}_{\boldsymbol{\lambda}}} v^{\ell(w)} T_w,$$

where $Q'_l = 1 + (v - v^{-1}) Q_l$ as in §1.1. The renormalization of $u_{\boldsymbol{\lambda}}$ by $1/Q'_{l+1}$ is not strictly necessary. When $Q'_{l+1} = 0$ this factor can be omitted from the definition of $u_{\boldsymbol{\lambda}}$, at the expense of some aesthetics in some of the formulas that follow. In the integral case, which is what we care most about, this problem does not arise because $Q'_l = v^{\kappa_l} \neq 0$ since $Q_l = [\kappa_l]$, for $1 \leq l \leq \ell$.

Using the relations in $\mathscr{H}_n^\Lambda$ it is not hard to show that $u_{\boldsymbol\lambda}$ and $x_{\boldsymbol\lambda}$ commute. Consequently, $m_{\mathsf{st}}^* = m_{\mathsf{ts}}$, for all $(\mathsf{s},\mathsf{t}) \in \operatorname{Std}^2(\mathcal{P}_n^\Lambda)$.

Theorem 1.5.1 ([31, Theorem 3.26] **and** [13, Theorem 6.3])**.** *The cyclotomic Hecke algebra* $\mathscr{H}_n^\Lambda$ *is free as a* $\mathcal{Z}$*-module with cellular basis* $\{\, m_{\mathsf{st}} \mid \mathsf{s},\mathsf{t} \in \operatorname{Std}(\boldsymbol\lambda)$ *for* $\boldsymbol\lambda \in \mathcal{P}_n^\Lambda \,\}$ *with respect to the poset* $(\mathcal{P}_n^\Lambda, \trianglerighteq)$.

Consequently, $\mathscr{H}_n^\Lambda$ is a cellular algebra so all of the theory in §1.3 applies. In particular, for each $\boldsymbol\lambda \in \mathcal{P}_n^\Lambda$ there exists a Specht module $\underline{S}^{\boldsymbol\lambda}$ with basis $\{\, m_{\mathsf{t}} \mid \mathsf{t} \in \operatorname{Std}(\boldsymbol\lambda) \,\}$. Concretely, we could take $m_{\mathsf{t}} = m_{\mathsf{t}^{\boldsymbol\lambda}\mathsf{t}} + \mathscr{H}_n^{\triangleright\boldsymbol\lambda}$, for $\mathsf{t} \in \operatorname{Std}(\boldsymbol\lambda)$.

Let $\underline{D}^{\boldsymbol\lambda} = \underline{S}^{\boldsymbol\lambda} / \operatorname{rad} \underline{S}^{\boldsymbol\lambda}$ be the quotient of $\underline{S}^{\boldsymbol\lambda}$ by the radical of its bilinear form. Set $\mathcal{K}_n^\Lambda = \{\, \boldsymbol\mu \in \mathcal{P}_n^\Lambda \mid \underline{D}^{\boldsymbol\mu} \neq 0 \,\}$. By Theorem 1.3.4 we obtain:

Corollary 1.5.2 ([13, 31, 48])**.** *Suppose that* $\mathcal{Z} = F$ *is a field. Then* $\{\, \underline{D}^{\boldsymbol\mu} \mid \boldsymbol\mu \in \mathcal{K}_n^\Lambda \,\}$ *is a complete set of pairwise non-isomorphic irreducible* $\mathscr{H}_n^\Lambda$*-modules.*

The set of multipartitions $\mathcal{K}_n^\Lambda$ has been determined by Ariki [4]; see also [8, 22]. We describe and recover his classification of the irreducible $\mathscr{H}_n^\Lambda$-modules in Corollary 3.5.28 below. When $e \neq 2$ and $\ell \geq 3$ the only known descriptions of $\mathcal{K}_n^\Lambda$ are recursive. See [11, 29] for $\ell \leq 2$ and [103] when $e = 2$.

1.6 *Semisimple cyclotomic Hecke algebras of type A*

We now explicitly describe the semisimple representation theory of $\mathscr{H}_n^\Lambda$ using the *seminormal coefficient systems* introduced in [57]. As we are ultimately interested in the cyclotomic quiver Hecke algebras, which are typically non-semisimple algebras even over $\mathbb{Q}$, it is a little surprising that we are interested in these results. We will see, however, that the semisimple representation theory of $\mathscr{H}_n^\Lambda$ and the KLR grading are closely intertwined.

The **Gelfand-Zetlin subalgebra** of $\mathscr{H}_n$ is the subalgebra $\mathscr{L}_n = \mathscr{L}_n(\mathcal{Z}) = \langle L_1, L_2, \dots, L_n \rangle$. We believe that understanding this subalgebra is crucial to understanding the representation theory of $\mathscr{H}_n$. To explain how $\mathscr{L}_n$ acts on $\mathscr{H}_n^\Lambda$ define two **content** functions for $\mathsf{t} \in \operatorname{Std}(\mathcal{P}_n^\Lambda)$ by

$$(1.6.1)\quad c_r^{\mathcal{Z}}(\mathsf{t}) = v^{2(c-b)} Q_l + [c-b]_v \in \mathcal{Z} \qquad \text{and} \qquad c_r^{\mathbb{Z}}(\mathsf{t}) = \kappa_l + c - b \in \mathbb{Z},$$

where $\mathsf{t}(l,b,c) = r$ and $1 \leq r \leq n$. In the special case of the integral parameters, where $Q_l = [\kappa_l]_v$ for $1 \leq l \leq \ell$, the reader can check that $c_r^{\mathcal{Z}}(\mathsf{t}) = [c_r^{\mathbb{Z}}(\mathsf{t})]_v$, for $1 \leq r \leq n$.

The next result is well-known and extremely useful.

Lemma 1.6.2 (James-Mathas [62, Proposition 3.7]**).** *Suppose that* $1 \le r \le n$ *and that* $\mathsf{s}, \mathsf{t} \in \operatorname{Std}(\boldsymbol{\lambda})$, *for* $\boldsymbol{\lambda} \in \mathcal{P}_n^\Lambda$. *Then*

$$m_{\mathsf{st}} L_r \equiv c_r^{\mathcal{Z}}(\mathsf{t}) m_{\mathsf{st}} + \sum_{\substack{\mathsf{v} \rhd \mathsf{t} \\ \mathsf{v} \in \operatorname{Std}(\boldsymbol{\lambda})}} a_{\mathsf{v}} m_{\mathsf{sv}} \pmod{\mathscr{H}_n^{\rhd \boldsymbol{\lambda}}},$$

for some $a_{\mathsf{v}} \in \mathcal{Z}$.

Proof. Let $(l, b, c) = \mathsf{t}^{-1}(r)$. Using our notation, [62, Proposition 3.7] says that $m_{\mathsf{st}} L'_r = Q'_l v^{2(c-b)} m_{\mathsf{st}}$ plus linear combination of more dominant terms, where $Q'_l = 1 + (v - v^{-1}) Q_l$. As $L_r = 1 + (v - v^{-1}) L'_r$ this easily implies the result when $v^2 \neq 1$. The case when $v^2 = 1$ now follows by specialization — or, see [13, Lemma 6.6]. □

In the integral case, $m_{\mathsf{st}} L_r \equiv [c_r^{\mathbb{Z}}(\mathsf{t})] m_{\mathsf{st}} + \sum_{\mathsf{v} \rhd \mathsf{t}} a_{\mathsf{v}} m_{\mathsf{st}} \pmod{\mathscr{H}_n^{\rhd \boldsymbol{\lambda}}}$. This agrees with [57, Lemma 2.9].

The Hecke algebra $\mathscr{H}_n$ is **content separated** if whenever $\mathsf{s}, \mathsf{t} \in \operatorname{Std}(\mathcal{P}_n^\Lambda)$ are standard tableaux, not necessarily of the same shape, then $\mathsf{s} = \mathsf{t}$ if and only if $c_r^{\mathcal{Z}}(\mathsf{s}) = c_r^{\mathcal{Z}}(\mathsf{t})$, for $1 \le r \le n$. The following is an immediate corollary of Lemma 1.6.2 using the theory of JM-elements developed in [108, Theorem 3.7].

Corollary 1.6.3 ([57, Proposition 3.4])**.** *Suppose that* $\mathcal{Z} = F$ *is a field and that* $\mathscr{H}_n$ *is content separated. Then, as an* $(\mathscr{L}_n, \mathscr{L}_n)$*-bimodule,*

$$\mathscr{H}_n = \bigoplus_{(\mathsf{s},\mathsf{t}) \in \operatorname{Std}^2(\mathcal{P}_n^\Lambda)} H_{\mathsf{st}},$$

where $H_{\mathsf{st}} = \{\, h \in \mathscr{H}_n \mid L_r h = c_r^{\mathcal{Z}}(\mathsf{s}) h$ *and* $h L_r = c_r^{\mathcal{Z}}(\mathsf{t}) h$, *for* $1 \le r \le n \,\}$.

For the rest of §1.6 we assume that $\mathscr{H}_n$ is content separated. Corollary 1.6.3 motivates the following definition.

Definition 1.6.4 (Hu-Mathas [57, Definition 3.7]**).** *Suppose that* $\mathcal{Z} = K$ *is a field. A* $*$*-seminormal basis of* $\mathscr{H}_n$ *is a basis of the form*

$$\{\, f_{\mathsf{st}} \mid 0 \neq f_{\mathsf{st}} \in H_{\mathsf{st}} \text{ and } f_{\mathsf{st}}^* = f_{\mathsf{ts}}, \text{ for } (\mathsf{s}, \mathsf{t}) \in \operatorname{Std}^2(\mathcal{P}_n^\Lambda) \,\}.$$

There is a vast literature on seminormal bases. This story started with Young's seminormal forms for the symmetric groups [137] and has now been extended to Hecke algebras and many other diagram algebras including the Brauer, BMW and partition algebras; see, for example, [105, 115, 118].

Suppose that $\{f_{\mathsf{st}}\}$ is a $*$-seminormal basis and that $(\mathsf{s},\mathsf{t}),(\mathsf{u},\mathsf{v}) \in \operatorname{Std}^2(\mathcal{P}_n^\Lambda)$. Let $\mathscr{C}_n = \{\, c_r^{\mathcal{Z}}(\mathsf{s}) \mid \mathsf{s} \in \operatorname{Std}(\mathcal{P}_n^\Lambda) \text{ for } 1 \le r \le n \,\}$ be the set of all possible contents for the tableaux in $\operatorname{Std}(\mathcal{P}_n^\Lambda)$. Following Murphy [108, 112], for a standard tableau $\mathsf{s} \in \operatorname{Std}(\mathcal{P}_n^\Lambda)$ define

$$F_{\mathsf{s}} = \prod_{r=1}^{n} \prod_{\substack{c \in \mathscr{C}_n \\ c \neq c_r^{\mathcal{Z}}(\mathsf{s})}} \frac{L_r - c}{c_r^{\mathcal{Z}}(\mathsf{s}) - c}.$$

By Definition 1.6.4, if $(\mathsf{s},\mathsf{t}),(\mathsf{u},\mathsf{v}) \in \operatorname{Std}^2(\mathcal{P}_n^\Lambda)$ then $\delta_{\mathsf{su}}\delta_{\mathsf{tv}} f_{\mathsf{st}} = F_{\mathsf{u}} f_{\mathsf{st}} F_{\mathsf{v}}$. In particular, F_{s} is a non-zero element of $\mathscr{H}_n$. It follows that F_{s} is a scalar multiple of f_{ss}, which implies that $\{\, F_{\mathsf{s}} \mid \mathsf{s} \in \operatorname{Std}(\mathcal{P}_n^\Lambda) \,\}$ is a complete set of pairwise orthogonal idempotents in $\mathscr{H}_n$. (In fact, in [108] these properties are used to establish Corollary 1.6.3.) Consequently, there exists a non-zero scalar $\gamma_{\mathsf{s}} \in F$ such that $F_{\mathsf{s}} = \frac{1}{\gamma_{\mathsf{s}}} f_{\mathsf{ss}}$. If $(\mathsf{s},\mathsf{t}),(\mathsf{u},\mathsf{v}) \in \operatorname{Std}^2(\mathcal{P}_n^\Lambda)$ then

$$f_{\mathsf{st}} f_{\mathsf{uv}} = f_{\mathsf{st}} F_{\mathsf{t}} F_{\mathsf{u}} f_{\mathsf{uv}} = \delta_{\mathsf{tu}} \gamma_{\mathsf{t}} f_{\mathsf{sv}}. \tag{1.6.5}$$

The next definition allows us to classify all seminormal bases and to describe how $\mathscr{H}_n^\Lambda$ acts on them.

Definition 1.6.6 (Hu-Mathas [57, §3]**).** *A* ***$*$-seminormal coefficient system*** *is a collection of scalars*

$$\boldsymbol{\alpha} = \{\, \alpha_r(\mathsf{t}) \mid \mathsf{t} \in \operatorname{Std}(\mathcal{P}_n^\Lambda) \text{ and } 1 \le r \le n \,\}$$

such that $\alpha_r(\mathsf{t}) = 0$ *if* $\mathsf{v} = \mathsf{t}(r, r+1)$ *is not standard, if* $\mathsf{v} \in \operatorname{Std}(\mathcal{P}_n^\Lambda)$ *then*

$$\alpha_r(\mathsf{v})\alpha_r(\mathsf{t}) = \frac{\big(1 - v^{-1}c_r^{\mathcal{Z}}(\mathsf{t}) + v c_r^{\mathcal{Z}}(\mathsf{v})\big)\big(1 + v c_r^{\mathcal{Z}}(\mathsf{t}) - v^{-1}c_r^{\mathcal{Z}}(\mathsf{v})\big)}{\big(c_r^{\mathcal{Z}}(\mathsf{t}) - c_r^{\mathcal{Z}}(\mathsf{v})\big)\big(c_r^{\mathcal{Z}}(\mathsf{v}) - c_r^{\mathcal{Z}}(\mathsf{t})\big)},$$

and $\alpha_r(\mathsf{t})\alpha_{r+1}(\mathsf{t}s_r)\alpha_r(\mathsf{t}s_r s_{r+1}) = \alpha_{r+1}(\mathsf{t})\alpha_r(\mathsf{t}s_{r+1})\alpha_{r+1}(\mathsf{t}s_{r+1}s_r)$, *and if* $|r - r'| > 1$ *then* $\alpha_r(\mathsf{t})\alpha_{r'}(\mathsf{t}s_r) = \alpha_{r'}(\mathsf{t})\alpha_r(\mathsf{t}s_{r'})$, *for* $1 \le r, r' < n$.

As the reader might guess, the conditions on the scalars $\alpha_r(\mathsf{t})$ in Definition 1.6.6 correspond to the quadratic relations $(T_r - v)(T_r + v^{-1}) = 0$ and the braid relations for $T_1, \dots, T_{n-1}$. The simplest example of a seminormal coefficient system is

$$\alpha_r(\mathsf{t}) = \frac{\big(1 - v^{-1}c_{r+1}^{\mathcal{Z}}(\mathsf{t}) + v c_r^{\mathcal{Z}}(\mathsf{t})\big)}{\big(c_{r+1}^{\mathcal{Z}}(\mathsf{t}) - c_r^{\mathcal{Z}}(\mathsf{t})\big)},$$

whenever $1 \le r < n$ and $\mathsf{t}, \mathsf{t}(r, r+1) \in \operatorname{Std}(\mathcal{P}_n^\Lambda)$. Another seminormal coefficient system is given in (1.7.1) below.

Seminormal coefficient systems arise because they describe the action of $\mathscr{H}_n$ on a seminormal basis. More precisely, we have the following:

Theorem 1.6.7 (Hu-Mathas [57]). *Suppose that $\mathcal{Z} = K$ is a field and that $\mathscr{H}_n$ is content separated and that $\{ f_{\mathsf{st}} \mid (\mathsf{s},\mathsf{t}) \in \operatorname{Std}^2(\mathcal{P}_n^\Lambda) \}$ is a seminormal basis of $\mathscr{H}_n$. Then $\{f_{\mathsf{st}}\}$ is a cellular basis of $\mathscr{H}_n$ and there exists a unique seminormal coefficient system $\boldsymbol{\alpha}$ such that*

$$f_{\mathsf{st}}T_r = \alpha_r(\mathsf{t})f_{\mathsf{sv}} + \frac{1+(v-v^{-1})c^{\mathcal{Z}}_{r+1}(\mathsf{t})}{c^{\mathcal{Z}}_{r+1}(\mathsf{t}) - c^{\mathcal{Z}}_r(\mathsf{t})} f_{\mathsf{st}},$$

where $\mathsf{v} = \mathsf{t}(r, r+1)$. Moreover, if $\mathsf{s} \in \operatorname{Std}(\boldsymbol{\lambda})$ then $F_{\mathsf{s}} = \frac{1}{\gamma_{\mathsf{s}}} f_{\mathsf{ss}}$ is a primitive idempotent and $\underline{S}^{\boldsymbol{\lambda}} \cong F_{\mathsf{s}}\mathscr{H}_n$ is irreducible for all $\boldsymbol{\lambda} \in \mathcal{P}_n^\Lambda$.

Proof. [Sketch of proof] By definition, $\{f_{\mathsf{st}}\}$ is a basis of $\mathscr{H}_n$ such that $f^*_{\mathsf{st}} = f_{\mathsf{ts}}$ for all $(\mathsf{s},\mathsf{t}) \in \operatorname{Std}^2(\mathcal{P}_n^\Lambda)$. Therefore, it follows from (1.6.5) that $\{f_{\mathsf{st}}\}$ is a cellular basis of $\mathscr{H}_n$ with cellular automorphism $*$.

It is an amusing application of the relations in Definition 1.1.1 to show that there exists a seminormal coefficient system that describes the action of T_r on the seminormal basis. See [57, Lemma 3.13] for details. The uniqueness of $\boldsymbol{\alpha}$ is clear.

We have already observed in (1.6.5) that $F_{\mathsf{s}} = \frac{1}{\gamma_{\mathsf{s}}} f_{\mathsf{ss}}$, for $\mathsf{s} \in \operatorname{Std}(\boldsymbol{\lambda})$, so it remains to show that F_{s} is primitive and that $\underline{S}^{\boldsymbol{\lambda}} \cong F_{\mathsf{s}}\mathscr{H}_n$. By what we have already shown, $F_{\mathsf{s}}\mathscr{H}_n$ is contained in the span of $\{ f_{\mathsf{st}} \mid \mathsf{t} \in \operatorname{Std}(\boldsymbol{\lambda}) \}$. On the other hand, if $f = \sum_{\mathsf{t}} r_{\mathsf{t}} f_{\mathsf{st}} \in F_{\mathsf{s}}\mathscr{H}_n$ and $r_{\mathsf{v}} \neq 0$ then $r_{\mathsf{v}} f_{\mathsf{sv}} = f F_{\mathsf{v}} \in F_{\mathsf{s}}\mathscr{H}_n$. It follows that $F_{\mathsf{s}}\mathscr{H}_n = \sum_{\mathsf{t}} K f_{\mathsf{st}}$, as a vector space. Consequently, $F_{\mathsf{s}}\mathscr{H}_n$ is irreducible and F_{s} is a primitive idempotent in $\mathscr{H}_n$. Finally, $\underline{S}^{\boldsymbol{\lambda}} \cong F_{\mathsf{s}}\mathscr{H}_n$ by Lemma 1.6.2 since $\mathscr{H}_n$ is content separated. □

Corollary 1.6.8 ([57, Corollary 3.17]). *Suppose that $\boldsymbol{\alpha}$ is a seminormal coefficient system and that $\mathsf{s} \rhd \mathsf{t} = \mathsf{s}(r, r+1)$, for tableaux $\mathsf{s},\mathsf{t} \in \operatorname{Std}(\mathcal{P}_n^\Lambda)$ and where $1 \le r < n$. Then $\alpha_r(\mathsf{s})\gamma_{\mathsf{t}} = \alpha_r(\mathsf{t})\gamma_{\mathsf{s}}$.*

Consequently, if the seminormal coefficient system $\boldsymbol{\alpha}$ is known then fixing γ_{t}, for some $\mathsf{t} \in \operatorname{Std}(\boldsymbol{\lambda})$, determines γ_{s} for all $\mathsf{s} \in \operatorname{Std}(\boldsymbol{\lambda})$. Conversely, these scalars, together with $\boldsymbol{\alpha}$, determines the seminormal basis.

Corollary 1.6.9 (Classifying seminormal bases [57, Theorem 3.14]**).** *There is a one-to-one correspondence between the $*$-seminormal bases of $\mathscr{H}_n$ and the pairs $(\boldsymbol{\alpha}, \boldsymbol{\gamma})$ where $\boldsymbol{\alpha} = \{ \alpha_r(\mathsf{s}) \mid 1 \le r < n \text{ and } \mathsf{s} \in \operatorname{Std}(\mathcal{P}_n^\Lambda) \}$ is a seminormal coefficient system and $\boldsymbol{\gamma} = \{ \gamma_{\mathsf{t}^{\boldsymbol{\lambda}}} \mid \boldsymbol{\lambda} \in \mathcal{P}_n^\Lambda \}$.*

Finally, the seminormal basis machinery in this section can be used to classify the semisimple cyclotomic Hecke algebras $\mathscr{H}_n$, thus re-proving Ariki's semisimplicity criterion [2], when $v^2 \neq 1$, and [13, Theorem 6.11], when $v^2 = 1$.

Theorem 1.6.10 (Ariki [2] and [13, Theorem 6.11]). *Suppose that F is a field. The following are equivalent:*

a) $\mathscr{H}_n = \mathscr{H}_n(F, v, Q_1, \dots, Q_\ell)$ *is semisimple.*

b) $\mathscr{H}_n$ *is content separated.*

c) $[1]_v[2]_v \dots [n]_v \prod_{1 \le r < s \le \ell} \prod_{-n<d<n} (v^{2d}Q_r + [d]_v - Q_s) \neq 0.$

We want to rephrase the semisimplicity criterion of Theorem 1.6.10 for the integral cyclotomic Hecke algebras $\mathscr{H}_n^\Lambda$, for $\Lambda \in P^+$. For each $i \in I$ define the **i-string of length** n to be $\alpha_{i,n} = \alpha_i + \alpha_{i+1} + \dots + \alpha_{i+n-1}$. Then $\alpha_{i,n} \in Q^+$.

Corollary 1.6.11. *Suppose that $\Lambda \in P^+$ and that $\mathcal{Z} = F$ is a field. Then $\mathscr{H}_n^\Lambda$ is semisimple if and only if $e > n$ and $(\Lambda, \alpha_{i,n}) \le 1$, for all $i \in I$.*

Proof. As $Q_r = [\kappa_r]$, for $1 \le r \le \ell$, the statement of Theorem 1.6.10(c) simplifies because $v^{2d}Q_r + [d]_v - Q_s = v^{-2\kappa_s}[d + \kappa_r - \kappa_s]_v$. Therefore, $\mathscr{H}_n^\Lambda$ is semisimple if and only if

$$[1]_v[2]_v \dots [n]_v \prod_{1 \le r < s \le \ell} \prod_{-n<d<n} [d + \kappa_r - \kappa_s]_v \neq 0.$$

On the other hand, $[1]_v[2]_v \dots [n]_v \neq 0$ if and only if $e > n$. Furthermore, $(\Lambda, \alpha_{i,n}) \le 1$, for all $i \in I$, if and only if $\kappa_r + d \neq \kappa_s$, for $1 \le r < s \le \ell$ and all $-n < d < n$. The result follows. □

1.7 *Gram determinants and the Jantzen sum formula*

For future use, we now recall the closed formula for the Gram determinants of the Specht modules $\underline{S}^{\boldsymbol{\lambda}}$ and the connection between these formulas and Jantzen filtrations. Throughout this section we assume that $\mathscr{H}_n$ is content separated over the field $K = \mathcal{Z}$.

For $\boldsymbol{\lambda} \in \mathcal{P}_n^\Lambda$ let $\underline{G}^{\boldsymbol{\lambda}} = \big(\langle m_{\mathsf{s}}, m_{\mathsf{t}} \rangle\big)_{\mathsf{s},\mathsf{t} \in \mathrm{Std}(\boldsymbol{\lambda})}$ be the **Gram matrix** of the Specht module $\underline{S}^{\boldsymbol{\lambda}}$, where we fix an arbitrary ordering of the rows and columns of $\underline{G}^{\boldsymbol{\lambda}}$.

For $(\mathsf{s},\mathsf{t}) \in \operatorname{Std}^2(\mathcal{P}_n^\Lambda)$ set $f_{\mathsf{st}} = F_{\mathsf{s}} m_{\mathsf{st}} F_{\mathsf{t}}$. By Lemma 1.6.2 and (1.6.5),

$$f_{\mathsf{st}} = m_{\mathsf{st}} + \sum_{(\mathsf{u},\mathsf{v}) \rhd (\mathsf{s},\mathsf{t})} r_{\mathsf{uv}} m_{\mathsf{uv}},$$

for some $r_{\mathsf{uv}} \in K$. By construction, $\{f_{\mathsf{st}}\}$ is a seminormal basis of $\mathscr{H}_n$. By [57, Proposition 3.18] this basis corresponds to the seminormal coefficient system given by

$$(1.7.1)\quad \alpha_r(\mathsf{t}) = \begin{cases} 1, & \text{if } \mathsf{t} \rhd \mathsf{t}(r, r+1), \\ \frac{\left(1 - v^{-1} c_r(\mathsf{t}) + v c_r(\mathsf{v})\right)\left(1 + v c_r(\mathsf{t}) - v^{-1} c_r(\mathsf{v})\right)}{\left(c_r(\mathsf{t}) - c_r(\mathsf{v})\right)\left(c_r(\mathsf{v}) - c_r(\mathsf{t})\right)}, & \text{otherwise,} \end{cases}$$

for $\mathsf{t} \in \operatorname{Std}(\mathcal{P}_n^\Lambda)$ and $1 \le r < n$ such that $\mathsf{t}s_r$ is standard. The γ-coefficients $\{\gamma_{\mathsf{t}}\}$ for this basis are explicitly known by [62, Corollary 3.29]. Moreover,

$$(1.7.2)\qquad \det \underline{G}^{\boldsymbol{\lambda}} = \prod_{\mathsf{t} \in \operatorname{Std}(\boldsymbol{\lambda})} \gamma_{\mathsf{t}}.$$

By explicitly computing the scalars γ_{t}, and using an intricate inductive argument based on the semisimple branching rules for the Specht modules, James and the author proved the following:

Theorem 1.7.3 (James-Mathas [62, Corollary 3.38]**).** *Suppose that $\mathscr{H}_n$ is content separated. Then there exist explicitly known scalars $g_{\boldsymbol{\lambda\mu}}$ and signs $\varepsilon_{\boldsymbol{\lambda\mu}} = \pm 1$ such that*

$$\det \underline{G}^{\boldsymbol{\lambda}} = \prod_{\substack{\boldsymbol{\mu} \in \mathcal{P}_n^\Lambda \\ \boldsymbol{\lambda} \rhd \boldsymbol{\mu}}} g_{\boldsymbol{\lambda\mu}}^{\varepsilon_{\boldsymbol{\lambda\mu}} \dim \underline{S}^{\boldsymbol{\mu}}}.$$

The scalars $g_{\boldsymbol{\lambda\mu}}$ are described combinatorially as the quotient of at most two *hook lengths*. The sign $\varepsilon_{\boldsymbol{\lambda\mu}}$ is the parity of the sum of the leg lengths of these hooks.

Theorem 1.7.3 gives a very pretty closed formula for the Gram determinant $\underline{G}^{\boldsymbol{\lambda}}$, generalizing a classical result of James and Murphy [64]. One problem with this formula is that $\det \underline{G}^{\boldsymbol{\lambda}}$ is a *polynomial* in $v, v^{-1}, Q_1, \dots, Q_\ell$ whereas Theorem 1.7.3 computes this determinant as a *rational function* in $v, Q_1, \dots, Q_\ell$. On the other hand, as we now recall, Theorem 1.7.3 has an impressive module theoretic application in the *Jantzen sum formula*.

Fix a modular system $(K, \mathcal{Z}, F)$, where $\mathcal{Z}$ is a discrete valuation ring with maximal ideal $\mathfrak{p}$ and such that $\mathcal{Z}$ contains $v, v^{-1}, Q_1, \dots, Q_\ell$. Let K be the field of fractions of $\mathcal{Z}$ and let $F = \mathcal{Z}/\mathfrak{p}$ be the residue field of $\mathcal{Z}$. Let $\mathscr{H}_n^{\mathcal{Z}}$, $\mathscr{H}_n^K \cong \mathscr{H}_n^{\mathcal{Z}} \otimes_{\mathcal{Z}} K$ and $\mathscr{H}_n^F = \mathscr{H}_n^{\mathcal{Z}} \otimes_{\mathcal{Z}} F$ be the corresponding

Hecke algebras. Therefore, $\mathscr{H}_n^F$ has Hecke parameter $v + \mathfrak{p}$ and cyclotomic parameters $Q_l + \mathfrak{p}$, for $1 \le l \le \ell$.

Let $\boldsymbol{\lambda} \in \mathcal{P}_n^\Lambda$ and let $\underline{S}_{\mathcal{Z}}^{\boldsymbol{\lambda}}$ and $\underline{S}_F^{\boldsymbol{\lambda}} \cong \underline{S}_{\mathcal{Z}}^{\boldsymbol{\lambda}} \otimes_{\mathcal{Z}} F$ be the corresponding Specht modules for $\mathscr{H}_n^{\mathcal{Z}}$ and $\mathscr{H}_n^F$, respectively. Define a filtration of the Specht module $\underline{S}_{\mathcal{Z}}^{\boldsymbol{\lambda}}$ by $J_k(\underline{S}_{\mathcal{Z}}^{\boldsymbol{\lambda}}) = \{\, x \in \underline{S}_{\mathcal{Z}}^{\boldsymbol{\lambda}} \mid \langle x, y\rangle_{\boldsymbol{\lambda}} \in \mathfrak{p}^k \text{ for all } y \in \underline{S}_{\mathcal{Z}}^{\boldsymbol{\lambda}}\,\}$, for $k \ge 0$. The **Jantzen filtration** of $\underline{S}_F^{\boldsymbol{\lambda}}$ is the filtration

$$\underline{S}_F^{\boldsymbol{\lambda}} = J_0(\underline{S}_F^{\boldsymbol{\lambda}}) \supseteq J_1(\underline{S}_F^{\boldsymbol{\lambda}}) \supseteq \cdots \supseteq J_z(\underline{S}_F^{\boldsymbol{\lambda}}) = 0,$$

where $J_k(\underline{S}_F^{\boldsymbol{\lambda}}) = \big(J_k(\underline{S}_{\mathcal{Z}}^{\boldsymbol{\lambda}}) + \mathfrak{p}\underline{S}_{\mathcal{Z}}^{\boldsymbol{\lambda}}\big)/\mathfrak{p}\underline{S}_{\mathcal{Z}}^{\boldsymbol{\lambda}}$ for $k \ge 0$. (As $\underline{S}_F^{\boldsymbol{\lambda}}$ is finite dimensional, $J_z(\underline{S}_F^{\boldsymbol{\lambda}}) = 0$ for $z \gg 0$.)

Let $\operatorname{Rep}(\mathscr{H}_n)$ be the category of finitely generated $\mathscr{H}_n$-modules and let $[\operatorname{Rep}(\mathscr{H}_n)]$ be its Grothendieck group. Let $[M]$ be the image of the $\mathscr{H}_n$-module M in $[\operatorname{Rep}(\mathscr{H}_n)]$. Let $\nu_{\mathfrak{p}}$ be the $\mathfrak{p}$-adic valuation map on $\mathcal{Z}^\times$.

Theorem 1.7.4 (James-Mathas [62, Theorem 4.6]**).** *Let $(K, \mathcal{Z}, F)$ be a modular system and suppose that $\boldsymbol{\lambda} \in \mathcal{P}_n^\Lambda$. Then, in $[\operatorname{Rep}(\mathscr{H}_n^F)]$,*

$$\sum_{k>0} [J_k(\underline{S}_F^{\boldsymbol{\lambda}})] = \sum_{\boldsymbol{\lambda} \rhd \boldsymbol{\mu}} \varepsilon_{\boldsymbol{\lambda}\boldsymbol{\mu}} \nu_{\mathfrak{p}}(g_{\boldsymbol{\lambda}\boldsymbol{\mu}}) [\underline{S}_F^{\boldsymbol{\mu}}].$$

Intuitively, the proof of Theorem 1.7.4 amounts to taking the $\mathfrak{p}$-adic valuation of the formula in Theorem 1.7.3. In fact, this is exactly how Theorem 1.7.4 is proved except that you need the corresponding formulas for the Gram determinants of the weight spaces of the Weyl modules of the cyclotomic Schur algebras of [31]. This is enough because the dimensions of the weight spaces of a module uniquely determine its image in the Grothendieck group of the Schur algebra. The proof given in [62] is stated only for the non-degenerate case $v^2 \ne 1$, however, the arguments apply equally well for the degenerate case when $v^2 = 1$.

The main point that we want to emphasize in this section is that the *rational* formula for $\det \underline{G}_K^{\boldsymbol{\lambda}}$ in Theorem 1.7.3 corresponds to writing the left-hand side of the Jantzen sum formula sum as a $\mathbb{Z}$-linear combination of Specht modules. Therefore, when the right-hand side of the sum formula is written as a linear combination of simple modules some of the terms must cancel. We give a cancellation free sum formula in §4.1.

Theorem 1.7.4 is a useful inductive tool because it gives an upper bound on the decomposition numbers of $\underline{S}_F^{\boldsymbol{\lambda}}$. Let $j_{\boldsymbol{\lambda}\boldsymbol{\mu}} = \varepsilon_{\boldsymbol{\lambda}\boldsymbol{\mu}} \nu_{\mathfrak{p}}(g_{\boldsymbol{\lambda}\boldsymbol{\mu}})$, for $\boldsymbol{\lambda}, \boldsymbol{\mu} \in \mathcal{P}_n^\Lambda$ and set $d_{\boldsymbol{\lambda}\boldsymbol{\mu}}^F = [\underline{S}_F^{\boldsymbol{\lambda}} : \underline{D}_F^{\boldsymbol{\mu}}]$. Using Theorem 1.7.4 to compute the multiplicity of $\underline{D}_F^{\boldsymbol{\mu}}$ in $\bigoplus_{k>0} J_k(\underline{S}_F^{\boldsymbol{\lambda}})$ yields the following.

Corollary 1.7.5. *Suppose that* $\boldsymbol{\lambda}, \boldsymbol{\mu} \in \mathcal{P}_n^\Lambda$. *Then* $0 \le d_{\boldsymbol{\lambda}\boldsymbol{\mu}}^F \le \sum_{\substack{\boldsymbol{\nu}\in\mathcal{P}_n^\Lambda \\ \boldsymbol{\lambda}\rhd\boldsymbol{\nu}\unrhd\boldsymbol{\mu}}} j_{\boldsymbol{\lambda}\boldsymbol{\nu}} d_{\boldsymbol{\nu}\boldsymbol{\mu}}^F$.

As a second application, Theorem 1.7.4 classifies the irreducible Specht modules $\underline{S}_F^{\boldsymbol{\lambda}}$, for $\boldsymbol{\lambda} \in \mathcal{K}_n^\Lambda$. Note that there are explicit combinatorial formulas for the integers $j_{\boldsymbol{\lambda}\boldsymbol{\mu}}$.

Corollary 1.7.6 (James-Mathas [62, Theorem 4.7]**).** *Suppose that F is a field and $\boldsymbol{\lambda} \in \mathcal{K}_n^\Lambda$. Then the Specht module $\underline{S}_F^{\boldsymbol{\lambda}}$ is irreducible if and only if $j_{\boldsymbol{\lambda}\boldsymbol{\mu}} = 0$ for all $\boldsymbol{\mu} \rhd \boldsymbol{\lambda}$.*

1.8 *The blocks of* $\mathscr{H}_n^F$

The most important application of the Jantzen Sum Formula (Theorem 1.7.4) is to the classification of the blocks of $\mathscr{H}_n^F$. The algebra $\mathscr{H}_n^F$, and in fact any finite-dimensional algebra over a field, can be written as a direct sum of indecomposable two-sided ideals: $\mathscr{H}_n^F = B_1 \oplus \cdots \oplus B_d$. The indecomposable two-sided ideals, or subalgebras $B_1, \dots, B_z$, are the **blocks** of $\mathscr{H}_n^F$. They are uniquely determined up to permutation. Any $\mathscr{H}_n^F$-module M splits into a direct sum of block components $M = MB_1 \oplus \cdots \oplus MB_d$, where we allow some of the summands to be zero. The module M **belongs** to the block B_r if $M = MB_r$. It is well-known that two simple modules $\underline{D}^{\boldsymbol{\lambda}}$ and $\underline{D}^{\boldsymbol{\mu}}$ belong to the same block if and only if they are in the same **linkage class**. That is, there exist indecomposable modules $\underline{M}_0, \dots, \underline{M}_z$ and a sequence of multipartitions $\boldsymbol{\nu}_0 = \boldsymbol{\lambda}$, $\boldsymbol{\nu}_1, \dots, \boldsymbol{\nu}_z = \boldsymbol{\mu}$ such that $[\underline{M}_r : \underline{D}^{\boldsymbol{\nu}_r}] \neq 0$ and $[\underline{M}_r : \underline{D}^{\boldsymbol{\nu}_{r+1}}] \neq 0$, for $0 \le r < z$. In fact, we can assume that the $\underline{M}_r$ are Specht modules, even though the Specht modules are not necessarily indecomposable.

We want an explicit combinatorial description of the blocks of $\mathscr{H}_n^F$. Define two equivalence relations $\sim_C$ and $\sim_J$ on $\mathcal{P}_n^\Lambda$ as follows. First, $\boldsymbol{\lambda} \sim_C \boldsymbol{\mu}$ if there is an equality of *multisets* $\{\, c_{\mathfrak{t}^{\boldsymbol{\lambda}}}^F(r) \mid 1 \le r \le n \,\} = \{\, c_{\mathfrak{t}^{\boldsymbol{\mu}}}^F(r) \mid 1 \le r \le n \,\}$. The second relation, **Jantzen equivalence**, is more involved: $\boldsymbol{\lambda} \sim_J \boldsymbol{\mu}$ if there exists a sequence $\boldsymbol{\nu}_0 = \boldsymbol{\lambda}, \boldsymbol{\nu}_1, \dots, \boldsymbol{\nu}_z = \boldsymbol{\mu}$ of multipartitions in $\mathcal{P}_n^\Lambda$ such that $j_{\boldsymbol{\nu}_r\boldsymbol{\nu}_{r+1}} \neq 0$ or $j_{\boldsymbol{\nu}_{r+1}\boldsymbol{\nu}_r} \neq 0$, for $0 \le r < z$.

Theorem 1.8.1 (Lyle-Mathas [96]**, Brundan** [19]**).** *Suppose that F is a field and that $\boldsymbol{\lambda}, \boldsymbol{\mu} \in \mathcal{P}_n^\Lambda$. Then the following are equivalent:*

a) $\underline{D}^{\boldsymbol{\lambda}}$ *and* $\underline{D}^{\boldsymbol{\mu}}$ *are in the same* $\mathscr{H}_n^F$*-block.*
b) $\underline{S}^{\boldsymbol{\lambda}}$ *and* $\underline{S}^{\boldsymbol{\mu}}$ *are in the same* $\mathscr{H}_n^F$*-block.*
c) $\boldsymbol{\lambda} \sim_J \boldsymbol{\mu}$.
d) $\boldsymbol{\lambda} \sim_C \boldsymbol{\mu}$.

Parts (a) and (b) are equivalent by the general theory of cellular algebras [48] whereas the equivalence of parts (b) and (c) is a general property of Jantzen filtrations from [96]. (In fact, part (c) is general property of the standard modules of a quasi-hereditary algebra.) In practice, part (d) is the most useful because it is easy to compute.

The hard part in proving Theorem 1.8.1 is in showing that parts (c) and (d) are equivalent. The argument is purely combinatorial with work of Fayers [36, 37] playing an important role.

When $\mathscr{H}_n^F = \mathscr{H}_n^\Lambda$ for some $\Lambda \in P^+$, the integral case, there is a nice reformulation of Theorem 1.8.1. The **residue sequence** of a standard tableau $\mathfrak{t}$ is $\mathbf{i}^{\mathfrak{t}} = (i_1^{\mathfrak{t}}, \dots, i_n^{\mathfrak{t}}) \in I^n$ where $i_r^{\mathfrak{t}} = c_r^{\mathbb{Z}}(\mathfrak{t}) + e\mathbb{Z}$. If $\boldsymbol{\lambda} \in \mathcal{P}_n^\Lambda$ define

$$\beta^{\boldsymbol{\lambda}} = \sum_{r=1}^{n} \alpha_{i_r^{\mathfrak{t}}} = \sum_{r=1}^{n} \alpha_{i_r^{\boldsymbol{\lambda}}} \in Q^+, \qquad \text{for some } \mathfrak{t} \in \operatorname{Std}(\boldsymbol{\lambda}).$$

By definition, $\beta^{\boldsymbol{\lambda}} \in Q^+$ depends only on $\boldsymbol{\lambda}$, and not on the choice of $\mathfrak{t}$. Moreover, $\boldsymbol{\lambda} \sim_C \boldsymbol{\mu}$ if and only if $\beta^{\boldsymbol{\lambda}} = \beta^{\boldsymbol{\mu}}$. Hence, we have the following:

Corollary 1.8.2. *Suppose that $\Lambda \in P^+$ and $\boldsymbol{\lambda}, \boldsymbol{\mu} \in \mathcal{P}_n^\Lambda$. Then $\underline{S}^{\boldsymbol{\lambda}}$ and $\underline{S}^{\boldsymbol{\mu}}$ are in the same $\mathscr{H}_n^\Lambda$-block if and only if $\beta^{\boldsymbol{\lambda}} = \beta^{\boldsymbol{\mu}}$.*

2 Cyclotomic Quiver Hecke Algebras of Type A

This section introduces the quiver Hecke algebras, and their cyclotomic quotients. We use the relations to reveal some of the properties of these algebras. The main aim of this section is to give the reader an appreciation of, and some familiarity with, the KLR relations without appealing to any general theory.

2.1 *Graded algebras*

In this section we quickly review the theory of graded (cellular) algebras. For more details the reader is referred to [15, 54, 114]. Throughout, $\mathcal{Z}$ is a commutative integral domain. Unless otherwise stated, all modules and algebras will be free and of finite rank as $\mathcal{Z}$-modules.

In this chapter a **graded module** will always mean a $\mathbb{Z}$-graded module. That is, a $\mathcal{Z}$-module M that has a decomposition $M = \bigoplus_{d\in\mathbb{Z}} M_d$ as a $\mathcal{Z}$-module. A **positively graded** module is a graded module $M = \bigoplus_d M_d$ such that $M_d = 0$ if $d < 0$.

A **graded algebra** is a unital associative $\mathcal{Z}$-algebra $A = \bigoplus_{d\in\mathbb{Z}} A_d$ that is a $\mathbb{Z}$-graded $\mathcal{Z}$-module such that $A_dA_e \subseteq A_{d+e}$, for all $d,e\in\mathbb{Z}$. It follows that $1\in A_0$ and that A_0 is a $\mathbb{Z}$-graded subalgebra of A. A graded (right) A-module is a graded $\mathcal{Z}$-module M such that $\underline{M}$ is an $\underline{A}$-module and $M_dA_e \subseteq M_{d+e}$, for all $d,e\in\mathbb{Z}$. Here $\underline{M}$ is the (ungraded) module, and $\underline{A}$ is the (ungraded) algebra, obtained by forgetting the $\mathbb{Z}$-gradings on M and A respectively. Graded submodules, graded left A-modules and so on are all defined in the obvious way.

Suppose that M is a graded A-module. If $m\in M_d$, for $d\in\mathbb{Z}$, then m is **homogeneous** of **degree** d and we set $\deg m = d$. Every element $m\in M$ can be written uniquely as a linear combination $m=\sum_d m_d$ of its **homogeneous components**, where $\deg m_d = d$ and $m_d\in M$.

A homomorphism of graded A-modules M and N is an $\underline{A}$-module homomorphism $f\colon \underline{M}\longrightarrow \underline{N}$ such that $\deg f(m) = \deg m$, whenever $m\in M$ is homogeneous. That is, f is a degree preserving $\underline{A}$-module homomorphism.

Let $\operatorname{Rep}(A)$ be the category of finitely generated graded A-modules together with degree preserving homomorphisms. Similarly, $\operatorname{Proj}(A)$ is the category of finitely generated projective A-modules with degree preserving maps. A **graded functor** between such categories is any functor that commutes with the grading shift functor that sends M to $M\langle 1\rangle$.

If M is a graded $\mathcal{Z}$-module and $s\in\mathbb{Z}$ let $M\langle s\rangle$ be the graded $\mathcal{Z}$-module obtained by shifting the grading on M up by s; that is, $M\langle s\rangle_d = M_{d-s}$, for $d\in\mathbb{Z}$. If $M\neq 0$ then $M\cong M\langle s\rangle$ as A-modules if and only if $s=0$. In contrast, $\underline{M}\cong\underline{M\langle s\rangle}$ as $\underline{A}$-modules, for all $s\in\mathbb{Z}$.

Let $\operatorname{Hom}_A(M,N)$ be the space of (degree preserving) A-module homomorphisms and set

$$\mathcal{H}om_A(M,N) = \bigoplus_{d\in\mathbb{Z}} \mathcal{H}om_A(M,N\langle d\rangle) \cong \bigoplus_{d\in\mathbb{Z}} \mathcal{H}om_A(M\langle -d\rangle, N).$$

The reader may check that $\mathcal{H}om_A(M,N)\cong \operatorname{Hom}_{\underline{A}}(\underline{M},\underline{N})$ as $\mathcal{Z}$-modules.

Suppose that q is an indeterminate and that M is a graded module. The **graded dimension** of M is the Laurent polynomial

$$\mathit{dim}_q\, M = \sum_{d\in\mathbb{Z}} (\dim M_d) q^d \in \mathbb{N}[q,q^{-1}].$$

If M is a graded A-module, and D is an irreducible graded A-module, then the **graded decomposition number** is the Laurent polynomial

$$[M:D]_q = \sum_{s\in\mathbb{Z}} [M:D\langle s\rangle]\, q^s \in \mathbb{N}[q,q^{-1}].$$

By definition, the (ungraded) decomposition multiplicity $[\underline{M}:\underline{D}]$ is given by evaluating $[M:D]_q$ at $q=1$.

Suppose that A is a graded algebra and that $\underline{m}$ is an (ungraded) $\underline{A}$-module. A graded **lift** of $\underline{m}$ is any graded A-module M such that $\underline{M}\cong\underline{m}$ as $\underline{A}$-modules. If M is a graded lift of $\underline{m}$ then so is $M\langle s\rangle$, for any $s\in\mathbb{Z}$, so graded lifts are not unique. By Fitting's Lemma, if $\underline{m}$ is indecomposable then its graded lift, if it exists, is unique up to grading shift [15, Lemma 2.5.3].

Following [54], the theory of cellular algebras from §1.3 extends to the graded setting in a natural way.

Definition 2.1.1 ([54, §2]). *Suppose that A is $\mathbb{Z}$-graded $\mathcal{Z}$-algebra that is free of finite rank over $\mathcal{Z}$. A* ***graded cell datum*** *for A is a cell datum $(\mathcal{P},T,C)$ together with a degree function*

$$\deg:\coprod_{\lambda\in\mathcal{P}}T(\lambda)\longrightarrow\mathbb{Z}$$

such that

(GC_d) *the element c_{st} is homogeneous of degree $\deg c_{\mathsf{st}}=\deg(\mathsf{s})+\deg(\mathsf{t})$, for all $\lambda\in\mathcal{P}$ and $\mathsf{s},\mathsf{t}\in T(\lambda)$.*

Then, A is a ***graded cellular algebra*** *with* ***graded cellular basis*** $\{c_{\mathsf{st}}\}$.

We use $\star$ for the *homogeneous* cellular algebra involution of A that is determined by $c_{\mathsf{st}}^{\star}=c_{\mathsf{ts}}$, for $\mathsf{s},\mathsf{t}\in T(\lambda)$.

2.1.2. **Example** (Toy example) The most basic example of a graded algebra is the truncated polynomial ring $A=F[x]/(x^{n+1})$, for some integer $n>0$, where $\deg x=2$. As an ungraded algebra, $\underline{A}$ has exactly one simple module, namely the field F with x acting as multiplication by zero. This algebra is a graded cellular algebra with $\mathcal{P}=\{0,1,\dots,n\}$, with its natural order, and $T(d)=\{d\}$ and $c_{dd}=x^d$. The irreducible graded A-modules are $F\langle d\rangle$, for $d\in\mathbb{Z}$, and $\dim_q A=1+q^2+\cdots+q^{2n}$. ◊

2.1.3. **Example** Let $A=\mathrm{Mat}_n(\mathcal{Z})$ be the $\mathcal{Z}$-algebra of $n\times n$-matrices. The basis of matrix units $\{\,e_{st}\mid 1\le s,t\le n\,\}$ is a cellular basis for A, where $\mathcal{P}=\{\heartsuit\}$ and $T(\heartsuit)=\{1,2,\dots,n\}$. We want to put a non-trivial grading on A. Let $\{d_1,\dots,d_n\}\subset\mathbb{Z}$ be a set of integers such that $d_s+d_{n-s+1}=0$, for $1\le s\le n$. Set $c_{st}=e_{s(n-t+1)}$ and define a degree function $\deg:T(\heartsuit)\longrightarrow\mathbb{Z}$ by $\deg s=d_s$. Then $\{\,c_{st}\mid 1\le s,t\le n\,\}$ is a graded cellular basis of A and $\dim_q A=\sum_{s=1}^{n}q^{d_s}$. Consequently, semisimple algebras can have non-trivial gradings. ◊

Exactly as in §1.3, for each $\lambda \in \mathcal{P}$ we obtain a *graded* cell module C^λ with homogeneous basis $\{ c_\mathfrak{t} \mid \mathfrak{t} \in T(\lambda) \}$ and $\deg c_\mathfrak{t} = \deg \mathfrak{t}$. Generalizing (1.3.2), the graded cell module C^λ comes equipped with a *homogeneous* symmetric bilinear form $\langle\ ,\ \rangle_\lambda$ of degree zero. Therefore, if x and y are homogeneous elements of C^λ then $\langle x, y\rangle_\lambda \neq 0$ only if $\deg x + \deg y = 0$. Moreover, $\langle xa, y\rangle_\lambda = \langle x, ya^\star\rangle_\lambda$, for all $x, y \in C^\lambda$ and all $a \in A$. Consequently,

$$\operatorname{rad} C^\lambda = \{ x \in C^\lambda \mid \langle x, y\rangle_\lambda = 0 \text{ for all } y \in C^\lambda \}$$

is a graded submodule of C^λ so that $D^\lambda = C^\lambda / \operatorname{rad} C^\lambda$ is a graded A-module.

If M is an A-module then its (graded) **dual** is the A-module

$$M^\circledast = \mathcal{H}om_\mathcal{Z}(M, \mathcal{Z}), \tag{2.1.4}$$

with A-action $(f \cdot a)(m) = f(ma^\star)$, for $f \in M^\circledast$, $a \in A$ and $m \in M$.

Theorem 2.1.5 ([54, Theorem 2.10]). *Suppose that $\mathcal{Z}$ is a field and that A is a graded cellular algebra. Then:*

a) *If $\lambda \in \mathcal{P}$ then D^λ is either 0 or an absolutely irreducible graded A-module. If $D^\lambda \neq 0$ then $(D^\lambda)^\circledast \cong D^\lambda$.*

b) *Let $\mathcal{K} = \{ \mu \in \mathcal{P} \mid D^\mu \neq 0 \}$. Then $\{ D^\lambda\langle s\rangle \mid \lambda \in \mathcal{K}$ and $s \in \mathbb{Z} \}$ is a complete set of pairwise non-isomorphic irreducible (graded) A-modules.*

c) *If $\lambda \in \mathcal{P}$ and $\mu \in \mathcal{K}$ then $[C^\lambda{:}D^\mu]_q \neq 0$ only if $\lambda \trianglerighteq \mu$. Moreover, $[C^\mu{:}D^\mu]_q = 1$.*

Forgetting the grading, the basis $\{c_{\mathfrak{s}\mathfrak{t}}\}$ is still a cellular basis of $\underline{A}$. Comparing Theorem 1.3.4 and Theorem 2.1.5 it follows that every (ungraded) irreducible $\underline{A}$-module has a graded lift that is unique up to shift. Conversely, if D is an irreducible graded A-module then $\underline{D}$ is an irreducible $\underline{A}$-module. (This holds more generally whenever a grading is put on a finite dimensional algebra; see [114, Theorem 4.4.4].) It is an instructive exercise to prove that if A is a finite dimensional graded algebra then every simple $\underline{A}$-module has a graded lift and, up to shift, every graded simple A-module is of this form.

By [47, Theorems 3.2 and 3.3] every projective indecomposable $\mathscr{H}_n^\Lambda$-module has a graded lift. More generally, as shown in [114, §4], if M is a finitely generated graded A-module then the Jacobson radical of $\underline{M}$ has a graded lift.

The matrix $\mathbf{D}_A(q) = ([C^\lambda : D^\mu]_q)_{\lambda\in\mathcal{P},\mu\in\mathcal{K}}$ is the **graded decomposition matrix** of A. For each $\mu \in \mathcal{K}$ let P^μ be the projective cover of D^μ in $\operatorname{Rep}(A)$. The matrix $\mathbf{C}_A(q) = ([P^\lambda : D^\mu]_q)_{\lambda,\mu\in\mathcal{K}}$ is the **graded Cartan matrix** of A.

An A-module M has a **cell filtration** if there exists a filtration $M = M_0 \supset M_1 \supset \cdots \supset M_z \supset 0$ such that each subquotient M_r/M_{r+1} is isomorphic, up to shift, to some graded cell module. Fixing isomorphisms $M_r/M_{r+1} \cong C^{\lambda_r}\langle d_r\rangle$, for some $\lambda_r \in \mathcal{P}$ and $d_r \in \mathbb{Z}$, define $(M : C^\lambda)_q = \sum_d m_d q^d$, where $m_d = \#\{\,1 \le r \le z \mid \lambda_r = \lambda \text{ and } d_r = d\,\}$. In general, the multiplicities $(M : C^\lambda)_q$ depend upon the choice of filtration *and* the labelling of the isomorphisms $M_r/M_{r+1} \cong C^{\lambda_r}\langle d_r\rangle$ because the cell modules are not guaranteed to be pairwise non-isomorphic, even up to shift.

Corollary 2.1.6 ([54, Theorem 2.17]). *Suppose that $\mathcal{Z} = F$ is a field. If $\mu \in \mathcal{K}$ then P^μ has a cell filtration such that $(P^\mu : C^\lambda)_q = [C^\lambda : D^\mu]_q$, for all $\lambda \in \mathcal{P}$. Consequently, $\mathbf{C}_A(q) = \mathbf{D}_A(q)^{tr}\mathbf{D}_A(q)$ is a symmetric matrix.*

2.2 *Cyclotomic quiver Hecke algebras*

We are now ready to define cyclotomic quiver Hecke algebras. We start by defining the affine versions of these algebras and then pass to the cyclotomic quotients. Throughout this section we will make extensive use of the Lie theoretic data that is attached to the quiver Γ_e in §1.2.

If $\beta \in Q^+$ let $I^\beta = \{\,\mathbf{i} \in I^n \mid \beta = \alpha_{i_1} + \cdots + \alpha_{i_n}\,\}$, a finite set.

Definition 2.2.1 (Khovanov and Lauda [74, 75] and Rouquier [121]). *Suppose that $n \ge 0$, $e \ge 1$, and $\beta \in Q^+$. The* ***quiver Hecke algebra****, or* ***Khovanov-Lauda–Rouquier algebra****, $\mathscr{R}_\beta = \mathscr{R}_\beta(\mathcal{Z})$ of type Γ_e is the unital associative $\mathcal{Z}$-algebra with generators*

$$\{\psi_1, \dots, \psi_{n-1}\} \cup \{y_1, \dots, y_n\} \cup \{\, e(\mathbf{i}) \mid \mathbf{i} \in I^\beta \,\}$$

and relations

$$e(\mathbf{i})e(\mathbf{j}) = \delta_{\mathbf{i}\mathbf{j}}e(\mathbf{i}), \qquad \textstyle\sum_{\mathbf{i}\in I^\beta} e(\mathbf{i}) = 1,$$

$$y_r e(\mathbf{i}) = e(\mathbf{i})y_r, \qquad \psi_r e(\mathbf{i}) = e(s_r{\cdot}\mathbf{i})\psi_r, \qquad y_r y_s = y_s y_r,$$

$$\psi_r\psi_s = \psi_s\psi_r, \qquad \text{if } |r-s| > 1,$$

$$\psi_r y_s = y_s\psi_r, \qquad \text{if } s \ne r, r+1,$$

$$(2.2.2)\qquad \begin{aligned} \psi_r y_{r+1} e(\mathbf{i}) &= (y_r\psi_r + \delta_{i_r i_{r+1}})e(\mathbf{i}),\\ y_{r+1}\psi_r e(\mathbf{i}) &= (\psi_r y_r + \delta_{i_r i_{r+1}})e(\mathbf{i}),\end{aligned}$$

$$(2.2.3) \qquad \psi_r^2 e(\mathbf{i}) = \begin{cases} (y_{r+1} - y_r)(y_r - y_{r+1})e(\mathbf{i}), & \text{if } i_r \rightleftarrows i_{r+1}, \\ (y_r - y_{r+1})e(\mathbf{i}), & \text{if } i_r \to i_{r+1}, \\ (y_{r+1} - y_r)e(\mathbf{i}), & \text{if } i_r \leftarrow i_{r+1}, \\ 0, & \text{if } i_r = i_{r+1}, \\ e(\mathbf{i}), & \text{otherwise}, \end{cases}$$

and $(\psi_r\psi_{r+1}\psi_r - \psi_{r+1}\psi_r\psi_{r+1})e(\mathbf{i})$ *is equal to*

$$(2.2.4) \qquad \begin{cases} (y_r + y_{r+2} - 2y_{r+1})e(\mathbf{i}), & \text{if } i_{r+2} = i_r \rightleftarrows i_{r+1}, \\ -e(\mathbf{i}), & \text{if } i_{r+2} = i_r \to i_{r+1}, \\ e(\mathbf{i}), & \text{if } i_{r+2} = i_r \leftarrow i_{r+1}, \\ 0, & \text{otherwise}, \end{cases}$$

for $\mathbf{i}, \mathbf{j} \in I^\beta$ *and all admissible* r *and* s.

Part of the point of these definitions is that $\mathscr{R}_\beta$ is a $\mathbb{Z}$-graded algebra with degree function determined by

$$\deg e(\mathbf{i}) = 0, \qquad \deg y_r = 2 \qquad \text{and} \qquad \deg \psi_s e(\mathbf{i}) = -c_{i_s, i_{s+1}},$$

for $1 \le r \le n$, $1 \le s < n$ and $\mathbf{i} \in I^n$.

Suppose that $n \ge 0$. Then $I^n = \bigsqcup_\beta I^\beta$ is the decomposition of I^n into a disjoint union of $\mathfrak{S}_n$-orbits. Define

$$(2.2.5) \qquad \mathscr{R}_n = \bigoplus_{\beta \in Q^+} \mathscr{R}_\beta.$$

Set $\beta = \sum_{\mathbf{i} \in I^\beta} e(\mathbf{i})$, for $\beta \in Q^+$. Then $\mathscr{R}_\beta = e_\beta \mathscr{R}_n e_\beta$ is a two-sided ideal of $\mathscr{R}_n$ and (2.2.5) is the decomposition of $\mathscr{R}_n$ into blocks. That is, $\mathscr{R}_\beta$ is indecomposable for all $\beta \in Q^+$.

Khovanov and Lauda [74, 75] and Rouquier [121] define quiver Hecke algebras for quivers of arbitrary type. In the short time since their inception a lot has been discovered about these algebras. The first important result is that these algebras categorify the negative part of the corresponding quantum group [22, 74, 122, 132].

Remark 2.2.6. We have defined only a special case of the quiver Hecke algebras from [74, 121]. In addition to allowing arbitrary quivers, Khovanov and Lauda allow a more general choice of signs. Rouquier's definition, which is the most general, defines the quiver Hecke algebras in terms of a matrix $Q = (Q_{ij})_{i,j \in I}$ with entries in a polynomial ring $\mathcal{Z}[u, v]$ with the properties that $Q_{ii} = 0$, Q_{ij} is not a zero divisor in $\mathcal{Z}[u, v]$ for $i \ne j$ and $Q_{ij}(u, v) = Q_{ji}(v, u)$, for $i, j \in I$. For an arbitrary quiver Γ,

Rouquier [121, Definition 3.2.1] defines $\mathscr{R}_\beta(\Gamma)$ to be the algebra generated by $\psi_r, y_s, e(\mathbf{i})$ subject to the relations above except that the quadratic and braid relations are replaced with $\psi_r^2 e(\mathbf{i}) = Q_{i_r,i_{r+1}}(y_r, y_{r+1})e(\mathbf{i})$ and $(\psi_r\psi_{r+1}\psi_r - \psi_{r+1}\psi_r\psi_{r+1})e(\mathbf{i})$ is equal to

$$\begin{cases} \frac{Q_{i_r,i_{r+1}}(y_r,y_{r+1})-Q_{i_r,i_{r+1}}(y_r,y_{r+1})}{y_{r+2}-y_r}, & \text{if } i_{r+2} = i_r, \\ 0, & \text{otherwise.} \end{cases}$$

The assumptions on Q ensure that the last expression is a polynomial in the generators. In general, $y_r e(\mathbf{i})$ is homogeneous of degree $(\alpha_{i_r}, \alpha_{i_r})$, for $1 \le r \le n$ and $\mathbf{i} \in I^n$. Under some mild assumptions, the isomorphism type of $\mathscr{R}_\beta$ is independent of the choice of Q by [121, Proposition 3.12]. We leave it to the reader to find a suitable matrix Q for Definition 2.2.1.

For the rest of these notes for $w \in \mathfrak{S}_n$ we arbitrarily fix a reduced expression $w = s_{r_1} \dots s_{r_k}$, with $1 \le r_j < n$. Using this fixed reduced expression for w define $\psi_w = \psi_{r_1} \dots \psi_{r_k}$.

2.2.7. **Example** As the ψ-generators of $\mathscr{R}_n$ do not satisfy the braid relations the element ψ_w will, in general, depend upon the choice of reduced expression for $w \in \mathfrak{S}_n$. For example, by (2.2.4) if $e \ne 2$, $n = 3$ and $w = s_1s_2s_1 = s_2s_1s_2$ then $\psi_1\psi_2\psi_1 e(0,2,0) = \psi_2\psi_1\psi_2 e(0,2,0) + e(0,2,0)$, by (2.2.4). Therefore, these two reduced expressions determine different elements of $\mathscr{R}_n$. $\diamond$

Khovanov and Lauda [74, Theorem 2.5] and Rouquier [121, Theorem 3.7] proved the following.

Theorem 2.2.8 (Khovanov-Lauda [74] and Rouquier [121]). *Suppose that $\beta \in Q^+$. Then $\mathscr{R}_\beta(\mathcal{Z})$ is free as an $\mathcal{Z}$-algebra with homogeneous basis $\{\,\psi_w y_1^{a_1} \dots y_n^{a_n} e(\mathbf{i}) \mid w \in \mathfrak{S}_n,\ a_1, \dots, a_n \in \mathbb{N}$ and $\mathbf{i} \in I^\beta\,\}$.*

Li [92, Theorem 4.3.10] has constructed a graded cellular basis of $\mathscr{R}_n$. In the special case when $e = \infty$, Kleshchev, Loubert and Miemietz [85] give a graded affine cellular basis of $\mathscr{R}_n$, in the sense of König and Xi [87].

In these notes we are not directly concerned with the quiver Hecke algebras $\mathscr{R}_n$. Rather, we are more interested in certain *cyclotomic quotients* of these algebras.

Definition 2.2.9 (Brundan-Kleshchev [21]). *Suppose that $\Lambda \in P^+$. The **cyclotomic quiver Hecke algebra** of type Γ_e and weight Λ is the quotient algebra $\mathscr{R}_n^\Lambda = \mathscr{R}_n / \langle y_1^{(\Lambda,\alpha_{i_1})} e(\mathbf{i}) \mid \mathbf{i} \in I^n \rangle$.*

We abuse notation and identify the KLR generators of $\mathscr{R}_n$ with their images in $\mathscr{R}_n^\Lambda$. That is, we consider the algebra $\mathscr{R}_n^\Lambda$ to be generated by $\psi_1, \dots, \psi_{n-1}, y_1, \dots, y_n$ and $e(\mathbf{i})$, for $\mathbf{i} \in I^n$, subject to the relations in Definition 2.2.1 and Definition 2.2.9. From this point onwards, fix $\Lambda \in P^+$.

When Λ is a weight of level 2, the algebras $\mathscr{R}_n^\Lambda$ first appeared in the work of Brundan and Stroppel [26] in their series of papers on the Khovanov diagram algebras. In full generality, the cyclotomic quotients of $\mathscr{R}_n$ were introduced by Khovanov-Lauda [74] and Rouquier [121]. Brundan and Kleshchev initiated the study of the cyclotomic quiver Hecke algebras $\mathscr{R}_n^\Lambda$, for any $\Lambda \in P^+$.

Although we will not use this here we note that, rather than working algebraically, it is often easier to work diagrammatically by identifying the elements of $\mathscr{R}_n^\Lambda$ with certain diagrams. In these diagrams, the end-points of the strings are labeled by $\{1, 2, \dots, n, 1', 2', \dots, n'\}$ and the strings themselves are coloured by I^n. For example, following [74], the KLR generators can be identified with the diagrams:

$e(\mathbf{i}) =$ 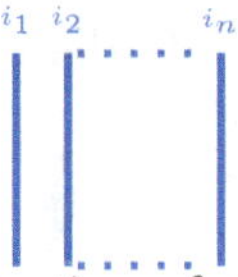

$\psi_r e(\mathbf{i}) =$ 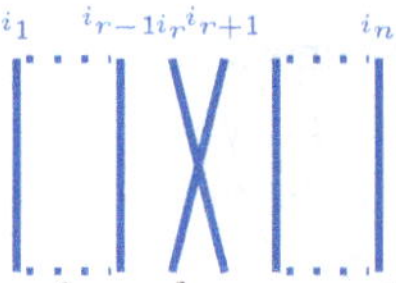

$y_s e(\mathbf{i}) =$ 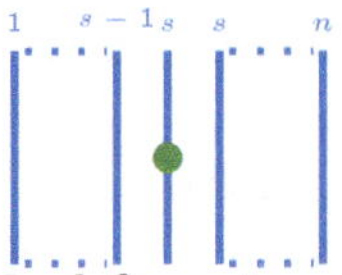

.

Multiplication of diagrams is given by concatenation, read from top to bottom, subject to the relations above that are also interpreted diagrammatically. As an exercise, we leave it to the reader to identify the two relations in Definition 2.2.1 that correspond to the following 'local' relations on strings inside braid diagrams:

$=$ 

$+ \delta_{ij}$ 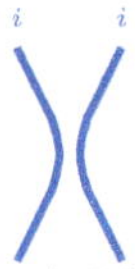

and 

$=$ 

$\pm$ 

.

(For the second relation, $e \neq 2$.) For more rigorous definitions of such diagrams, and non-trivial examples of their application, we refer the reader to the papers [53, 81, 92, 97].

2.2.10. **Example** (Rank one algebras) Suppose that $n = 1$ and $\Lambda \in P^+$. Then $\mathscr{R}_1^\Lambda = \langle y_1, e(i) \mid y_1 e(i) = e(i) y_1 \text{ and } y_1^{\langle \Lambda, \alpha_i \rangle} e(i) = 0, \text{ for } i \in I \rangle$, with $\deg y_1 = 2$ and $\deg e(i) = 0$, for $i \in I$. Therefore, there is an isomorphism of graded algebras

$$\mathscr{R}_1^\Lambda \cong \bigoplus_{\substack{i \in I \\ (\Lambda, \alpha_i) > 0}} \mathcal{Z}[y] / y^{(\Lambda, \alpha_i)} \mathcal{Z}[y],$$

where $y = y_1$ is in degree 2. Armed with this description of $\mathscr{R}_n^\Lambda$ it is now straightforward to show that $\mathscr{H}_n^\Lambda \cong \mathscr{R}_n^\Lambda$ when $\mathcal{Z}$ is a field and $n = 1$. $\Diamond$

2.3 *Nilpotence and small representations*

In this section and the next we use the KLR relations to prove some results about the cyclotomic quiver Hecke algebras $\mathscr{R}_n^\Lambda$ for particular Λ and n.

By Theorem 2.2.8 the algebra $\mathscr{R}_n$ is infinite dimensional, so it is not obvious from the relations that the cyclotomic Hecke algebra $\mathscr{R}_n^\Lambda$ is finite dimensional — or even that $\mathscr{R}_n^\Lambda$ is non-zero. The following result shows that y_r is nilpotent, for $1 \le r \le n$, which implies that $\mathscr{R}_n^\Lambda$ is finite dimensional.

Lemma 2.3.1 (Brundan and Kleshchev [21, Lemma 2.1]**).** *Suppose that $1 \le r \le n$ and $\mathbf{i} \in I^n$. Then $y_r^N e(\mathbf{i}) = 0$ for $N \gg 0$.*

Proof. We argue by induction on r. If $r = 1$ then $y_1^{(\Lambda,\alpha_{i_1})} e(\mathbf{i}) = 0$ by Definition 2.2.9, proving the base step of the induction. Now consider $y_{r+1}e(\mathbf{i})$. By induction, we may assume that there exists $N \gg 0$ such that $y_r^N e(\mathbf{j}) = 0$, for all $\mathbf{j} \in I^n$. There are three cases to consider.

Case 1. $i_{r+1} \not\to i_r$.
By (2.2.3) and (2.2.2),

$$y_{r+1}^N e(\mathbf{i}) = y_{r+1}^N \psi_r^2 e(\mathbf{i}) = \psi_r y_r^N \psi_r e(\mathbf{i}) = \psi_r y_r^N e(s_r \cdot \mathbf{i}) \psi_r = 0,$$

where the last equality follows by induction.

Case 2. $i_{r+1} = i_r \pm 1$.
Suppose first that $e \neq 2$. This is a variation on the previous case, with a twist. By (2.2.3) and (2.2.2), again

$$\begin{aligned}
y_{r+1}^{2N} e(\mathbf{i}) &= y_{r+1}^{2N-1} y_r e(\mathbf{i}) + y_{r+1}^{2N-1}(y_{r+1} - y_r) e(\mathbf{i}) \\
&= y_r y_{r+1}^{2N-1} e(\mathbf{i}) \pm y_{r+1}^{2N-1} \psi_r^2 e(\mathbf{i}) \\
&= y_r y_{r+1}^{2N-1} e(\mathbf{i}) \pm \psi_r y_r^{2N-1} e(s_r \cdot \mathbf{i}) \psi_r \\
&= y_r y_{r+1}^{2N-1} e(\mathbf{i}) = \cdots = y_r^N y_{r+1}^N e(\mathbf{i}) = 0.
\end{aligned}$$

The case when $e = 2$ is similar. First, observe that $y_{r+1}^2 e(\mathbf{i}) = (2y_r y_{r+1} - y_r^2 - \psi_r^2) e(\mathbf{i})$ by (2.2.3). Therefore, arguing as before,

$$y_{r+1}^{3N} e(\mathbf{i}) = y_r(2y_{r+1} - y_r) y_{r+1}^{3N-2} e(\mathbf{i}) = \cdots = y_r^N (2y_{r+1} - y_r)^N y_{r+1}^N e(\mathbf{i}) = 0.$$

Case 3. $i_{r+1} = i_r$.
Let $\phi_r = \psi_r(y_r - y_{r+1})$. Then $\phi_r \psi_r e(\mathbf{i}) = -2\psi_r e(\mathbf{i})$ by (2.2.2), so that $(1 + \phi_r)^2 e(\mathbf{i}) = e(\mathbf{i})$. Moreover,

$$(1 + \phi_r) y_r (1 + \phi_r) e(\mathbf{i}) = (y_r + \phi_r y_r + y_r \phi_r + \phi_r y_r \phi_r) e(\mathbf{i}) = y_{r+1} e(\mathbf{i}),$$

where the last equality uses (2.2.2). Now we are done because

$$y_{r+1}^N e(\mathbf{i}) = \big((1+\phi_r)y_r(1+\phi_r)\big)^N e(\mathbf{i}) = (1+\phi_r)y_r^N(1+\phi_r)e(\mathbf{i}) = 0,$$

since ϕ_r commutes with $e(\mathbf{i})$ and $y_r^N e(\mathbf{i}) = 0$ by induction. □

We have marginally improved on Brundan and Kleshchev's original proof of Lemma 2.3.1 because the argument above gives an upper bound for the nilpotency index of y_r. In general, this bound is far from sharp. For a better estimate of the nilpotency index of y_r see [57, Corollary 4.6] (and [53] when $e = \infty$). See [67, Lemma 4.4] for another argument that applies to cyclotomic quiver Hecke algebras of arbitrary type.

Combining Theorem 2.2.8 and Lemma 2.3.1 we have:

Corollary 2.3.2 (Brundan and Kleshchev [21, Corollary 2.2]**).** *Suppose $\mathcal{Z}$ is an integral domain. Then $\mathscr{R}_n^\Lambda$ is finite dimensional.*

As our next exercise we classify the one dimensional representations of $\mathscr{R}_n^\Lambda$ when $\mathcal{Z} = F$ is a field. For $i \in I$ let $\mathbf{i}_n^+ = (i, i+1, \dots, i+n-1)$ and $\mathbf{i}_n^- = (i, i-1, \dots, i-n+1)$. Then $\mathbf{i}_n^\pm \in I^n$. If $(\Lambda, \alpha_i) = 0$ then $e(\mathbf{i}_n^\pm) = 0$ by Definition 2.2.9. However, if $(\Lambda, \alpha_i) \neq 0$ then using the relations it is easy to see that $\mathscr{R}_n^\Lambda$ has unique one dimensional representations $D_{i,n}^+ = Fd_{i,n}^+$ and $D_{i,n}^- = Fd_{i,n}^-$ such that

$$d_{i,n}^\pm e(\mathbf{i}) = \delta_{\mathbf{i},\mathbf{i}_n^\pm} d_{i,n}^\pm \quad \text{and} \quad d_{i,n}^+ y_r = 0 = d_{i,n}^\pm \psi_s,$$

for $\mathbf{i} \in I^n$, $1 \le r \le n$ and $1 \le s < n$ and such that $\deg d_{i,n}^\pm = 0$. In particular, this shows that $e(\mathbf{i}_n^\pm) \neq 0$ and hence that $\mathscr{R}_n^\Lambda \neq 0$. If $e \neq 2$ then $\{\, D_{i,n}^\pm \mid i \in I \text{ and } (\Lambda, \alpha_i) \neq 0 \,\}$ are pairwise non-isomorphic irreducible representations of $\mathscr{R}_n^\Lambda$. If $e = 2$ then $\mathbf{i}_n^+ = \mathbf{i}_n^-$ so that $D_{i,n}^+ = D_{i,n}^-$.

Proposition 2.3.3. *Suppose that $\mathcal{Z} = F$ is a field and that D is a one dimensional graded $\mathscr{R}_n^\Lambda$-module. Then $D \cong D_{i,n}^\pm\langle k\rangle$, for some $k \in \mathbb{Z}$ and $i \in I$ such that $(\Lambda, \alpha_i) \neq 0$.*

Proof. Let d be a non-zero element of D so that $D = Fd$. Then $d = \sum_{\mathbf{j} \in I^n} de(\mathbf{j})$ so that $de(\mathbf{i}) \neq 0$ for some $\mathbf{i} \in I^n$. Moreover, $de(\mathbf{j}) = 0$ if and only if $\mathbf{j} = \mathbf{i}$ since otherwise $de(\mathbf{i})$ and $de(\mathbf{j})$ are linearly independent elements of D, contradicting assumption that D is one dimensional. Now, $\deg dy_r = 2 + \deg d$, so $dy_r = 0$, for $1 \le r \le n$, since D is one dimensional. Similarly, $d\psi_r = de(\mathbf{i})\psi_r = 0$ if $i_r = i_{r+1}$ or $i_r = i_{r+1} \pm 1$ since in these cases $\deg e(\mathbf{i})\psi_r \neq 0$.

It remains to show that $\mathbf{i} = \mathbf{i}_n^\pm$ and that $(\Lambda, \alpha_{i_1}) \neq 0$. First, since $0 \neq d = d\,e(\mathbf{i})$ we have that $e(\mathbf{i}) \neq 0$ so that $(\Lambda, \alpha_{i_1}) \neq 0$ by Definition 2.2.9.

To complete the proof we show that if $\mathbf{i} \neq \mathbf{i}_n^{\pm}$ then $d = 0$, which will give a contradiction. First, suppose that $i_r = i_{r+1}$ for some r, with $1 \le r < n$. Then $d = de(\mathbf{i}) = d(\psi_r y_{r+1} - y_r\psi_r)e(\mathbf{i}) = 0$ by (2.2.2), which is not possible, so $i_r \neq i_{r+1}$. Next, suppose that $i_{r+1} \neq i_r \pm 1$. Then $d = de(\mathbf{i}) = d\psi_r^2 e(\mathbf{i}) = d\psi_r e(s_r \cdot \mathbf{i})\psi_r = 0$ because D is one dimensional and $de(\mathbf{j}) = 0$ if $\mathbf{j} \neq \mathbf{i}$. This is another contradiction, so we must have $i_{r+1} = i_r \pm 1$ for $1 \le r < n$. Therefore, if $\mathbf{i} \neq \mathbf{i}_n^{\pm}$ then $e \neq 2$, $n > 2$ and $i_r = i_{r+2} = i_{r+1} \pm 1$ for some r. Applying the braid relation (2.2.4),

$$d = de(\mathbf{i}) = \pm d \cdot (\psi_r\psi_{r+1}\psi_r - \psi_{r+1}\psi_r\psi_{r+1})e(\mathbf{i}) = 0,$$

a contradiction. Hence, $D \cong D_{i,n}^{\pm}\langle \deg d\rangle$, completing the proof. □

2.4 *Semisimple KLR algebras*

Now that we understand the one dimensional representations of $\mathscr{R}_n^\Lambda$ we consider the semisimple representation theory of the cyclotomic quiver Hecke algebras. These results do not appear in the literature, but there are few surprises here because everything we do can be easily deduced from results that are known. The main idea is to show by example how to use the quiver Hecke algebra relations.

In this section we fix $e > n$ and $\Lambda \in P^+$ such that $(\Lambda, \alpha_{i,n}) \le 1$, for all $i \in I$, and we study the algebras $\mathscr{R}_n^\Lambda$. Notice that these conditions ensure that $\mathscr{H}_n^\Lambda$ is semisimple by Corollary 1.6.11.

Recall from §1.8 that $\mathbf{i}^{\mathsf{t}} = (i_1^{\mathsf{t}}, \dots, i_n^{\mathsf{t}})$ is the residue sequence of $\mathsf{t} \in \operatorname{Std}(\mathcal{P}_n^\Lambda)$, where $i_r^{\mathsf{t}} = c_r^{\mathbb{Z}}(\mathsf{t}) + e\mathbb{Z}$. We caution the reader that if t is a standard tableau then the contents $c_r^{\mathbb{Z}}(\mathsf{t}) \in \mathbb{Z}$ and the residues $i_r^{\mathsf{t}} \in I$ are in general different.

If $i \in I$ then a node $A = (l, r, c)$ is an i-**node** if $i = \kappa_l + c - r + e\mathbb{Z}$. Therefore, extending the definitions of §1.4, we can now talk of addable and removable i-nodes.

Lemma 2.4.1. *Suppose that $e > n$ and $(\Lambda, \alpha_{i,n}) \le 1$, for all $i \in I$. Let $\mathsf{s}, \mathsf{t} \in \operatorname{Std}(\mathcal{P}_n^\Lambda)$. Then $\mathsf{s} = \mathsf{t}$ if and only if $\mathbf{i}^{\mathsf{s}} = \mathbf{i}^{\mathsf{t}}$.*

Proof. Observe that if $i \in I$ and $\boldsymbol{\mu} \in \mathcal{P}_m^\Lambda$, where $0 \le m < n$, then $\boldsymbol{\mu}$ has at most one addable i-node since $(\Lambda, \alpha_{i,n}) \le 1$. Hence, it follows easily by induction on n that $\mathsf{s} = \mathsf{t}$ if and only if $\mathbf{i}^{s} = \mathbf{i}^{\mathsf{t}}$. □

Lemma 2.4.1 also follows from Theorem 1.6.10 and Corollary 1.6.11.

Let $I_\Lambda^n = \{\, \mathbf{i}^{\mathsf{t}} \mid \mathsf{t} \in \operatorname{Std}(\mathcal{P}_n^\Lambda) \,\}$ be the set of residue sequences of all of the standard tableaux in $\operatorname{Std}(\mathcal{P}_n^\Lambda)$. By the proof of Lemma 2.4.1, if $\mathbf{i} = \mathbf{i}^{\mathsf{t}} \in I_\Lambda^n$

and $i_{r+1} = i_r \pm 1$ then r and $r+1$ must be in either in the same row or in the same column of t. Hence, we have the following useful fact.

Corollary 2.4.2. *Suppose that $e > n$ and that $(\Lambda, \alpha_{i,n}) \le 1$, for all $i \in I$, and that $\mathbf{i} \in I^n_\Lambda$ with $i_{r+1} = i_r \pm 1$. Then $s_r \cdot \mathbf{i} \notin I^n_\Lambda$.*

When $\Lambda = \Lambda_0$ the next result is due to Brundan and Kleshchev [21, §5.5]. More generally, Kleshchev and Ram [83, Theorem 3.4] prove similar results for quiver Hecke algebras of simply laced type.

Proposition 2.4.3 (Seminormal representations of $\mathscr{R}^\Lambda_n$). *Suppose that $\mathcal{Z} = F$ is a field, $e > n$ and that $\Lambda \in P^+$ with $(\Lambda, \alpha_{i,n}) \le 1$, for all $i \in I$. Then for each $\boldsymbol{\lambda} \in \mathcal{P}^\Lambda_n$ there is a unique irreducible graded $\mathscr{R}^\Lambda_n$-module $S^{\boldsymbol{\lambda}}$ with homogeneous basis $\{\, \psi_\mathsf{t} \mid \mathsf{t} \in \operatorname{Std}(\boldsymbol{\lambda}) \,\}$ such that $\deg \psi_\mathsf{t} = 0$, for all $\mathsf{t} \in \operatorname{Std}(\boldsymbol{\lambda})$, and where the $\mathscr{R}^\Lambda_n$-action is given by*

$$\psi_\mathsf{t} e(\mathbf{i}) = \delta_{\mathbf{i},\mathbf{i}^\mathsf{t}} \psi_\mathsf{t}, \qquad \psi_\mathsf{t} y_r = 0 \qquad \text{and} \qquad \psi_\mathsf{t}\psi_r = \psi_{\mathsf{t}(r,r+1)},$$

where we set $\psi_{\mathsf{t}(r,r+1)} = 0$ if $\mathsf{t}(r, r+1)$ is not standard.

Proof. By Lemma 2.4.1, if $\mathsf{s}, \mathsf{t} \in \operatorname{Std}(\boldsymbol{\lambda})$ then $\mathsf{s} = \mathsf{t}$ if and only if $\mathbf{i}^\mathsf{s} = \mathbf{i}^\mathsf{t}$. Moreover, $i^\mathsf{t}_{r+1} = i^\mathsf{t}_r \pm 1$ if and only if r and $r+1$ are in the same row or in the same column of t. Similarly, $i^\mathsf{t}_r \ne i^\mathsf{t}_{r+1}$ for any r. Consequently, since $\psi_\mathsf{t} = \psi_\mathsf{t} e(\mathbf{i}^\mathsf{t})$ almost all of the relations in Definition 2.2.1 are trivially satisfied. In fact, all that we need to check is that $\psi_1, \dots, \psi_{n-1}$ satisfy the braid relations of the symmetric group $\mathfrak{S}_n$ with ψ_r^2 acting as zero when $i^\mathsf{t}_{r+1} = i^\mathsf{t}_r \pm 1$, which follows automatically by Corollary 2.4.2. By the same reasoning if $\mathsf{t}(r, r+1)$ is standard then $\deg e(\mathbf{i}^\mathsf{t})\psi_r = 0$. Hence, we can set $\deg \psi_\mathsf{t} = 0$, for all $\mathsf{t} \in \operatorname{Std}(\boldsymbol{\lambda})$. This proves that $S^{\boldsymbol{\lambda}}$ is a graded $\mathscr{R}^\Lambda_n$-module.

It remains to show that $S^{\boldsymbol{\lambda}}$ is irreducible. If $\mathsf{s}, \mathsf{t} \in \operatorname{Std}(\boldsymbol{\lambda})$ then $\mathsf{s} = \mathsf{t}^{\boldsymbol{\lambda}} d(\mathsf{s}) = \mathsf{t} d(\mathsf{t})^{-1} d(\mathsf{s})$, so $\psi_\mathsf{s} = \psi_\mathsf{t} \psi_{d(\mathsf{t})^{-1}} \psi_{d(\mathsf{s})}$. Suppose that $x = \sum_\mathsf{t} r_\mathsf{t} \psi_\mathsf{t}$ is a non-zero element of $S^{\boldsymbol{\lambda}}$. If $r_\mathsf{t} \ne 0$ then $\psi_\mathsf{t} = \frac{1}{r_\mathsf{t}} x e(\mathbf{i}^\mathsf{t})$, so it follows that $\psi_\mathsf{s} \in x\mathscr{R}^\Lambda_n$, for any $\mathsf{s} \in \operatorname{Std}(\boldsymbol{\lambda})$. Therefore, $S^{\boldsymbol{\lambda}} = x\mathscr{R}^\Lambda_n$ so that $S^{\boldsymbol{\lambda}}$ is irreducible as claimed. □

Hence, $e(\mathbf{i}) \ne 0$ in $\mathscr{R}^\Lambda_n$, for all $\mathbf{i} \in I^\Lambda_n$. This was not clear until now.

We want to show that Proposition 2.4.3 describes all of the graded irreducible representations of $\mathscr{R}^\Lambda_n$, up to degree shift. To do this we need a better understanding of the set I^n_Λ. Okounkov and Vershik [117, Theorem 6.7] explicitly described the set of all *content sequences* $(c^\mathbb{Z}_1(\mathsf{t}), \dots, c^\mathbb{Z}_n(\mathsf{t}))$ when $\ell = 1$. This combinatorial result easily extends to higher levels and so suggests a description of I^n_Λ.

If $\mathbf{i} \in I^n$ and $1 \le m \le n$ set $\mathbf{i}_{\downarrow m} = (i_1, \dots, i_m)$. Then $\mathbf{i}_{\downarrow m} \in I^m$ and $I^m_\Lambda = \{\, \mathbf{i}_{\downarrow m} \mid \mathbf{i} \in I^n_\Lambda \,\}$.

Lemma 2.4.4 (*cf.* Ogievetsky-d'Andecy [116, Proposition 5]**).** *Suppose that $e > n$ and $(\Lambda, \alpha_{i,n}) \le 1$, for all $i \in I$. Let $\mathbf{i} \in I^n$. Then $\mathbf{i} \in I^n_\Lambda$ if and only if it satisfies the following three conditions:*

a) $(\Lambda, \alpha_{i_1}) \neq 0$.
b) If $1 < r \le n$ and $(\Lambda, \alpha_{i_r}) = 0$ then $\{i_r - 1, i_r + 1\} \cap \{i_1, \dots, i_{r-1}\} \neq \emptyset$.
c) If $1 \le s < r \le n$ and $i_r = i_s$ then $\{i_r - 1, i_r + 1\} \subseteq \{i_{s+1}, \dots, i_{r-1}\}$.

Proof. Suppose that $\mathsf{t} \in \operatorname{Std}(\mathcal{P}^\Lambda_n)$ and let $\mathbf{i} = \mathbf{i}^{\mathsf{t}}$. We prove by induction on r that $\mathbf{i}_{\downarrow r} \in I^r_\Lambda$. By definition, $i_1 = \kappa_t + e\mathbb{Z}$ for some t with $1 \le t \le \ell$, so (a) holds. By induction we may assume that the subsequence $(i_1, \dots, i_{r-1})$ satisfies properties (a)–(c). If $(\Lambda, \alpha_{i_r}) = 0$ then r cannot be in both the first row and in the first column of any component of t, so t has an entry in the row directly above r or in the column immediately to the left of r — or both! Hence, there exists an integer s with $1 \le s < r$ such that $i^{\mathsf{t}}_s = i^{\mathsf{t}}_r \pm 1$. Hence, (b) holds. Finally, suppose that $i_r = i_s$ as in (c). As the residues of the nodes in different components of t are disjoint it follows that s and r are in same component of t and on the same diagonal. In particular, r is not in the first row or in the first column of its component in t. As t is standard, the entries in t that are immediately above or to the left of r are both larger than s and smaller than r. Hence, (c) holds.

Conversely, suppose that $\mathbf{i} \in I^n$ satisfies properties (a)–(c). We show by induction on m that $\mathbf{i}_{\downarrow m} \in I^m_\Lambda$, for $1 \le m \le n$. If $m = 1$ then $\mathbf{i}_{\downarrow 1} \in I^1_\Lambda$ by property (a). Now suppose that $1 < m < n$ and that $\mathbf{i}_{\downarrow m} \in I^m_\Lambda$. By induction $\mathbf{i}_{\downarrow m} = \mathbf{i}^{\mathsf{s}}$, for some $\mathsf{s} \in \operatorname{Std}(\mathcal{P}^\Lambda_m)$. Let $\boldsymbol{\nu} = \operatorname{Shape}(\mathsf{s})$. If $i \in I$ then $(\Lambda, \alpha_{i,n}) \le 1$, so the multipartition $\boldsymbol{\nu}$ can have at most one addable i-node. On the other hand, reversing the argument of the last paragraph, using properties (b) and (c) with $r = m+1$, shows that $\boldsymbol{\nu}$ has at least one addable i_{m+1}-node. Let A be the unique addable i_{m+1}-node of $\boldsymbol{\nu}$. Then $\mathbf{i}_{\downarrow(m+1)} = \mathbf{i}^{\mathsf{t}}$ where $\mathsf{t} \in \operatorname{Std}(\mathcal{P}^\Lambda_{m+1})$ is the unique standard tableau such that $\mathsf{t}_{\downarrow m} = \mathsf{s}$ and $\mathsf{t}(A) = m + 1$. Hence, $\mathbf{i} \in I^{m+1}_\Lambda$ as required. $\square$

By Proposition 2.4.3, if $\mathbf{i} \in I^n_\Lambda$ then $e(\mathbf{i}) \neq 0$. We use Lemma 2.4.4 to show that $e(\mathbf{i}) = 0$ if $\mathbf{i} \notin I^n_\Lambda$. First, a result that holds for all $\Lambda \in P^+$.

Lemma 2.4.5. *Suppose that $\Lambda \in P^+$ and that $e(\mathbf{i}) \neq 0$, for $\mathbf{i} \in I^n$. Then $(\Lambda, \alpha_{i_1}) \neq 0$. Moreover, $\{i_r - 1, i_r + 1\} \cap \{i_1, \dots, i_{r-1}\} \neq \emptyset$ whenever $(\Lambda, \alpha_{i_r}) = 0$ for some $1 < r \le n$.*

Proof. By Definition 2.2.9, $e(\mathbf{i}) = 0$ whenever $(\Lambda, \alpha_{i_1}) = 0$. To prove the second claim suppose that $(\Lambda, \alpha_{i_r}) = 0$ and $i_r \pm 1 \notin \{i_1, \dots, i_{r-1}\}$. We may assume that $i_r \neq i_s$ for $1 \le s < r$. Applying (2.2.3) r-times,

$$\begin{aligned} e(\mathbf{i}) &= \psi_{r-1}^2 e(\mathbf{i}) = \psi_{r-1} e(i_1, \dots, i_r, i_{r-1}, i_{r+1}, \dots, i_n)\psi_{r-1} \\ &= \cdots = \psi_{r-1} \dots \psi_1 e(i_r, i_1, \dots, i_{r-1}, i_{r+1}, \dots, i_n)\psi_1 \dots \psi_{r-1} = 0, \end{aligned}$$

where the last equality follows because $(\Lambda, \alpha_{i_r}) = 0$. □

Proposition 2.4.6. *Suppose that $1 \le m \le n$ and that $(\Lambda, \alpha_{i,m}) \le 1$, for all $i \in I$. Then $y_1 = \cdots = y_m = 0$. Moreover, if $\mathbf{i} \in I^n$ then $e(\mathbf{i}) \neq 0$ only if $\mathbf{i}_{\downarrow m} \in I^m_\Lambda$.*

Proof. We argue by induction on r to show that $y_r = 0$ and $e(\mathbf{i}) = 0$ if $\mathbf{i}_{\downarrow r} \notin I^r_\Lambda$, for $1 \le r \le m$. If $r = 1$ this is immediate because $y_1^{(\Lambda,\alpha_{i_1})} e(\mathbf{i}) = 0$ by Definition 2.2.9 and $(\Lambda, \alpha_{i_1}) \le 1$ by assumption. Suppose then that $1 < r \le m$.

We first show that $e(\mathbf{i}) = 0$ if $\mathbf{i}_{\downarrow r} \notin I^r_\Lambda$. By induction, Lemma 2.4.4 and Lemma 2.4.5, it is enough to show that $e(\mathbf{i}) = 0$ whenever there exists an integer $1 \le s < r$ such that $i_s = i_r$ and $\{i_r - 1, i_r + 1\} \nsubseteq \{i_{s+1}, \dots, i_{r-1}\}$. We may assume that s is maximal such that $i_s = i_r$ and $1 \le s < r$. There are three cases to consider.

Case 1. $r = s + 1$.
By (2.2.2), $e(\mathbf{i}) = (y_{s+1}\psi_s - \psi_s y_s)e(\mathbf{i}) = y_{s+1}\psi_s e(\mathbf{i})$, since $y_s = 0$ by induction. Using this identity twice, reveals that $e(\mathbf{i}) = y_{s+1}\psi_s e(\mathbf{i}) = y_{s+1} e(\mathbf{i})\psi_s = y_{s+1}^2 \psi_s e(\mathbf{i})\psi_s = y_{s+1}^2 \psi_s^2 e(\mathbf{i}) = 0$, where the last equality comes from (2.2.3). Therefore, $e(\mathbf{i}) = 0$ as we wanted to show.

Case 2. $s < r - 1$ *and* $\{i_r - 1, i_r + 1\} \cap \{i_{s+1}, \dots, i_{r-1}\} = \emptyset$.
By the maximality of s, $i_r \notin \{i_{s+1}, \dots, i_{r-1}\}$. Therefore, arguing as in the proof of Lemma 2.4.5, there exists a permutation $w \in \mathfrak{S}_r$ such that $e(\mathbf{i}) = \psi_w e(i_1, \dots, i_s, i_r, i_{s+1}, \dots, i_{r-1}, i_{r+1}, \dots, i_n)\psi_w$. Hence, $e(\mathbf{i}) = 0$ by Case 1.

Case 3. $s < r - 1$ *and* $\{i_r - 1, i_r + 1\} \cap \{i_{s+1}, \dots, i_{r-1}\} = \{j\}$, *where* $j = i_r \pm 1$.
Let t be an index such that $i_t = j = i_r \pm 1$ and $s < t < r$. Note that if there exists an integer t' such that $i_t = i_{t'}$ and $s < t < t' < r$ then we may assume that $i_s \in \{i_{t+1}, \dots, i_{t'-1}\}$ by Lemma 2.4.4(c) and induction. Therefore, since s was chosen to be maximal, t is the unique integer such

that $i_t = j$ and $s < t < r$. Hence, arguing as in Case 2, there exists a permutation $w \in \mathfrak{S}_r$ such that

$$e(\mathbf{i}) = \psi_w e(\dots, i_{s-1}, i_{s+1}, \dots, i_{t-1}, i_s, i_t, i_r, i_{t+1}, \dots, i_{r-1}, i_{r+1}, \dots)\psi_w.$$

For convenience, we identify $e(i_1, \dots, i_s, i_t, i_r, \dots, i_n)$ with $e(i, j, i)$, where $i = i_s = i_r$ and $j = i \pm 1$. Then, by (2.2.4),

$$\begin{aligned} e(i,j,i) &= \pm(\psi_1\psi_2\psi_1 - \psi_2\psi_1\psi_2)e(i,j,i) \\ &= \pm\psi_1\psi_2 e(j,i,i)\psi_1 \mp \psi_2\psi_1 e(i,i,j)\psi_2 = 0, \end{aligned}$$

where the last equality follows by Case 1.

Combining Cases 1–3, if $e(\mathbf{i}) \neq 0$ then $\{i_r - 1, i_r + 1\} \subseteq \{i_{s+1}, \dots, i_{r-1}\}$ whenever there exists an integer s such that $i_s = i_r$ and $1 \le s < r$. Hence, as remarked above, induction, Lemma 2.4.5 and Lemma 2.4.4 show that $e(\mathbf{i}) \neq 0$ only if $\mathbf{i}_{\downarrow r} \in I^r_\Lambda$.

To complete the proof of the inductive step (and of the proposition), it remains to show that $y_r = 0$. Using what we have just proved, it is enough to show that $y_r e(\mathbf{i}) = 0$ whenever $\mathbf{i}_{\downarrow r} \in I^r_\Lambda$. If $i_{r-1} = i_r \pm 1$ then, by induction and (2.2.3),

$$y_r e(\mathbf{i}) = (y_r - y_{r-1})e(\mathbf{i}) = \pm\psi_{r-1}^2 e(\mathbf{i}) = \pm\psi_{r-1} e(s_{r-1}\cdot\mathbf{i})\psi_{r-1} = 0,$$

where the last equality follows because $(s_{r-1}\cdot\mathbf{i})_{\downarrow r} \notin I^r_\Lambda$ by Corollary 2.4.2. If $i_{r-1} \neq i_r \pm 1$ then $i_{r-1} \not\leftarrow i_r$ by Lemma 2.4.4 since $\mathbf{i}_{\downarrow r} \in I^r_\Lambda$. Therefore, $y_r e(\mathbf{i}) = y_r\psi_{r-1}^2 e(\mathbf{i}) = \psi_{r-1}y_{r-1}\psi_{r-1}e(\mathbf{i}) = 0$ since $y_{r-1} = 0$ by induction. This completes the proof. □

Before giving our main application of Proposition 2.4.6 we interpret this result for the cyclotomic quiver Hecke algebras of the symmetric groups.

2.4.7. **Example** (Symmetric groups) Suppose that $\Lambda = \Lambda_0$, $n \ge 0$ and set $f = \min\{e, n\}$. Then $(\Lambda, \alpha_{i,f-1}) \le 1$ for all $i \in I$. Therefore, Proposition 2.4.6 shows that $y_r = 0$ for $1 \le r < f$ and that $e(\mathbf{i}) \neq 0$ only if $\mathbf{i}_{\downarrow(f-1)} \in I^{f-1}_\Lambda$. In addition, we also have $\psi_1 = 0$ because if $\mathbf{i} \in I^n$ then $\psi_1 e(\mathbf{i}) = e(s_1\cdot\mathbf{i})\psi_1 = 0$ because if $\mathbf{i}_{\downarrow(f-1)} \in I^{f-1}_\Lambda$ then $(s_1\cdot\mathbf{i})_{f-1} \notin I^{f-1}_\Lambda$.

Translating the proof of Proposition 2.4.6 back to Lemma 2.4.1, the reason why $\psi_1 = 0$ is that if $\mathbf{i} = \mathbf{i}^\mathfrak{t}$ is the residue sequence of some standard tableau $\mathfrak{t} \in \operatorname{Std}(\mathcal{P}^\Lambda_n)$ then $i_1 = 0$ and $i_2 \neq 0$, so that $s_1\cdot\mathbf{i} \notin I^\Lambda_n$ is not a residue sequence and, consequently, $\psi_1 e(\mathbf{i}) = e(s_1\cdot\mathbf{i})\psi_1 = 0$. By the same reasoning, $\psi_1 \neq 0$ if Λ has level $\ell > 1$. ◊

We now completely describe the structure of the KLR algebras $\mathscr{R}_n^\Lambda$ when $e > n$ and $\Lambda \in P^+$ such that $(\Lambda, \alpha_{i,n}) \le 1$, for all $i \in I$. For $(\mathsf{s},\mathsf{t}) \in \operatorname{Std}^2(\mathcal{P}_n^\Lambda)$ define $e_{\mathsf{st}} = \psi_{d(\mathsf{s})^{-1}} e(\mathbf{i}^{\boldsymbol{\lambda}}) \psi_{d(\mathsf{t})}$, where $\mathbf{i}^{\boldsymbol{\lambda}} = \mathbf{i}^{\mathsf{t}^{\boldsymbol{\lambda}}}$.

Theorem 2.4.8. *Suppose that $e > n$ and $\Lambda \in P^+$ with $(\Lambda, \alpha_{i,n}) \le 1$, for all $i \in I$. Then $\mathscr{R}_n^\Lambda$ is a graded cellular algebra with graded cellular basis $\{ e_{\mathsf{st}} \mid (\mathsf{s},\mathsf{t}) \in \operatorname{Std}^2(\mathcal{P}_n^\Lambda) \}$ with $\deg e_{\mathsf{st}} = 0$ for all $(\mathsf{s},\mathsf{t}) \in \operatorname{Std}^2(\mathcal{P}_n^\Lambda)$.*

Proof. By Proposition 2.4.6, $y_r = 0$ for $1 \le r \le n$ and $e(\mathbf{i}) = 0$ if $\mathbf{i} \notin I_n^\Lambda$. In particular, this implies that $\psi_1, \dots, \psi_{n-1}$ satisfy the braid relations for the symmetric group $\mathfrak{S}_n$ because, by Lemma 2.4.4, if $\mathbf{i} \in I_\Lambda^n$ then $(i, i \pm 1, i)$ is not a subsequence of $\mathbf{i}$, for any $i \in I$. Therefore, $\mathscr{R}_n^\Lambda$ is spanned by the elements $\psi_v e(\mathbf{i}) \psi_w$, where $v, w \in \mathfrak{S}_n$ and $\mathbf{i} \in I_\Lambda^n$. Moreover, if $\mathbf{j} \in I^n$ then $e(\mathbf{j}) \psi_v e(\mathbf{i}) \psi_w = 0$ unless $\mathbf{j} = v \cdot \mathbf{i} \in I_\Lambda^n$. Therefore, $\mathscr{R}_n^\Lambda$ is spanned by the elements $\{ e_{\mathsf{st}} \mid (\mathsf{s},\mathsf{t}) \in \operatorname{Std}^2(\mathcal{P}_n^\Lambda) \}$ as required by the statement of the theorem. (Note that s and t must have the same shape because $\mathbf{i}^{\mathsf{s}}$ and $\mathbf{i}^{\mathsf{t}}$ are in the same $\mathfrak{S}_n$-orbit and the multiset of contents determines the shape; compare with Theorem 1.8.1) Hence, $\mathscr{R}_n^\Lambda$ has rank at most $\sum_{\boldsymbol{\lambda} \in \mathcal{P}_n^\Lambda} |\operatorname{Std}(\boldsymbol{\lambda})|^2 = \ell^n n!$, where this combinatorial identity comes from Theorem 1.6.7.

Let K be the algebraic closure of the field of fractions of $\mathcal{Z}$. Then $\mathscr{R}_n^\Lambda(K) \cong \mathscr{R}_n^\Lambda(\mathcal{Z}) \otimes_{\mathcal{Z}} K$. By the last paragraph, the dimension of $\mathscr{R}_n^\Lambda$ is at most $\ell^n n!$. Let $\operatorname{rad} \mathscr{R}_n^\Lambda(K)$ be the Jacobson radical of $\mathscr{R}_n^\Lambda(K)$. For each multipartition $\boldsymbol{\lambda} \in \mathcal{P}_n^\Lambda$, Proposition 2.4.3 constructs an irreducible graded Specht module $S^{\boldsymbol{\lambda}}$. By Lemma 2.4.1, if $\boldsymbol{\lambda}, \boldsymbol{\mu} \in \mathcal{P}_n^\Lambda$ and $d \in \mathbb{Z}$ then $S^{\boldsymbol{\lambda}} \cong S^{\boldsymbol{\mu}}\langle d \rangle$ if and only if $\boldsymbol{\lambda} = \boldsymbol{\mu}$ and $d = 0$. By the Wedderburn theorem,

$$\ell^n n! \ge \dim \mathscr{R}_n^\Lambda(K)/\operatorname{rad} \mathscr{R}_n^\Lambda(K) \ge \sum_{\boldsymbol{\lambda} \in \mathcal{P}_n^\Lambda} (\dim S^{\boldsymbol{\lambda}})^2$$

$$= \sum_{\boldsymbol{\lambda} \in \mathcal{P}_n^\Lambda} |\operatorname{Std}(\boldsymbol{\lambda})|^2 = \ell^n n!.$$

Hence, we have equality throughout, so $\{ e_{\mathsf{st}} \mid (\mathsf{s},\mathsf{t}) \in \operatorname{Std}^2(\mathcal{P}_n^\Lambda) \}$ is a basis of $\mathscr{R}_n^\Lambda(K)$. As the elements $\{e_{\mathsf{st}}\}$ span $\mathscr{R}_n^\Lambda(\mathcal{Z})$, and their images in $\mathscr{R}_n^\Lambda(K)$ are linearly independent, so $\{e_{\mathsf{st}}\}$ is a basis of $\mathscr{R}_n^\Lambda(\mathcal{Z})$.

It remains to prove that $\{e_{\mathsf{st}}\}$ is a graded cellular basis of $\mathscr{R}_n^\Lambda$. The orthogonality of the KLR idempotents implies that $e_{\mathsf{st}} e_{\mathsf{uv}} = \delta_{\mathsf{tu}} e_{\mathsf{sv}}$. Therefore, $\{e_{\mathsf{st}}\}$ is a basis of matrix units for $\mathscr{R}_n^\Lambda$. Consequently, $\mathscr{R}_n^\Lambda$ is a direct sum of matrix rings, for any integral domain $\mathcal{Z}$, and $\{e_{\mathsf{st}}\}$ is a cellular basis of $\mathscr{R}_n^\Lambda$.

Finally, we need to show that e_{st} is homogeneous of degree zero. This will follow if we show that $\deg \psi_r e(\mathbf{i}) = 0$, for $1 \le r < n$ and $\mathbf{i} \in I_\Lambda^n$. In fact, this is already clear because if $\mathbf{i} \in I^n_\Lambda$ then $i_r \neq i_{r+1}$, by Lemma 2.4.4, and if $i_{r+1}=i_r \pm 1$ then $\psi_r e(\mathbf{i})=0$ by Corollary 2.4.2 and Proposition 2.4.6. □

By definition, $e_{\mathsf{st}} e_{\mathsf{uv}} = \delta_{\mathsf{tv}} e_{\mathsf{sv}}$. Let $\mathrm{Mat}_d(\mathcal{Z})$ be the ring of $d \times d$ matrices over $\mathcal{Z}$. Hence, the proof of Theorem 2.4.8 yields the following.

Corollary 2.4.9. *Suppose that $\mathcal{Z}$ is an integral domain $e > n$ and that $\Lambda \in P^+$ with $(\Lambda, \alpha_{i,n}) \leq 1$, for all $i \in I$. Then*

$$\mathscr{R}^\Lambda_n(\mathcal{Z}) \cong \bigoplus_{\boldsymbol{\lambda} \in \mathcal{P}^\Lambda_n} \mathrm{Mat}_{s_{\boldsymbol{\lambda}}}(\mathcal{Z}),$$

where $s_{\boldsymbol{\lambda}} = \#\,\mathrm{Std}(\boldsymbol{\lambda})$ for $\boldsymbol{\lambda} \in \mathcal{P}^\Lambda_n$.

Another consequence of Theorem 2.4.8 is that the KLR relations simplify in the semisimple case — giving a non-standard presentation for a direct sum of matrix rings.

Corollary 2.4.10. *Suppose that $\mathcal{Z}$ is an integral domain, $e > n$ and that $\Lambda \in P^+$ with $(\Lambda, \alpha_{i,n}) \leq 1$, for all $i \in I$. Then $\mathscr{R}^\Lambda_n$ is the unital associative $\mathbb{Z}$-graded algebra generated by $\psi_1, \dots, \psi_{n-1}$ and $e(\mathbf{i})$, for $\mathbf{i} \in I^n$, subject to the relations*

$$e(\mathbf{i})^{(\Lambda,\alpha_{i_1})} = 0 \qquad \textstyle\sum_{\mathbf{i} \in I^n} e(\mathbf{i}) = 1, \qquad e(\mathbf{i})e(\mathbf{j}) = \delta_{\mathbf{ij}} e(\mathbf{i}),$$

$$\psi_r e(\mathbf{i}) = e(s_r \cdot \mathbf{i})\psi_r \qquad e(\mathbf{i}) = 0 \textit{ if } i_r = i_{r+1}, \qquad \psi_r^2 e(\mathbf{i}) = e(\mathbf{i})$$

$$\psi_r \psi_s = \psi_s \psi_r, \qquad \textit{if } |r-s| > 1,$$

$$\psi_r \psi_{r+1} \psi_r e(\mathbf{i}) = \begin{cases} (\psi_{r+1}\psi_r\psi_{r+1} - 1)e(\mathbf{i}), & \textit{if } i_{r+2} = i_r \to i_{r+1}, \\ (\psi_{r+1}\psi_r\psi_{r+1} + 1)e(\mathbf{i}), & \textit{if } i_{r+2} = i_r \leftarrow i_{r+1}, \\ \psi_{r+1}\psi_r\psi_{r+1} e(\mathbf{i}), & \textit{otherwise}, \end{cases}$$

for all $\mathbf{i}, \mathbf{j} \in I^n$ and admissible r and s. Moreover, $\mathscr{R}^\Lambda_n$ is concentrated in degree zero.

The reader is encouraged to check the details here. Note that these relations, together with the argument of Proposition 2.4.6, imply that $e(\mathbf{i}) \neq 0$ if only if $\mathbf{i} \in I^n_\Lambda$. In particular, the combinatorics of tableau content sequences is partially encoded in the failure of the braid relations for the ψ_r.

As a final application, we prove Brundan and Kleshchev's graded isomorphism theorem in this special case.

Corollary 2.4.11. *Suppose that $\mathcal{Z} = K$ is a field, $e > n$, and that $\Lambda \in P^+$ with $(\Lambda, \alpha_{i,n}) \leq 1$, for all $i \in I$. Then $\mathscr{R}^\Lambda_n \cong \mathscr{H}^\Lambda_n$.*

Proof. By Corollary 2.4.10 and Theorem 1.6.7, there is a well-defined homomorphism $\Theta\colon \mathscr{R}_n^\Lambda \longrightarrow \mathscr{H}_n^\Lambda$ determined by $e(\mathbf{i}^{\mathsf{s}}) \mapsto F_{\mathsf{s}}$ and

$$\psi_r e(\mathbf{i}^{\mathsf{s}}) \mapsto \begin{cases} \frac{1}{\alpha_r(\mathsf{s})}\Big(T_r - \frac{c^{\mathbb{Z}}_{r+1}(\mathsf{s})-c^{\mathbb{Z}}_r(\mathsf{s})}{1+(v-v^{-1})c^{\mathbb{Z}}_{r+1}(\mathsf{s})}\Big)F_{\mathsf{s}}, & \text{if } \alpha_r(\mathsf{s}) \neq 0,\\ 0, & \text{otherwise,}\end{cases}$$

for $\mathsf{s} \in \operatorname{Std}(\mathcal{P}_n^\Lambda)$ and $1 \le r < n$. Using Theorem 2.4.8, or Proposition 2.4.3, it follows that Θ is an isomorphism. □

We emphasize that it is essential to work over a field in Corollary 2.4.11 because Corollary 2.4.9 says that $\mathscr{R}_n^\Lambda$ is always a direct sum of matrix rings. If $n > 1$ this is only true of $\mathscr{H}_n^\Lambda$ when it is defined over a field.

These results suggest that $\mathscr{R}_n^\Lambda$ should be considered as the "idempotent completion" of the algebra $\mathscr{H}_n^\Lambda$ obtained by adjoining idempotents $e(\mathbf{i})$, for $\mathbf{i} \in I^n$. We will see how to make sense of the idempotents $e(\mathbf{i}) \in \mathscr{H}_n^\Lambda$ for any $\mathbf{i} \in I^n$ in Theorem 3.1.1 and Lemma 4.2.2 below.

2.5 *The nil-Hecke algebra*

Still working just with the relations we now consider the shadow of the nil-Hecke algebra in the cyclotomic KLR setting. For the affine KLR algebras the nil-Hecke algebras case has been well-studied [74,121]. For the cyclotomic quotients (in type A) the story is similar.

For this section fix $i \in I$ and set $\beta = n\alpha_i$ and $\Lambda = n\Lambda_i$. Following (2.2.5), set $\mathscr{R}_\beta^\Lambda = e(\mathbf{i})\mathscr{R}_n^\Lambda e(\mathbf{i})$, where $\mathbf{i} = \mathbf{i}^\beta = (i^n)$. Then $\mathscr{R}_\beta^\Lambda$ is a direct summand of $\mathscr{R}_n^\Lambda$ and, moreover, it is a non-unital subalgebra with identity element $e(\mathbf{i})$. As $e(\mathbf{i})$ is the unique non-zero KLR idempotent in $\mathscr{R}_\beta^\Lambda$, $\psi_r = \psi_r e(\mathbf{i})$ and $y_s = y_s e(\mathbf{i})$. Therefore, $\mathscr{R}_\beta^\Lambda$ is the unital associative graded algebra generated by ψ_r and y_s, for $1 \le r < n$ and $1 \le s \le n$, with relations

$$\begin{gathered} y_1^n = 0, \qquad \psi_r^2 = 0, \qquad y_r y_s = y_s y_r,\\ \psi_r y_{r+1} = y_r \psi_r + 1, \qquad y_{r+1}\psi_r = \psi_r y_r + 1,\\ \psi_r\psi_s = \psi_s\psi_r \quad \text{if } |r-s|>1, \qquad \psi_r y_s = y_s\psi_r \quad \text{if } s \neq r, r+1,\\ \psi_r\psi_{r+1}\psi_r = \psi_{r+1}\psi_r\psi_{r+1}. \end{gathered}$$

The grading on $\mathscr{R}_\beta^\Lambda$ is determined by $\deg \psi_r = -2$ and $\deg y_s = 2$. Some readers will recognize this presentation as defining as a cyclotomic quotient of the **nil-Hecke algebra** of type A [88]. Note that the argument from Case 3 of Lemma 2.3.1 shows that $y_r^\ell = 0$ for $1 \le r \le \ell$.

Let $\boldsymbol{\lambda} = (1|1|\dots|1) \in \mathcal{P}_\beta^\Lambda$. Then the map $\mathsf{t} \mapsto d(\mathsf{t})$ defines a bijection between the set of standard $\boldsymbol{\lambda}$-tableaux and the symmetric group $\mathfrak{S}_n$.

For convenience, we identify the standard $\boldsymbol{\lambda}$-tableaux with the set of (non-standard) tableaux of partition shape (n) by concatenating their components. In other words, if $d = d(\mathsf{t})$ then $\mathsf{t} = \boxed{d_1|d_2|\cdots|d_n}$, where $d = d_1 \dots d_n$ is the permutation written in one-line notation.

If $\mathsf{v}, \mathsf{s} \in \operatorname{Std}(\boldsymbol{\lambda})$ then write $\mathsf{s} \mathrel{\triangleright\!\!\!>} \mathsf{v}$ if $\mathsf{s} \triangleright \mathsf{v}$ and $\ell(d(\mathsf{v})) = \ell(d(\mathsf{s})) + 1$. To make this more explicit write $t \prec_{\mathsf{v}} m$ if t is in an earlier component of v than m — that is, t is to the left of m in v. The reader can check that $\mathsf{s} \mathrel{\triangleright\!\!\!>} \mathsf{v}$ if and only if there exist integers $1 \le m < t \le n$ such that $\mathsf{s} = \mathsf{v}(m,t)$, $m \prec_{\mathsf{v}} t$ and if $m < l < t$ then either $l \prec_{\mathsf{v}} m$ or $t \prec_{\mathsf{v}} l$.

2.5.1. **Example** Suppose that $n = 6$. Let $\mathsf{v} = \boxed{4|6|5|3|1|2}$ and take $t = 3$. Then

$$\left\{ \boxed{3|6|5|4|1|2},\ \boxed{4|6|3|5|1|2},\ \boxed{4|6|5|2|1|3},\ \boxed{4|6|5|1|3|2} \right\}$$

is the set of $\boldsymbol{\lambda}$-tableaux $\{\, \mathsf{s} \mid \mathsf{s} = \mathsf{v}(3,r) \mathrel{\triangleright\!\!\!>} \mathsf{v} \text{ for } 1 \le r \le n \,\}$. ◊

We can now state the main result of the section.

Proposition 2.5.2. *Suppose that $\beta = n\alpha_i$ and $\Lambda = n\Lambda_i$, for $i \in I$. Then there is a unique graded $\mathscr{R}^\Lambda_\beta$-module $S^{\boldsymbol{\lambda}}$ with homogeneous basis $\{\, \psi_{\mathsf{s}} \mid \mathsf{s} \in \operatorname{Std}(\boldsymbol{\lambda}) \,\}$ such that $\deg \psi_{\mathsf{s}} = \binom{n}{2} - 2\ell(d(\mathsf{s}))$ and*

$$\psi_{\mathsf{s}}\psi_r = \begin{cases} \psi_{\mathsf{s}(r,r+1)}, & \text{if } \mathsf{s} \triangleright \mathsf{s}(r,r+1) \in \operatorname{Std}(\boldsymbol{\lambda}), \\ 0, & \text{otherwise,} \end{cases}$$

$$\psi_{\mathsf{v}} y_t = \sum_{\substack{1 \le k < t \\ \mathsf{u} = \mathsf{v}(k,t) \mathrel{\triangleright\!\!\!>} \mathsf{u}}} \psi_{\mathsf{u}} \quad - \sum_{\substack{t < k \le n \\ \mathsf{u} = \mathsf{v}(k,t) \mathrel{\triangleright\!\!\!>} \mathsf{v}}} \psi_{\mathsf{u}},$$

for $\mathsf{s}, \mathsf{v} \in \operatorname{Std}(\boldsymbol{\lambda})$, $1 \le r < n$ and $1 \le t \le n$. Moreover, if $\mathcal{Z}$ is a field then $S^{\boldsymbol{\lambda}}$ is irreducible.

Proof. The uniqueness is clear. To show that $S^{\boldsymbol{\lambda}}$ is a $\mathscr{R}^\Lambda_\beta$-module we check that the action respects the relations of $\mathscr{R}^\Lambda_\beta$. By definition, if $\mathsf{v} \in \operatorname{Std}(\boldsymbol{\lambda})$ then $\psi_{\mathsf{v}} = \psi_{\mathsf{t}^{\boldsymbol{\lambda}}}\psi_{d(\mathsf{v})}$ and $\psi_{\mathsf{v}}\psi_r^2 = 0$ since $\psi_{\mathsf{v}}\psi_r = 0$ if $\mathsf{v}(r,r+1) \triangleright \mathsf{v}$. In particular, this implies that the action of $\psi_1 \dots, \psi_{n-1}$ on $S^{\boldsymbol{\lambda}}$ respects the braid relations of $\mathfrak{S}_n$ and that ψ_{v} has the specified degree. Further, note that if $\mathsf{u} \mathrel{\triangleright\!\!\!>} \mathsf{v}$ then $\ell(d(\mathsf{v})) = \ell(d(\mathsf{u})) + 1$ so that $\deg \psi_{\mathsf{u}} = \deg \psi_{\mathsf{v}} + 2$.

By the last paragraph, the action of $\mathscr{R}^\Lambda_\beta$ is compatible with the grading on $S^{\boldsymbol{\lambda}}$, but we still need to check the relations involving $y_1, \dots, y_n$. First consider $\psi_{\mathsf{v}} y_r y_t = \psi_{\mathsf{v}} y_t y_r$, for $1 \le r, t \le n$ and $\mathsf{v} \in \operatorname{Std}(\boldsymbol{\lambda})$. If $r = t$ there is nothing to prove so suppose $r \ne t$. By definition,

$$\psi_{\mathsf{v}} y_t y_r = \sum_{\mathsf{u} \mathrel{\triangleright\!\!\!>} \mathsf{v}} \sum_{\mathsf{s} \mathrel{\triangleright\!\!\!>} \mathsf{u}} \varepsilon_t(\mathsf{v},\mathsf{u})\varepsilon_r(\mathsf{u},\mathsf{s})\psi_{\mathsf{s}},$$

for appropriate choices of the signs $\varepsilon_t(\mathsf{v},\mathsf{u})$ and $\varepsilon_r(\mathsf{u},\mathsf{s})$. Suppose that ψ_s appears with non-zero coefficient in this sum. Then we can write $\mathsf{u} = \mathsf{v}(m,t)$ and $\mathsf{s} = \mathsf{v}(l,r)$, for some l,m such that $\mathsf{s} \rhd \mathsf{u} \rhd \mathsf{v}$. Suppose first that $l \neq m$. Then the permutations (m,t) and (l,r) commute and, as their lengths add, we have $\mathsf{s} \rhd \mathsf{v}(l,r) \rhd \mathsf{v}$. Therefore, ψ_s appears with the same coefficient in $\psi_\mathsf{v} y_t y_r$ and $\psi_\mathsf{v} y_r y_t$. If $l = m$ then $\mathsf{s} \rhd \mathsf{u} \unrhd \mathsf{v}$ only if m is in between r and t in v. That is, either $r \prec_\mathsf{v} m \prec_\mathsf{v} t$ or $t \prec_\mathsf{v} m \prec_\mathsf{v} r$. However, this implies that either $\mathsf{s} \not\rhd \mathsf{u}$ or $\mathsf{u} \not\rhd \mathsf{v}$, so that ψ_s does not appear in either $\psi_\mathsf{v} y_r y_t$ or in $\psi_\mathsf{v} y_t y_r$. Hence, the actions of y_r and y_t on S^λ commute.

Similar, but easier, calculations with tableaux show that the action defined on S^λ respects the three relations $\psi_r y_{r+1} = y_r \psi_r + 1$, $y_{r+1}\psi_r = \psi_r y_r + 1$ and $\psi_r y_t = y_t \psi_r$ when $t \neq r, r+1$. To complete the verification of the relations in $\mathscr{R}^\Lambda_\beta$ it remains to show that $\psi_\mathsf{v} y_1^n = 0$, for all $\mathsf{v} \in \operatorname{Std}(\lambda)$. This is clear, however, because $\psi_\mathsf{v} y_1$ is equal to a linear combination of terms ψ_s where 1 appears in an earlier component of s than it does in v.

Finally, it remains to prove that S^λ is irreducible over a field. First we need some more notation. Let $\mathsf{t}_\lambda = \boxed{n}\boxed{\cdots}\boxed{\cdots}\boxed{2}\boxed{1}$ and set $w_\lambda = d(\mathsf{t}_\lambda)$. Then w_λ is the unique element of maximal length in $\mathfrak{S}_n$. Recall from §1.4, that $d'(\mathsf{t})$ is the unique permutation such that $\mathsf{t} = \mathsf{t}_\lambda d'(\mathsf{t})$ and, moreover, $d(\mathsf{t})d'(\mathsf{t})^{-1} = w_\lambda$ with the lengths adding. Therefore, if $\ell(d(\mathsf{s})) \geq \ell(d(\mathsf{t}))$ then $\psi_\mathsf{s}\psi^\star_{d'(\mathsf{t})} = \delta_{\mathsf{st}}\psi_{\mathsf{t}_\lambda}$.

We are now ready to show that S^λ is irreducible. Suppose that $x = \sum_\mathsf{s} r_\mathsf{s}\psi_\mathsf{s}$ is a non-zero element of S^λ. Let t be any tableau such that $r_\mathsf{t} \neq 0$ and $\ell(d(\mathsf{t}))$ is minimal. Then, by the last paragraph, $x\psi^\star_{d'(\mathsf{t})} = r_\mathsf{t}\psi_{\mathsf{t}_\lambda}$, so $\psi_{\mathsf{t}_\lambda} \in x\mathscr{R}^\Lambda_\beta$. We have already observed that y_1 acts by moving 1 to an earlier component. Therefore, $\psi_{\mathsf{t}_\lambda} y_1^{n-1} = (-1)^{n-1}\psi_{\mathsf{t}_{\lambda,1}}$, where $\mathsf{t}_{\lambda,1} = \boxed{1}\boxed{n}\boxed{\cdots}\boxed{3}\boxed{2}$. Similarly, $\psi_{\mathsf{t}_\lambda} y_1^{n-1}y_2^{n-2} = (-1)^{2n-3}\psi_{\mathsf{t}_{\lambda,2}}$, where $\mathsf{t}_{\lambda,2} = \boxed{1}\boxed{2}\boxed{n}\boxed{\cdots}\boxed{3}$. Continuing in this way shows that $\psi_{\mathsf{t}_\lambda} y_1^{n-1}y_2^{n-2}\dots y_{n-1} = (-1)^{\frac12 n(n-1)}\psi_{\mathsf{t}_\lambda}$. Hence, $x\mathscr{R}^\Lambda_\beta = S^\lambda$, so that S^λ is irreducible as claimed.□

The proof of Proposition 2.5.2 shows that $y_1^{n-1}y_2^{n-2}\dots y_{n-1}$ is a non-zero element of $\mathscr{R}^\Lambda_\beta$. Using the relations, and a bit of ingenuity, it is possible to show that $\{\,\psi_w y_1^{a_1}\dots y_n^{a_n} \mid w \in \mathfrak{S}_n \text{ and } 0 \leq a_r \leq n-r, \text{ for } 1 \leq r \leq n\,\}$ is a basis of $\mathscr{R}^\Lambda_\beta$. Alternatively, it follows from [22, Theorem 4.20] that $\dim \mathscr{R}^\Lambda_\beta = (n!)^2$. Hence, we obtain the following.

Corollary 2.5.3. *Suppose that* $\beta = n\alpha_i$ *and* $\Lambda = n\Lambda_i$, *for* $i \in I$. *Let* $\lambda = (1|1|\dots|1)$ *and for* $\mathsf{s},\mathsf{t} \in \operatorname{Std}(\lambda)$ *define* $\psi_{\mathsf{st}} = \psi^\star_{d(\mathsf{s})}e(\mathbf{i}^\lambda)y^\lambda\psi_{d(\mathsf{t})}$, *where* $\mathbf{i}^\lambda = \mathbf{i}^{\mathsf{t}^\lambda}$ *and* $y^\lambda = y_1^{n-1}y_2^{n-2}\dots y_{n-1}$. *Then* $\{\,\psi_{\mathsf{st}} \mid \mathsf{s},\mathsf{t} \in \operatorname{Std}(\lambda)\,\}$ *is a* *graded cellular basis of* $\mathscr{R}^\Lambda_\beta$.

The basis of the Specht module $S^{\boldsymbol{\lambda}}$ in Proposition 2.5.2 is well-known because it is really a disguised version of the basis of *Schubert polynomials* of the coinvariant algebra of the symmetric group $\mathfrak{S}_n$ [90, 101]. The **coinvariant algebra** $\mathscr{C}_n$ is the quotient of the polynomial ring $\mathbb{Z}[\mathbf{x}] = \mathbb{Z}[x_1, \dots, x_n]$ by the deal generated by the symmetric polynomials in $x_1, \dots, x_n$ of *positive degree*. Then $\mathscr{C}_n$ is free of rank $n!$. As we have quotiented out by a homogeneous ideal, $\mathscr{C}_n$ inherits a grading from $\mathbb{Z}[\mathbf{x}]$, where we set $\deg x_r = 2$ for $1 \le r \le n$. Identify x_r with its image in $\mathscr{C}_n$, for $1 \le r \le n$. There is a well-defined action of $\mathscr{R}^\Lambda_\beta$ on $\mathscr{C}_n$ where y_r acts as multiplication by x_r, and ψ_r acts from the right as a *divided difference operator*:

$$f(\mathbf{x})\psi_r = \partial_r f(\mathbf{x}) = \frac{f(\mathbf{x}) - f(s_r \cdot \mathbf{x})}{x_r - x_{r+1}},$$

where $\mathbf{x} = (x_1, \dots, x_n)$ and $s_r \cdot \mathbf{x} = (x_1, \dots, x_{r+1}, x_r, \dots, x_n)$ for $1 \le r < n$. Here we are secretly thinking of $\mathscr{R}^\Lambda_\beta$ as being a quotient of the nil-Hecke algebra, where this action is well-known.

For $d \in \mathfrak{S}_n$ define $\sigma_d = (x_1^{n-1} x_2^{n-2} \dots x_{n-1})\psi_{w_0 d}$. Then $\{\, \sigma_d \mid d \in \mathfrak{S}_n \,\}$ is the basis of **Schubert polynomials** of $\mathscr{C}_n$. The Specht module is isomorphic to $\mathscr{C}_n$ as a $\mathscr{R}^\Lambda_\beta$-module, where an isomorphism is given by $\psi_{\mathsf{t}} \mapsto \sigma_{d'(\mathsf{t})}$. To see this it is enough to know that the Schubert polynomials satisfy the identity

$$\partial_r \sigma_d = \begin{cases} \sigma_{s_r d}, & \text{if } \ell(s_r d) = \ell(d) - 1, \\ 0, & \text{otherwise.} \end{cases}$$

By the last paragraph of the proof of Proposition 2.5.2, if $\mathsf{t} \in \mathfrak{S}_n$ then

$$\psi_{\mathsf{t}} = \psi_{\mathsf{t}^{\boldsymbol{\lambda}}} \psi_{d(\mathsf{t})} = \psi_{\mathsf{t}_{\boldsymbol{\lambda}}} y_1^{n-1} y_2^{n-1} \dots y_{n-1} \psi_{d(\mathsf{t})}.$$

Therefore, our claim follows by identifying $\psi_{\mathsf{t}_{\boldsymbol{\lambda}}}$ with the polynomial $1 \in \mathscr{C}_n$.

Finally, we remark that the formula for the action of $y_1, \dots, y_n$ in Proposition 2.5.2 is a well-known corollary of Monk's rule; see, for example, [101, Exercise 2.7.3].

3 Isomorphisms, Specht modules and categorification

In the last section we proved that the algebras $\mathscr{R}^\Lambda_n$ and $\mathscr{H}^\Lambda_n$ are isomorphic when $e > n$ and $(\Lambda, \alpha_{i,n}) \le 1$, for all $i \in I$. In this section we state Brundan and Kleshchev's Graded Isomorphism Theorem, which says that $\mathscr{R}^\Lambda_n \cong \mathscr{H}^\Lambda_n$, and we start to investigate the consequences of this result for both algebras.

3.1 *The Graded Isomorphism Theorem*

One of the most fundamental results for the cyclotomic Hecke algebras $\mathscr{H}_n^\Lambda$ is Brundan and Kleshchev's spectacular isomorphism theorem.

Theorem 3.1.1 (Graded Isomorphism Theorem [21, 121]**).** *Suppose that $\mathcal{Z} = F$ is a field, $v \in F$ has quantum characteristic e and that $\Lambda \in P^+$. Then there is an isomorphism of algebras $\mathscr{R}_n^\Lambda \cong \mathscr{H}_n^\Lambda$.*

Suppose that F is a field of characteristic $p > 0$ and that $e = pf$, where $f > 1$. Then F cannot contain an element v of quantum characteristic e, so Theorem 3.1.1 says nothing about the quiver Hecke algebra $\mathscr{R}_n^\Lambda(F)$.

As a first consequence of Theorem 3.1.1, by identifying $\mathscr{H}_n^\Lambda$ and $\mathscr{R}_n^\Lambda$ we can consider $\mathscr{H}_n^\Lambda$ as a graded algebra.

Corollary 3.1.2. *Suppose that $\Lambda \in P^+$ and $\mathcal{Z} = F$ is a field. Then there is a unique grading on $\mathscr{H}_n^\Lambda$ such that $\deg e(\mathbf{i}) = 0$, $\deg y_r = 2$ and $\deg \psi_s e(\mathbf{i}) = -c_{i_s,i_{s+1}}$, for $1 \le r \le n$, $1 \le s < n$ and $\mathbf{i} \in I^n$.*

Brundan and Kleshchev prove Theorem 3.1.1 by constructing family of isomorphisms $\mathscr{R}_n^\Lambda \longrightarrow \mathscr{H}_n^\Lambda$, together with their inverses, and then painstakingly checking that these isomorphisms respect the relations of both algebras. Their argument starts with the well-known fact that $\mathscr{H}_n^\Lambda$ decomposes into a direct sum of simultaneous generalized eigenspaces for the Jucys-Murphy elements $L_1, \dots, L_n$. These eigenspaces are indexed by I^n, so for each $\mathbf{i} \in I^n$ there is an element $e(\mathbf{i}) \in \mathscr{H}_n^\Lambda$, possibly zero, such that $e(\mathbf{i})e(\mathbf{j}) = \delta_{\mathbf{ij}} e(\mathbf{i})$. We describe these idempotents explicitly in Lemma 4.2.2 below.

Translating through Definition 1.1.1, Brundan and Kleshchev's isomorphism is given by $e(\mathbf{i}) \mapsto e(\mathbf{i})$ and

$$y_r \mapsto \sum_{\mathbf{i} \in I^n} v^{-i_r}\big(L_r - [i_r]_v\big)e(\mathbf{i}), \quad \text{and} \quad \psi_s \mapsto \sum_{\mathbf{i} \in I^n} \big(T_s + P_s(\mathbf{i})\big)\frac{1}{Q_s(\mathbf{i})}e(\mathbf{i}),$$

for $1 \le r \le n$, $1 \le s < n$, $\mathbf{i} \in I^n$ and where $P_r(\mathbf{i})$ and $Q_r(\mathbf{i})$ are certain rational functions in y_r and y_{r+1} that are well-defined because $(L_t - [i_t]_v)e(\mathbf{i})$ is nilpotent in $\mathscr{H}_n^\Lambda$, for $1 \le t \le n$; see [21, §3.3 and §4.3]. Here we are abusing notation by identifying the KLR generators with their images in $\mathscr{H}_n^\Lambda$. The inverse isomorphism is given by $e(\mathbf{i}) \mapsto e(\mathbf{i})$,

$$L_r \mapsto \sum_{\mathbf{i} \in I^n} \big(v^{i_r} y_r + [i_r]_v\big)e(\mathbf{i}) \quad \text{and} \quad T_s \mapsto \sum_{\mathbf{i} \in I^n} \big(\psi_s Q_s(\mathbf{i}) - P_s(\mathbf{i})\big)e(\mathbf{i}),$$

for $1 \le r \le n$, $1 \le s < n$ and $\mathbf{i} \in I^n$.

Rouquier [121, Corollary 3.20] has given a more direct proof of Theorem 3.1.1 by first showing that the (non-cyclotomic) quiver Hecke algebra $\mathscr{R}_n$ is isomorphic to the (extended) affine Hecke algebra of type A. Following [57], we sketch another approach to Theorem 3.1.1 in §4.2 below.

The following easy but important application of Theorem 3.1.1 was a surprise (at least to the author!).

Corollary 3.1.3. *Suppose that $\mathcal{Z} = F$ is a field and that $v, v' \in F$ are two elements of quantum characteristic e. Then $\mathscr{H}_n^\Lambda(F, v) \cong \mathscr{H}_n^\Lambda(F, v')$.*

Proof. By Theorem 3.1.1, $\mathscr{H}_n^\Lambda(F, v) \cong \mathscr{R}_n^\Lambda(F) \cong \mathscr{H}_n^\Lambda(F, v')$. □

Consequently, up to isomorphism, the algebra $\mathscr{H}_n^\Lambda$ depends only on e, Λ and the field F. Therefore, because $\mathscr{H}_n^\Lambda$ is cellular, the decomposition matrices of $\mathscr{H}_n^\Lambda$ depend only on e, Λ and p, where p is the characteristic of F. In the special case of the symmetric group, when $\Lambda = \Lambda_0$, this weaker statement for the decomposition matrices was conjectured in [104, Conjecture 6.38].

When $F = \mathbb{C}$ it is easy to prove Corollary 3.1.3 because there is a Galois automorphism of $\mathbb{Q}(v)$, as an extension of $\mathbb{Q}$, that interchanges v and v'. It is not difficult to see that this automorphism induces an isomorphism $\mathscr{H}_n^\Lambda(F, v) \cong \mathscr{H}_n^\Lambda(F, v')$. This argument fails for fields of positive characteristic because such fields have fewer automorphisms.

3.2 *Graded Specht modules*

As we noted in §2.1, if we impose a grading on an algebra $\underline{A}$ then it is not true that every (ungraded) $\underline{A}$-module has a graded lift, so there is no reason to expect that graded lifts of Specht modules $\underline{S}^{\boldsymbol{\lambda}}$ exist. Of course, graded Specht modules do exist and this section describes one way to define them.

Recall from §1.5 that the ungraded Specht module $\underline{S}^{\boldsymbol{\lambda}}$, for $\boldsymbol{\lambda} \in \mathcal{P}_n^\Lambda$, has basis $\{ m_{\mathfrak{t}} \mid \mathfrak{t} \in \mathrm{Std}(\boldsymbol{\lambda}) \}$. By construction, $\underline{S}^{\boldsymbol{\lambda}} = m_{\mathfrak{t}^{\boldsymbol{\lambda}}} \mathscr{H}_n^\Lambda$. Brundan, Kleshchev and Wang [25] proved that $\underline{S}^{\boldsymbol{\lambda}}$ has a graded lift essentially by declaring that $m_{\mathfrak{t}^{\boldsymbol{\lambda}}}$ should be homogeneous and then showing that this induces a grading on the Specht module $\underline{S}^{\boldsymbol{\lambda}} = m_{\mathfrak{t}^{\boldsymbol{\lambda}}} \mathscr{R}_n^\Lambda$.

Partly inspired by [25], Jun Hu and the author [54] showed that $\mathscr{H}_n^\Lambda$ is a graded cellular algebra. The graded cell modules constructed from this cellular basis coincide exactly with those of [25]. Perhaps most significantly, the construction of the graded Specht modules using cellular algebra techniques endows the graded Specht modules with a homogeneous bilinear form of degree zero.

Following Brundan, Kleshchev and Wang [25, §3.5] we now define the degree of a standard tableau. Suppose that $\boldsymbol{\mu} \in \mathcal{P}_n^\Lambda$. For $i \in I$ let $\operatorname{Add}_i(\boldsymbol{\mu})$ be the set of addable i-nodes of $\boldsymbol{\mu}$ and let $\operatorname{Rem}_i(\boldsymbol{\mu})$ be its set of removable i-nodes.

Definition 3.2.1. If A is an addable or removable i-node of $\boldsymbol{\mu}$ define:

$$d^A(\boldsymbol{\mu}) = \#\{\, B \in \operatorname{Add}_i(\boldsymbol{\mu}) \mid A > B \,\} - \#\{\, B \in \operatorname{Rem}_i(\boldsymbol{\mu}) \mid A > B \,\},$$
$$d_A(\boldsymbol{\mu}) = \#\{\, B \in \operatorname{Add}_i(\boldsymbol{\mu}) \mid A < B \,\} - \#\{\, B \in \operatorname{Rem}_i(\boldsymbol{\mu}) \mid A < B \,\},$$
$$d_i(\boldsymbol{\mu}) = \#\operatorname{Add}_i(\boldsymbol{\mu}) - \#\operatorname{Rem}_i(\boldsymbol{\mu}).$$

If t is a standard $\boldsymbol{\mu}$-tableau then its **codegree** and **degree** are defined inductively by setting $\operatorname{codeg}_e \mathsf{t} = 0 = \deg_e \mathsf{t}$ if $n = 0$ and if $n > 0$ then

$$\operatorname{codeg}_e \mathsf{t} = \operatorname{codeg}_e \mathsf{t}_{\downarrow(n-1)} + d^A(\boldsymbol{\mu}) \quad \text{and} \quad \deg_e \mathsf{t} = \deg_e \mathsf{t}_{\downarrow(n-1)} + d_A(\boldsymbol{\mu}),$$

where $A = \mathsf{t}^{-1}(n)$. If e is fixed write $\operatorname{codeg} \mathsf{t} = \operatorname{codeg}_e \mathsf{t}$ and $\deg \mathsf{t} = \deg_e \mathsf{t}$.

Implicitly, all of these definitions depend on the choice of multicharge $\boldsymbol{\kappa}$. The definition of the degree and codegree of a standard tableau is due to Brundan, Kleshchev and Wang [25], however, the underlying combinatorics dates back to Misra and Miwa [111] and their work on the crystal graph and Fock space representations of $U_q(\widehat{\mathfrak{sl}}_e)$.

Recall that we have fixed an arbitrary reduced expression for each permutation $w \in \mathfrak{S}_n$. In §1.4 for each standard tableau $\mathsf{t} \in \operatorname{Std}(\boldsymbol{\lambda})$ we have defined permutations $d'(\mathsf{t}), d(\mathsf{t}) \in \mathfrak{S}_n$ by $\mathsf{t}_{\boldsymbol{\lambda}} d'(\mathsf{t}) = \mathsf{t} = \mathsf{t}^{\boldsymbol{\lambda}} d(\mathsf{t})$.

Definition 3.2.2 ([54, Definitions 4.9 and 5.1]). *Suppose that $\boldsymbol{\mu} \in \mathcal{P}_n^\Lambda$. Define non-negative integers $d_1^{\boldsymbol{\mu}}, \dots, d_n^{\boldsymbol{\mu}}$ and $d_{\boldsymbol{\mu}}^1, \dots, d_{\boldsymbol{\mu}}^n$ recursively by requiring that $d_{\boldsymbol{\mu}}^1 + \dots + d_{\boldsymbol{\mu}}^k = \operatorname{codeg}(\mathsf{t}_{\downarrow k}^{\boldsymbol{\mu}})$ and $d_1^{\boldsymbol{\mu}} + \dots + d_k^{\boldsymbol{\mu}} = \deg(\mathsf{t}_{\downarrow k}^{\boldsymbol{\mu}})$, for $1 \le k \le n$. Now set $\mathbf{i}_{\boldsymbol{\mu}} = \mathbf{i}^{\mathsf{t}_{\boldsymbol{\mu}}}$, $\mathbf{i}^{\boldsymbol{\mu}} = \mathbf{i}^{\mathsf{t}^{\boldsymbol{\mu}}}$, $y_{\boldsymbol{\mu}} = y_1^{d_{\boldsymbol{\mu}}^1} \dots y_n^{d_{\boldsymbol{\mu}}^n}$ and $y^{\boldsymbol{\mu}} = y_1^{d_1^{\boldsymbol{\mu}}} \dots y_n^{d_n^{\boldsymbol{\mu}}}$. For $(\mathsf{s}, \mathsf{t}) \in \operatorname{Std}^2(\boldsymbol{\mu})$ define*

$$\psi'_{\mathsf{st}} = \psi^\star_{d'(\mathsf{s})} e(\mathbf{i}_{\boldsymbol{\mu}}) y_{\boldsymbol{\mu}} \psi_{d'(\mathsf{t})} \quad \text{and} \quad \psi_{\mathsf{st}} = \psi^\star_{d(\mathsf{s})} e(\mathbf{i}^{\boldsymbol{\mu}}) y^{\boldsymbol{\mu}} \psi_{d(\mathsf{t})},$$

where $\star$ is the unique (homogeneous) anti-isomorphism of $\mathscr{R}_n^\Lambda$ that fixes the KLR generators.

3.2.3. **Example** Suppose that $e = 3$, $\Lambda = \Lambda_0 + \Lambda_2$ and $\boldsymbol{\mu} = (7,6,3,2|4,3,1)$, with multicharge $\boldsymbol{\kappa} = (0,2)$. Then

$$\mathsf{t}^{\boldsymbol{\mu}} = \left(\begin{array}{ccccccc} 1 & 2 & 3 & 4 & 5 & 6 & 7 \\ 8 & 9 & 10 & 11 & 12 & 13 & \\ 14 & 15 & 16 & & & & \\ 17 & 18 & & & & & \end{array} \middle| \begin{array}{cccc} 19 & 20 & 21 & 22 \\ 23 & 24 & 25 & \\ 26 & & & \\ & & & \end{array} \right)$$

The reader may check that $e(\mathbf{i}^{\boldsymbol{\mu}}) = e(0120120201201120012012\,1200)$. We have shaded the nodes in $\mathsf{t}^{\boldsymbol{\mu}}$ when they have column index divisible by e and when they have residue $2 = \operatorname{res}_{\mathsf{t}^{\boldsymbol{\mu}}}(19)$. This should convince the reader that $y^{\boldsymbol{\mu}} = y_3^2 y_6^2 y_8 y_{10} y_{11} y_{13} y_{15} y_{16} y_{21} y_{25}$. With analogous shadings,

$$\mathsf{t}_{\boldsymbol{\mu}} = \left(\begin{array}{|c|c|c|c|c|c|c|} \hline 9 & 13 & 17 & 20 & 22 & 24 & 26 \\ \hline 10 & 14 & 18 & 21 & 23 & 25 \\ \cline{1-6} 11 & 15 & 19 \\ \cline{1-3} 12 & 16 \\ \cline{1-2} \end{array} \;\middle|\; \begin{array}{|c|c|c|c|} \hline 1 & 4 & 6 & 8 \\ \hline 2 & 5 & 7 \\ \cline{1-3} 3 \\ \cline{1-1} \end{array} \right).$$

Hence, reading right to left, $y_{\boldsymbol{\mu}} = y_3^2 y_4 y_7 y_{11} y_{15} y_{19}$. Note that $\operatorname{res}_{\mathsf{t}_{\boldsymbol{\mu}}}(9) = 0$. ◊

Theorem 3.2.4 (Hu-Mathas [54, Theorem 5.8]**).** *Suppose that $\mathcal{Z} = F$ is a field. Then $\{\, \psi_{\mathsf{st}} \mid (\mathsf{s}, \mathsf{t}) \in \operatorname{Std}^2(\mathcal{P}_n^\Lambda) \,\}$ is a graded cellular basis of $\mathscr{R}_n^\Lambda$ with $\psi_{\mathsf{st}}^\star = \psi_{\mathsf{ts}}$ and $\deg \psi_{\mathsf{st}} = \deg \mathsf{s} + \deg \mathsf{t}$, for $(\mathsf{s}, \mathsf{t}) \in \operatorname{Std}^2(\mathcal{P}_n^\Lambda)$.*

3.2.5. **Example** Let $\beta = n\alpha_i$ and $\Lambda = n\Lambda_i$, for some $i \in I$, so that $\mathscr{R}_\beta^\Lambda$ is the nil-Hecke algebra $\mathscr{R}_\beta^\Lambda$ of §2.5. Let $\boldsymbol{\lambda} = (1|1|\dots|1)$. Then the definitions give $y^{\boldsymbol{\lambda}} = y_1^{n-1} \dots y_{n-2}^2 y_{n-1}$. Hence, the basis $\{\psi_{\mathsf{st}}\}$ of $\mathscr{R}_\beta^\Lambda$ coincides with that of Corollary 2.5.3. ◊

3.2.6. **Example** As in Example 2.2.7, in general, the basis element ψ_{st} depends on the choices of reduced expressions that we have fixed for the permutations $d(\mathsf{s})$ and $d(\mathsf{t})$. For example, let $\Lambda = 2\Lambda_0 + \Lambda_1$, $\boldsymbol{\kappa} = (0, 1, 0)$ and $\boldsymbol{\mu} = (1|1|1)$ and consider the standard $\boldsymbol{\mu}$-tableaux $\mathsf{t}^{\boldsymbol{\mu}} = (\boxed{1} \,|\, \boxed{2} \,|\, \boxed{3})$ and $\mathsf{t}_{\boldsymbol{\mu}} = (\boxed{3} \,|\, \boxed{2} \,|\, \boxed{1})$. Then $d(\mathsf{t}^{\boldsymbol{\mu}}) = 1$ and $d(\mathsf{t}_{\boldsymbol{\mu}}) = (1, 3) = s_1 s_2 s_1 = s_2 s_1 s_2$ has two different reduced expressions. Let $\psi_{\mathsf{t}_{\boldsymbol{\mu}}\mathsf{t}_{\boldsymbol{\mu}}} = \psi_1 \psi_2 \psi_1 e(\mathbf{i}^{\boldsymbol{\mu}}) y^{\boldsymbol{\mu}} \psi_1 \psi_2 \psi_1$ and $\hat{\psi}_{\mathsf{t}_{\boldsymbol{\mu}}\mathsf{t}_{\boldsymbol{\mu}}} = \psi_2 \psi_1 \psi_2 e(\mathbf{i}^{\boldsymbol{\mu}}) y^{\boldsymbol{\mu}} \psi_2 \psi_1 \psi_2$. Then the calculation in Example 2.2.7 implies that

$$\hat{\psi}_{\mathsf{t}_{\boldsymbol{\mu}}\mathsf{t}_{\boldsymbol{\mu}}} = \psi_{\mathsf{t}_{\boldsymbol{\mu}}\mathsf{t}_{\boldsymbol{\mu}}} + \psi_{\mathsf{t}_{\boldsymbol{\mu}}\mathsf{t}^{\boldsymbol{\mu}}} + \psi_{\mathsf{t}^{\boldsymbol{\mu}}\mathsf{t}_{\boldsymbol{\mu}}} + \psi_{\mathsf{t}^{\boldsymbol{\mu}}\mathsf{t}^{\boldsymbol{\mu}}}.$$

This is probably the simplest example where different reduced expressions lead to different ψ-basis elements, but examples occur for almost all $\mathscr{R}_n^\Lambda$. This said, in view of Proposition 2.4.3, ψ_{st} is independent of the choice of reduced expressions for $d(\mathsf{s})$ and $d(\mathsf{t})$ whenever $e > n$ and $(\Lambda, \alpha_{i,n}) \le 1$, for all $i \in I$. The ψ-basis can be independent of the choice of reduced expressions even when $\mathscr{R}_n^\Lambda$ is not semisimple. For example, this is always the case when $e > n$ and $\ell = 2$ by [55, Appendix], yet these algebras are typically not semisimple. ◊

Using the theory of graded cellular algebras from §2.1, Theorem 3.2.4 allows us to construct a family $\{ S_F^{\boldsymbol{\lambda}} \mid \boldsymbol{\lambda} \in \mathcal{P}_n^\Lambda \}$ of graded Specht modules for $\mathscr{H}_n^\Lambda$. By [54, Corollary 5.10] the graded Specht modules attached to the ψ-basis coincide with those constructed by Brundan, Kleshchev and Wang [25]. When $e > n$ and $(\Lambda, \alpha_{i,n}) \leq 1$, for $i \in I$, it is not hard to show that these Specht modules coincide with those we constructed in Proposition 2.4.3 above. Similarly, for the nil-Hecke algebra considered in §2.5, the graded Specht module $S_F^{\boldsymbol{\lambda}}$, with $\boldsymbol{\lambda} = (1|1|\dots|1)$, is isomorphic to the graded module constructed in Proposition 2.5.2. Moreover, on forgetting the grading $S_F^{\boldsymbol{\lambda}}$ coincides exactly with the ungraded Specht module $\underline{S}_F^{\boldsymbol{\lambda}}$ constructed in §1.5, for $\boldsymbol{\lambda} \in \mathcal{P}_n^\Lambda$.

If $\boldsymbol{\lambda} \in \mathcal{P}_n^\Lambda$ the graded Specht module $S_F^{\boldsymbol{\lambda}}$ has basis $\{ \psi_{\mathsf{t}} \mid \mathsf{t} \in \operatorname{Std}(\boldsymbol{\lambda}) \}$, with $\deg \psi_{\mathsf{t}} = \deg \mathsf{t}$. The reader should be careful not to confuse $\psi_{\mathsf{t}} \in S_F^{\boldsymbol{\lambda}}$ with $\psi_{d(\mathsf{t})} \in \mathscr{R}_n^\Lambda$! By Theorem 3.2.4 we recover [22, Theorem 4.20]:

$$\dim_q \mathscr{H}_n^\Lambda = \sum_{(\mathsf{s},\mathsf{t}) \in \operatorname{Std}(\boldsymbol{\lambda})} q^{\deg \mathsf{s} + \deg \mathsf{t}} = \sum_{\boldsymbol{\lambda} \in \mathcal{P}_n^\Lambda} \big(\dim_q S_F^{\boldsymbol{\lambda}} \big)^2.$$

In essence, Theorem 3.2.4 is proved in much the same way that Brundan, Kleshchev and Wang [25] constructed a grading on the Specht modules: we proved that the transition matrix between the ψ-basis and the Murphy basis of Theorem 1.5.1 is triangular. In order to do this we needed the correct definition of the elements $y^{\boldsymbol{\mu}}$, which we discovered by first looking at the one dimensional two-sided ideals of $\mathscr{H}_n^\Lambda$ (which are necessarily homogeneous). We then used Brundan and Kleshchev's Graded Isomorphism Theorem 3.1.1, together with the seminormal forms (Theorem 1.6.7), to show that $e(\mathbf{i}^{\boldsymbol{\mu}})y^{\boldsymbol{\mu}} \neq 0$. This established that the basis of Theorem 3.2.4 is a graded cellular basis. Finally, the combinatorial results of [25] are used to determine the degree of ψ-basis elements.

Following the recipe in §2.1, for $\boldsymbol{\mu} \in \mathcal{P}_n^\Lambda$ define $D_F^{\boldsymbol{\mu}} = S_F^{\boldsymbol{\mu}} / \operatorname{rad} S_F^{\boldsymbol{\mu}}$, where $\operatorname{rad} S_F^{\boldsymbol{\mu}}$ is the radical of the homogeneous bilinear form on $S_F^{\boldsymbol{\mu}}$. This yields the classification of the graded irreducible $\mathscr{H}_n^\Lambda$-modules. The main point of the next result is that the labelling of the graded irreducible $\mathscr{H}_n^\Lambda$-modules agrees with Corollary 1.5.2.

Corollary 3.2.7 ([22, Theorem 5.13], [54, Corollary 5.11]). *Suppose that $\Lambda \in P^+$ and that $\mathcal{Z} = F$ is a field. Then $\{ D_F^{\boldsymbol{\mu}}\langle d \rangle \mid \boldsymbol{\mu} \in \mathcal{K}_n^\Lambda \text{ and } d \in \mathbb{Z} \}$ is a complete set of pairwise non-isomorphic graded $\mathscr{H}_n^\Lambda$-modules. Moreover, $(D_F^{\boldsymbol{\mu}})^{\circledast} \cong D_F^{\boldsymbol{\mu}}$ and $D_F^{\boldsymbol{\mu}}$ is absolutely irreducible, for all $\boldsymbol{\mu} \in \mathcal{K}_n^\Lambda$.*

The **graded decomposition numbers** are the Laurent polynomials

$$d^F_{\boldsymbol{\lambda\mu}}(q) = [S^{\boldsymbol{\lambda}}_F : D^{\boldsymbol{\mu}}_F]_q = \sum_{d\in\mathbb{Z}} [S^{\boldsymbol{\lambda}}_F : D^{\boldsymbol{\mu}}_F\langle d\rangle]\, q^d, \tag{3.2.8}$$

for $\boldsymbol{\lambda} \in \mathcal{P}^\Lambda_n$ and $\boldsymbol{\mu} \in \mathcal{K}^\Lambda_n$. Write $S^{\boldsymbol{\lambda}} = S^{\boldsymbol{\lambda}}_F$, $D^{\boldsymbol{\mu}} = D^{\boldsymbol{\mu}}_F$ and $d_{\boldsymbol{\lambda\mu}}(q) = d^F_{\boldsymbol{\lambda\mu}}(q)$ when F is understood. By definition, $d_{\boldsymbol{\lambda\mu}}(q) \in \mathbb{N}[q, q^{-1}]$ is a Laurent polynomial with non-negative coefficients. Let $\mathbf{d}_q = (d_{\boldsymbol{\lambda\mu}}(q))_{\boldsymbol{\lambda}\in\mathcal{P}^\Lambda_n, \boldsymbol{\mu}\in\mathcal{K}^\Lambda_n}$ be the **graded decomposition matrix** of $\mathscr{H}^\Lambda_n$.

The KLR algebra $\mathscr{R}_n$ is always $\mathbb{Z}$-free, however, it is not clear whether the same is true for the cyclotomic KLR algebra $\mathscr{R}^\Lambda_n$. To prove this you cannot use the Graded Isomorphism Theorem 3.1.1 because this result holds only over a field. Using extremely sophisticated diagram calculus, Li [92] proved the following.

Theorem 3.2.9 (Li [92]). *Suppose that $\Lambda \in P^+$. Then the quiver Hecke algebra $\mathscr{R}^\Lambda_n(\mathbb{Z})$ is free as a $\mathbb{Z}$-module of rank $\ell^n n!$. Moreover, $\mathscr{R}^\Lambda_n(\mathbb{Z})$ is a graded cellular algebra with graded cellular basis $\{\, \psi_{\mathsf{st}} \mid (\mathsf{s},\mathsf{t}) \in \operatorname{Std}^2(\mathcal{P}^\Lambda_n)\,\}$.*

Therefore, $\mathscr{R}^\Lambda_n$ is free over any commutative ring and any field is a splitting field for $\mathscr{R}^\Lambda_n$. Moreover, the graded Specht modules, together with their homogeneous bilinear forms, are defined over $\mathbb{Z}$. The integrality of the graded Specht modules can also be proved using Theorem 3.6.2 below.

The next result lists some important properties of the ψ-basis.

Proposition 3.2.10. *Suppose that $(\mathsf{s},\mathsf{t}) \in \operatorname{Std}^2(\mathcal{P}^\Lambda_n)$ and that $\mathcal{Z}$ is an integral domain. Then:*

a) [54, Lemma 5.2] *If $\mathbf{i},\mathbf{j} \in I^n$ then $\psi_{\mathsf{st}} = \delta_{\mathbf{i},\mathbf{i}^{\mathsf{s}}}\delta_{\mathbf{j},\mathbf{i}^{\mathsf{t}}} e(\mathbf{i})\psi_{\mathsf{st}} e(\mathbf{j})$.*

b) [55, Lemma 3.17] *Suppose that ψ_{st} and $\hat\psi_{\mathsf{st}}$ are defined using different reduced expressions for the permutations $d(\mathsf{s}), d(\mathsf{t}) \in \mathfrak{S}_n$. Then there exist $a_{\mathsf{uv}} \in \mathcal{Z}$ such that*

$$\hat\psi_{\mathsf{st}} = \psi_{\mathsf{st}} + \sum_{(\mathsf{u},\mathsf{v})\blacktriangleright(\mathsf{s},\mathsf{t})} a_{\mathsf{uv}}\psi_{\mathsf{uv}},$$

where $a_{\mathsf{uv}} \neq 0$ only if $\mathbf{i}^{\mathsf{u}} = \mathbf{i}^{\mathsf{s}}$, $\mathbf{i}^{\mathsf{v}} = \mathbf{i}^{\mathsf{t}}$ and $\deg\mathsf{u}+\deg\mathsf{v} = \deg\mathsf{s}+\deg\mathsf{t}$.

c) [56, Corollary 3.11] *If $1 \le r \le n$ then there exist $b_{\mathsf{uv}} \in \mathcal{Z}$ such that*

$$\psi_{\mathsf{st}} y_r = \sum_{(\mathsf{u},\mathsf{v})\blacktriangleright(\mathsf{s},\mathsf{t})} b_{\mathsf{uv}}\psi_{\mathsf{uv}},$$

where $b_{\mathsf{uv}} \neq 0$ only if $\mathbf{i}^{\mathsf{u}} = \mathbf{i}^{\mathsf{s}}$, $\mathbf{i}^{\mathsf{v}} = \mathbf{i}^{\mathsf{t}}$ and $\deg\mathsf{u} + \deg\mathsf{v} = \deg\mathsf{s} + \deg\mathsf{t} + 2$.

Part (a) follows quickly using the relations in Definition 2.2.1 and the definition of the ψ-basis. In contrast, parts (b) and (c) are proved by using

Theorem 3.1.1 to reduce to the seminormal basis. With part (c), it is fairly easy to show that $b_{\mathsf{uv}} \neq 0$ only if $\mathsf{u} \trianglerighteq \mathsf{s}$. The difficult part is showing that $b_{\mathsf{uv}} \neq 0$ only if $\mathsf{v} \trianglerighteq \mathsf{t}$. Again, this is done using seminormal bases.

Finally, we note that Theorem 3.2.9 implies that $e(\mathbf{i}) \neq 0$ in $\mathscr{R}_n^\Lambda$ if and only if $\mathbf{i} \in I_\Lambda^n = \{\, \mathbf{i}^{\mathsf{t}} \mid \mathsf{t} \in \operatorname{Std}(\mathcal{P}_n^\Lambda) \,\}$, generalizing Proposition 2.4.6. In fact, if F is a field and $\mathscr{H}_n^\Lambda(F) \cong \mathscr{R}_n^\Lambda(F)$ then it is shown in [54, Lemma 4.1] that the non-zero KLR idempotents are a complete set of primitive (central) idempotents in the Gelfand-Zetlin algebra $\mathscr{L}_n(F)$ and that $\mathscr{L}_n(F) = \langle y_1, \dots, y_n, e(\mathbf{i}) \mid \mathbf{i} \in I^n \rangle$. It follows that $\mathscr{L}_n(F)$ is a positively graded commutative algebra with one dimensional irreducible modules indexed by I_Λ^n, up to shift. It would be interesting to find a (homogeneous) basis of $\mathscr{L}_n(F)$. The author would also like to know whether $\mathscr{R}_n^\Lambda$ is projective as a graded $\mathscr{L}_n$-module.

3.3 *Blocks and dual Specht modules*

This section shows that the blocks of $\mathscr{H}_n^\Lambda$ are graded symmetric algebras and it sketches the proof of an analogous statement that relates the graded Specht modules and their graded duals.

Theorem 1.8.1 describes the block decomposition of $\mathscr{H}_n^\Lambda$ so, by Theorem 3.1.1, it gives the block decomposition of $\mathscr{R}_n^\Lambda$. As in (2.2.5), set

$$\mathscr{R}_\beta^\Lambda = \mathscr{R}_n^\Lambda e_\beta, \qquad \text{where } e_\beta = \sum_{\mathbf{i} \in I^\beta} e(\mathbf{i}).$$

It follows from Definition 2.2.1 that e_β is central in $\mathscr{R}_n^\Lambda$, so $\mathscr{R}_\beta^\Lambda = e_\beta \mathscr{R}_n^\Lambda e_\beta$ is a two-sided ideal of $\mathscr{R}_n^\Lambda$. Let $Q_n^+ = Q_n^+(\Lambda) = \{\, \beta \in Q^+ \mid e_\beta \neq 0 \,\}$ in $\mathscr{R}_n^\Lambda$. Similarly, let $\mathcal{P}_\beta^\Lambda = \{\, \boldsymbol{\lambda} \in \mathcal{P}_n^\Lambda \mid \mathbf{i}^{\boldsymbol{\lambda}} \in I^\beta \,\} = \{\, \boldsymbol{\lambda} \in \mathcal{P}_n^\Lambda \mid \beta^{\boldsymbol{\lambda}} = \beta \,\}$.

Combining Theorem 3.2.9, Theorem 3.1.1 and Corollary 1.8.2 we obtain the following.

Theorem 3.3.1. *Suppose that $\Lambda \in P^+$. Then $\mathscr{R}_n^\Lambda = \bigoplus_{\beta \in Q_n^+} \mathscr{R}_\beta^\Lambda$ is the decomposition of $\mathscr{R}_n^\Lambda$ into indecomposable two-sided ideals. Moreover, $\mathscr{R}_\beta^\Lambda$ is a graded cellular algebra with cellular basis $\{\, \psi_{\mathsf{st}} \mid (\mathsf{s}, \mathsf{t}) \in \operatorname{Std}^2(\mathcal{P}_\beta^\Lambda) \,\}$ and weight poset $\mathcal{P}_\beta^\Lambda$.*

By virtue of Theorem 3.2.9, the block decomposition of $\mathscr{R}_n^\Lambda$ holds over $\mathbb{Z}$, even though we cannot think about the blocks as linkage classes of simple modules in this case. Compare with Theorem 2.4.8 in the semisimple case.

Suppose that A is a graded $\mathcal{Z}$-algebra. Then A is a **graded symmetric algebra** if there exists a homogeneous non-degenerate trace form $\tau : A \longrightarrow \mathcal{Z}$, where $\mathcal{Z}$ is in degree zero. That is, $\tau(ab) = \tau(ba)$ and if $0 \neq a \in A$ then there exists $b \in A$ such that $\tau(ab) \neq 0$. The map τ is **homogeneous of degree** d if $\tau(a) \neq 0$ only if $\deg a = -d$.

Fix $\beta \in Q^+$. The **defect** of β is the non-negative integer

$$\operatorname{def}\beta = (\Lambda, \beta) - \frac{1}{2}(\beta, \beta) = \frac{1}{2}\Big((\Lambda, \Lambda) - (\Lambda - \beta, \Lambda - \beta)\Big).$$

Notice that $\operatorname{def}\beta = \operatorname{def}_\Lambda \beta$ depends on Λ. If $\boldsymbol{\lambda} \in \mathcal{P}_n^\Lambda$ set $\operatorname{def}\boldsymbol{\lambda} = \operatorname{def}\beta^{\boldsymbol{\lambda}}$ (see Corollary 1.8.2). If $\lambda \in \mathcal{P}_{1,n}$ is a partition then $\operatorname{def}\lambda$ is equal to its e-weight; see, for example, [36, Proposition 2.1] or the proof of [89, Lemma 7.6].

Definition 3.2.1, and the definition of defect, readily implies the following combinatorial relationships between degrees, codegrees and defects.

Lemma 3.3.2. *Suppose that* $\boldsymbol{\lambda} \in \mathcal{P}_n^\Lambda$.

a) [25, Lemma 3.11] *If* $A \in \operatorname{Add}_i(\boldsymbol{\lambda})$ *then* $d_A(\boldsymbol{\lambda}) + 1 + d^A(\boldsymbol{\lambda}) = d_i(\boldsymbol{\lambda})$ *and* $\operatorname{def}(\boldsymbol{\lambda}+A) = \operatorname{def}\boldsymbol{\lambda} + d_i(\boldsymbol{\lambda}) - 1$.

b) [25, Lemma 3.12] *If* $\mathsf{s} \in \operatorname{Std}(\boldsymbol{\lambda})$ *then* $\deg \mathsf{s} + \operatorname{codeg} \mathsf{s} = \operatorname{def}\boldsymbol{\lambda}$.

In Definition 3.2.2 we defined two sets of elements $\{\psi_{\mathsf{st}}\}$ and $\{\psi'_{\mathsf{st}}\}$ in $\mathscr{R}_n^\Lambda$. Just as there are two versions of the Murphy basis, $\{m_{\mathsf{st}}\}$ and $\{m'_{\mathsf{st}}\}$, that are built from the trivial and sign representations of $\mathscr{H}_n^\Lambda$ [106], respectively, there are two versions of the ψ-basis. By [54, Theorem 6.17], $\{\, \psi'_{\mathsf{st}} \mid (\mathsf{s},\mathsf{t}) \in \operatorname{Std}^2(\mathcal{P}_n^\Lambda)\,\}$ is a second graded cellular basis of $\mathscr{H}_n^\Lambda$ with weight poset $(\mathcal{P}_n^\Lambda, \trianglelefteq)$ and with $\deg \psi'_{\mathsf{st}} = \operatorname{codeg}\mathsf{s} + \operatorname{codeg}\mathsf{t}$. We warn the reader that we are following the conventions of [55], rather than the notation of [54]. See [55, Lemma 3.15 and Remark 3.12] for the translation.

The bases $\{\psi_{\mathsf{st}}\}$ and $\{\psi'_{\mathsf{uv}}\}$ of $\mathscr{R}_n^\Lambda$ are dual in the sense that if $(\mathsf{s},\mathsf{t}), (\mathsf{u},\mathsf{v}) \in \operatorname{Std}^2(\mathcal{P}_\beta^\Lambda)$ then, by [56, Theorem 6.17],

$$\psi_{\mathsf{st}}\psi'_{\mathsf{ts}} \neq 0 \quad\text{and}\quad \psi_{\mathsf{st}}\psi'_{\mathsf{uv}} \neq 0 \qquad \text{only if}\quad \mathbf{i}^{\mathsf{t}} = \mathbf{i}^{\mathsf{u}} \text{ and } \mathsf{u} \trianglerighteq \mathsf{t}. \tag{3.3.3}$$

Let τ be the usual non-degenerate trace form on $\mathscr{H}_n^\Lambda$ [20,100]. We can write $\tau = \sum_d \tau_d$, where τ_d is homogeneous of degree $d \in \mathbb{Z}$. Let $\tau_\beta = \tau_{-2\operatorname{def}\beta}$ be the homogeneous component of τ of degree $-2\operatorname{def}\beta$. By [56, Theorem 6.17], if $(\mathsf{s},\mathsf{t}) \in \operatorname{Std}^2(\mathcal{P}_n^\Lambda)$ then $\tau_\beta(\psi_{\mathsf{st}}\psi'_{\mathsf{ts}}) \neq 0$, so (3.3.3) implies the following.

Theorem 3.3.4 (Hu-Mathas [54, Corollary 6.18]**).** *Let* $\beta \in Q_n^+$. *Then* $\mathscr{R}_\beta^\Lambda$ *a graded symmetric algebra with homogeneous trace form* τ_β *of degree* $-2\operatorname{def}\beta$.

It would be better to have an intrinsic definition of τ_β for $\mathscr{R}_n^\Lambda(\mathbb{Z})$. Webster [134, Remark 2.27] has given a diagrammatic description of a trace form on an arbitrary cyclotomic KLR algebra. It is unclear to the author how these two forms on $\mathscr{R}_n^\Lambda$ are related.

The ψ'-basis is a graded cellular basis of $\mathscr{H}_n^\Lambda$ so it defines a collection of graded cell modules. For $\boldsymbol{\lambda} \in \mathcal{P}_\beta^\Lambda$, the **dual graded Specht module** $S_{\boldsymbol{\lambda}}$ is the corresponding graded cell module determined by the ψ'-basis. The dual Specht module $S_{\boldsymbol{\lambda}}$ has basis $\{\, \psi'_{\mathsf{t}} \mid \mathsf{t} \in \operatorname{Std}(\boldsymbol{\lambda}) \,\}$, with $\deg \psi'_{\mathsf{t}} = \operatorname{codeg} \mathsf{t}$, and

$$\dim_q S_{\boldsymbol{\lambda}} = \sum_{\mathsf{t} \in \operatorname{Std}(\boldsymbol{\lambda})} q^{\operatorname{codeg} \mathsf{t}}.$$

We can identify $S_{\boldsymbol{\lambda}}\langle \operatorname{codeg} \mathsf{t}_{\boldsymbol{\lambda}} \rangle$ with $(\psi'_{\mathsf{t}_{\boldsymbol{\lambda}}\mathsf{t}_{\boldsymbol{\lambda}}} + \mathscr{H}_n^{\prime \lhd \boldsymbol{\lambda}})\mathscr{H}_n^\Lambda$, where $\mathscr{H}_n^{\prime \lhd \boldsymbol{\lambda}}$ is the two-sided ideal of $\mathscr{H}_n^\Lambda$ spanned by ψ'_{st} where $(\mathsf{s}, \mathsf{t}) \in \operatorname{Std}^2(\boldsymbol{\mu})$ for some multipartition $\boldsymbol{\mu}$ such that $\boldsymbol{\lambda} \rhd \boldsymbol{\mu}$. Similarly, we can identify $S^{\boldsymbol{\lambda}}\langle \deg \mathsf{t}^{\boldsymbol{\lambda}} \rangle$ with $(\psi_{\mathsf{t}^{\boldsymbol{\lambda}}\mathsf{t}^{\boldsymbol{\lambda}}} + \mathscr{H}_n^{\rhd \boldsymbol{\lambda}})\mathscr{H}_n^\Lambda$. By (3.3.3) there is a non-degenerate pairing

$$\{\ ,\ \} : S^{\boldsymbol{\lambda}}\langle \deg \mathsf{t}^{\boldsymbol{\lambda}} \rangle \times S_{\boldsymbol{\lambda}}\langle \operatorname{codeg} \mathsf{t}_{\boldsymbol{\lambda}} \rangle \longrightarrow \mathbb{Z}$$

given by $\{a + \mathscr{H}_n^{\rhd \boldsymbol{\lambda}}, b + \mathscr{H}_n^{\prime \lhd \boldsymbol{\lambda}}\} = \tau_\beta(ab^\star)$. Hence, Lemma 3.3.2 implies:

Corollary 3.3.5 (Hu-Mathas [54, Proposition 6.19]**).** *Suppose that* $\boldsymbol{\lambda} \in \mathcal{P}_n^\Lambda$. *Then* $S^{\boldsymbol{\lambda}} \cong S_{\boldsymbol{\lambda}}^{\circledast}\langle \operatorname{def} \boldsymbol{\lambda} \rangle$ *and* $S_{\boldsymbol{\lambda}} = (S^{\boldsymbol{\lambda}})^{\circledast}\langle \operatorname{def} \boldsymbol{\lambda} \rangle$.

This result holds for the Specht modules defined over $\mathbb{Z}$ by Theorem 3.2.9 or by [81, Theorem 7.25].

There is an interesting byproduct of the proof of Corollary 3.3.5. In the ungraded setting the Specht module $\underline{S}^{\boldsymbol{\lambda}}$ is isomorphic to the submodule of $\mathscr{H}_n^\Lambda$ generated by an element $m_{\boldsymbol{\lambda}} T_{w_{\boldsymbol{\lambda}}} m'_{\boldsymbol{\lambda}}$; see [33, Definition 2.1 and Theorem 2.9]. By [54, Corollary 6.21], $m_{\boldsymbol{\lambda}} T_{w_{\boldsymbol{\lambda}}} m'_{\boldsymbol{\lambda}}$ is homogeneous. In fact, $\psi_{\mathsf{t}^{\boldsymbol{\lambda}}\mathsf{t}^{\boldsymbol{\lambda}}} \psi_{w_{\boldsymbol{\lambda}}} \psi'_{\mathsf{t}_{\boldsymbol{\lambda}}\mathsf{t}_{\boldsymbol{\lambda}}} = m_{\boldsymbol{\lambda}} T_{w_{\boldsymbol{\lambda}}} m'_{\boldsymbol{\lambda}}$ and $\psi_{\mathsf{t}^{\boldsymbol{\lambda}}\mathsf{t}^{\boldsymbol{\lambda}}} \psi_{w_{\boldsymbol{\lambda}}} \psi'_{\mathsf{t}_{\boldsymbol{\lambda}}\mathsf{t}_{\boldsymbol{\lambda}}} \mathscr{R}_n^\Lambda \cong S^{\boldsymbol{\lambda}}\langle \operatorname{def} \boldsymbol{\lambda} + \operatorname{codeg} \mathsf{t}_{\boldsymbol{\lambda}} \rangle$.

3.4 *Induction and restriction*

The cyclotomic Hecke algebra $\mathscr{H}_n^\Lambda$ is naturally a subalgebra of $\mathscr{H}_{n+1}^\Lambda$, and $\mathscr{H}_{n+1}^\Lambda$ is free as a $\mathscr{H}_n^\Lambda$-module by (1.1.2). This gives rise to the usual induction and restriction functors. These functors can be decomposed into the "classical" i-induction and i-restriction functors, for $i \in I$, by projecting onto the blocks of these two algebras. As we will see, these functors are implicitly built into the graded setting.

Recall that $I = \mathbb{Z}/e\mathbb{Z}$ and $\Lambda \in P^+$. For each $i \in I$ define

$$e_{n,i} = \sum_{\mathbf{j} \in I^n} e(\mathbf{j} \vee i) \in \mathscr{R}_{n+1}^\Lambda.$$

The relations for $\mathscr{R}^\Lambda_{n+1}$ in Definition 2.2.1 imply that $e_{n,i}$ is an idempotent and that $\sum_{i\in I} e_{n,i} = \sum_{\mathbf{i}\in I^{n+1}} e(\mathbf{i})$ is the identity element of $\mathscr{R}^\Lambda_{n+1}$.

Let $\mathrm{Rep}(\mathscr{R}^\Lambda_n)$, and $\mathrm{Rep}(\mathscr{R}^\Lambda_\beta)$ for $\beta \in Q^+$, be the category of finite dimensional (graded) $\mathscr{R}^\Lambda_n$-modules, respectively, $\mathscr{R}^\Lambda_\beta$-modules. Similarly, let $\mathrm{Proj}(\mathscr{R}^\Lambda_n)$ and $\mathrm{Proj}(\mathscr{R}^\Lambda_\beta)$ be the categories of finitely generated projective modules for these algebras.

Lemma 3.4.1. *Suppose that $i \in I$ and that $\mathcal{Z}$ is an integral domain. Then there is a (non-unital) embedding of graded algebras $\mathscr{R}^\Lambda_n \hookrightarrow \mathscr{R}^\Lambda_{n+1}$ given by*

$$e(\mathbf{j}) \mapsto e(\mathbf{j} \vee i), \quad y_r \mapsto e_{n,i} y_r \quad \text{and} \quad \psi_s \mapsto e_{n,i}\psi_s,$$

for $\mathbf{j} \in I^n$, $1 \le r \le n$ and $1 \le s < n$. This map induces an exact functor

$$i\text{-Ind} : \mathrm{Rep}(\mathscr{R}^\Lambda_n) \longrightarrow \mathrm{Rep}(\mathscr{R}^\Lambda_{n+1}); M \mapsto M \otimes_{\mathscr{R}^\Lambda_n} e_{n,i}\mathscr{R}^\Lambda_{n+1}.$$

Moreover, $\mathrm{Ind} = \bigoplus_{i\in I} i\text{-Ind}$ is the graded induction functor from $\mathrm{Rep}(\mathscr{R}^\Lambda_n)$ to $\mathrm{Rep}(\mathscr{R}^\Lambda_{n+1})$.

Proof. The images of the homogeneous generators of $\mathscr{R}^\Lambda_n$ under this embedding commute with $e_{n,i}$, so this map defines a non-unital degree preserving homomorphism from $\mathscr{R}^\Lambda_n$ to $\mathscr{R}^\Lambda_{n+1}$. This map is an embedding by Theorem 3.2.9. The remaining claims follow because, by definition, $e_{n,i}$ is an idempotent and $\sum_{i\in I} e_{n,i}$ is the identity element of $\mathscr{R}^\Lambda_{n+1}$. □

The i**-induction functor** i-Ind functor is obviously a left adjoint to the i**-restriction** functor i-Res, which sends a $\mathscr{R}^\Lambda_{n+1}$-module M to

$$i\text{-Res}\, M = M e_{n,i} \cong \mathcal{H}om_{\mathscr{R}^\Lambda_n}(e_{n,i}\mathscr{R}^\Lambda_n, M).$$

A much harder fact is that these functors are two-sided adjoints.

Theorem 3.4.2 (Kashiwara [71, Theorem 3.5]**).** *Suppose $i \in I$. Then $(i\text{-Res}, i\text{-Ind})$ is a biadjoint pair.*

Kashiwara proves this theorem by constructing explicit homogeneous adjunctions. He does this for the i-induction and i-restriction functors for any cyclotomic quiver Hecke algebra defined by a symmetrizable Cartan matrix. As we do not need this result we feel justified in stating it now, even though its proof builds upon Kang and Kashiwara's result that the cyclotomic quiver Hecke algebras of arbitrary type categorify the integrable highest weight modules of the corresponding quantum group [67]; compare with Proposition 3.5.12 and Corollary 3.5.27 below. This biadjointness property is also a consequence of Rouquier's Kac-Moody categorification axioms [121, Theorem 5.16]. Theorem 3.4.2 was conjectured by Khovanov-Lauda [74].

Recall from (2.1.4) that $\circledast$ defines a graded duality on $\operatorname{Rep}(\mathscr{R}_n^\Lambda)$. Similarly, define $\#$ to be the graded functor given by

$$(3.4.3)\qquad M^{\#} = \mathcal{H}om_{\mathscr{R}_n^\Lambda}(M, \mathscr{R}_n^\Lambda), \qquad \text{for } M \in \operatorname{Rep}(\mathscr{R}_n^\Lambda),$$

where the action of $\mathscr{R}_n^\Lambda$ on $M^{\#}$ is given by

$$(f \cdot h)(m) = h^\star f(m), \qquad \text{for } f \in M^{\#},\ h \in \mathscr{R}_n^\Lambda \text{ and } m \in M.$$

We consider $\circledast$ and $\#$ as endofunctors of $\operatorname{Rep}(\mathscr{R}_n^\Lambda) = \bigoplus_\beta \operatorname{Rep}(\mathscr{R}_\beta^\Lambda)$ and $\operatorname{Proj}(\mathscr{R}_n^\Lambda) = \bigoplus_\beta \operatorname{Proj}(\mathscr{R}_\beta^\Lambda)$. As noted in [22, Remark 4.7], Theorem 3.3.4 implies that these two functors agree up to shift.

Lemma 3.4.4. *As endofunctors of* $\operatorname{Rep}(\mathscr{R}_\beta^\Lambda)$, *there is an isomorphism of functors* $\# \cong \langle 2 \operatorname{def} \beta\rangle \circ \circledast$.

Proof. By Theorem 3.3.4, $\mathscr{R}_\beta^\Lambda \cong (\mathscr{R}_\beta^\Lambda)^\circledast \langle 2 \operatorname{def}\beta\rangle$. If $M \in \operatorname{Rep}(\mathscr{R}_\beta^\Lambda)$ then

$$\begin{aligned} M^{\#} &= \mathcal{H}om_{\mathscr{R}_\beta^\Lambda}(M, \mathscr{R}_\beta^\Lambda) = \mathcal{H}om_{\mathscr{R}_\beta^\Lambda}\big(M, (\mathscr{R}_\beta^\Lambda)^\circledast \langle 2 \operatorname{def}\beta\rangle\big) \\ &\cong \mathcal{H}om_{\mathscr{R}_\beta^\Lambda}\big(M, \mathcal{H}om_{\mathcal{Z}}(\mathscr{R}_\beta^\Lambda, \mathcal{Z})\big)\langle 2 \operatorname{def}\beta\rangle \\ &\cong \mathcal{H}om_{\mathcal{Z}}\big(M \otimes_{\mathscr{R}_\beta^\Lambda} \mathscr{R}_\beta^\Lambda, \mathcal{Z}\big)\langle 2 \operatorname{def}\beta\rangle \\ &\cong M^\circledast \langle 2 \operatorname{def}\beta\rangle, \end{aligned}$$

where the third isomorphism is the standard adjointness of tensor and hom. As all of these isomorphisms are functorial, the lemma follows. □

As M is finite dimensional, $(M^\circledast)^\circledast \cong M$ for all $M \in \operatorname{Rep}(\mathscr{H}_n^\Lambda)$. Hence, $(M^{\#})^{\#} \cong M$ by Lemma 3.4.4. Therefore, $\circledast$ and $\#$ define self-dual equivalences on the module categories $\operatorname{Rep}(\mathscr{R}_n^\Lambda)$ and $\operatorname{Proj}(\mathscr{R}_n^\Lambda)$.

Proposition 3.4.5. *Suppose that* $\beta \in Q^+$ *and* $i \in I$. *Then there are functorial isomorphisms*

$$\begin{aligned} \circledast \circ i\text{-Res} &\cong i\text{-Res} \circ \circledast : \operatorname{Rep}(\mathscr{R}_{n+1}^\Lambda) \longrightarrow \operatorname{Rep}(\mathscr{R}_n^\Lambda), \\ \# \circ i\text{-Ind} &\cong i\text{-Ind} \circ \# : \operatorname{Proj}(\mathscr{R}_n^\Lambda) \longrightarrow \operatorname{Proj}(\mathscr{R}_{n+1}^\Lambda). \end{aligned}$$

Proof. The isomorphism $\circledast \circ i\text{-Res} \cong i\text{-Res} \circ \circledast$ is immediate from the definitions. For the second isomorphism, recall that if $P \in \operatorname{Proj}(\mathscr{R}_\beta^\Lambda)$ then $\mathcal{H}om_{\mathscr{R}_n^\Lambda}(P, M) \cong \mathcal{H}om_{\mathscr{R}_n^\Lambda}(M, \mathscr{R}_n^\Lambda) \otimes_{\mathscr{R}_n^\Lambda} M$, for any $\mathscr{R}_n^\Lambda$-module M. Now,

$$(e_{n,i}\mathscr{R}_{n+1}^\Lambda)^{\#} = \mathcal{H}om_{\mathscr{R}_{n+1}^\Lambda}(e_{n,i}\mathscr{R}_{n+1}^\Lambda, \mathscr{R}_{n+1}^\Lambda) \cong e_{n,i}\mathscr{R}_{n+1}^\Lambda,$$

the last isomorphism following because $e^\star_{n,i} = e_{n,i}$. Therefore,

$$\begin{aligned} i\text{-Ind}(P^\#) &= \mathcal{H}om_{\mathscr{R}^\Lambda_n}(P, \mathscr{R}^\Lambda_n) \otimes_{\mathscr{R}^\Lambda_n} e_{n,i}\mathscr{R}^\Lambda_{n+1} \cong \mathcal{H}om_{\mathscr{R}^\Lambda_n}(P, e_{n,i}\mathscr{R}^\Lambda_{n+1}) \\ &\cong \mathcal{H}om_{\mathscr{R}^\Lambda_n}\left(P, \mathcal{H}om_{\mathscr{R}^\Lambda_{n+1}}(e_{n,1}\mathscr{R}^\Lambda_{n+1}, \mathscr{R}^\Lambda_{n+1})\right) \\ &\cong \mathcal{H}om_{\mathscr{R}^\Lambda_{n+1}}(P \otimes_{\mathscr{R}^\Lambda_n} e_{n,i}\mathscr{R}^\Lambda_{n+1}, \mathscr{R}^\Lambda_{n+1}) \\ &\cong (i\text{-Ind}\, P)^\#, \end{aligned}$$

where the second last isomorphism is the usual tensor-hom adjointness. □

It follows from Proposition 3.4.5 and Lemma 3.4.4 that the functors ⊛ and i-Ind, and # and i-Res, commute only up to shift. This difference in degree shift is what makes Lemma 3.5.13 work below.

We next describe the effect of the i-induction and i-restriction functors on the graded Specht modules, for $i \in I$. This result generalizes the well-known (ungraded) branching rules for the symmetric group [59, Example 17.16] and the cyclotomic Hecke algebras [12, 109, 125].

Recall the integers $d^A(\boldsymbol{\lambda})$ and $d_A(\boldsymbol{\lambda})$ from Definition 3.2.1.

Theorem 3.4.6. *Suppose that $\mathcal{Z}$ is an integral domain and $\boldsymbol{\lambda} \in \mathcal{P}^\Lambda_n$.*

a) [56, Main theorem] *Let $A_1 < A_2 \cdots < A_z$ be the addable i-nodes of $\boldsymbol{\lambda}$. Then i-Ind $S^{\boldsymbol{\lambda}}$ has a graded Specht filtration*

$$0 = I_0 \subset I_1 \subset \cdots \subset I_z = i\text{-Ind}\, S^{\boldsymbol{\lambda}},$$

such that $I_j/I_{j-1} \cong S^{\boldsymbol{\lambda}+A_j}\langle d_{A_j}(\boldsymbol{\lambda})\rangle$,

b) [25, Theorem 4.11] *Let $B_1 > B_2 > \cdots > B_y$ be the removable i-nodes of $\boldsymbol{\lambda}$. Then i-Res $S^{\boldsymbol{\lambda}}$ has a graded Specht filtration*

$$0 = R_0 \subset R_1 \subset \cdots \subset R_y = i\text{-Res}\, S^{\boldsymbol{\lambda}},$$

such that $R_j/R_{j-1} \cong S^{\boldsymbol{\lambda}-B_j}\langle d_{B_j}(\boldsymbol{\lambda})\rangle$, for $1 \le j \le y$.

c) [56, Corollary 4.7] *Let $A_1 > A_2 > \cdots > A_z$ be the addable i-nodes of $\boldsymbol{\lambda}$. Then i-Ind $S_{\boldsymbol{\lambda}}$ has a graded Specht filtration*

$$0 = I_0 \subset I_1 \subset \cdots \subset I_z = i\text{-Ind}\, S_{\boldsymbol{\lambda}},$$

such that $I_j/I_{j-1} \cong S_{\boldsymbol{\lambda}+A_j}\langle d^{A_j}(\boldsymbol{\lambda})\rangle$, for $1 \le j \le z$.

d) Let $B_1 < B_2 < \cdots < B_y$ be the removable i-nodes of $\boldsymbol{\lambda}$. Then i-Res $S^{\boldsymbol{\lambda}}$ has a graded Specht filtration

$$0 = R_0 \subset R_1 \subset \cdots \subset R_y = i\text{-Res}\, S_{\boldsymbol{\lambda}},$$

such that $R_j/R_{j-1} \cong S_{\boldsymbol{\lambda}-B_j}\langle d^{B_j}(\boldsymbol{\lambda})\rangle$, for $1 \le j \le y$.

Observe that parts (a) and (c), and parts (b) and (d), are equivalent by Corollary 3.3.5 (and Lemma 3.3.2).

Part (b) is proved using the fact that the action of $\mathscr{H}_n^\Lambda$ on the ψ-basis is compatible with restriction. Part (a), which was conjectured by Brundan, Kleshchev and Wang [25, Remark 4.12], is proved by extending elegant ideas of Ryom-Hansen [125] to the graded setting using [54].

3.5 *Grading Ariki's Categorification Theorem*

The aim of this section is to prove the Ariki-Brundan-Kleshchev Categorification Theorem [3] that connects the canonical bases of $U_q(\widehat{\mathfrak{sl}}_e)$-modules with the simple and projective indecomposable $\mathscr{R}_n^\Lambda$-modules in characteristic zero. Our argument runs parallel to Brundan and Kleshchev's with the key difference being that we use the representation theory of $\mathscr{H}_n^\Lambda$, and in particular the graded branching rules, to construct a *bar involution* on the Fock space. In this way we are able to show that the canonical basis is categorified by the basis of simple $\mathscr{H}_n^\Lambda$-modules if and only if the graded decomposition numbers are polynomials. As a consequence, Ariki's categorification theorem [3] lifts to the graded setting.

Throughout this section we assume that the Hecke algebra $\mathscr{H}_n^\Lambda$ is defined over a field $\mathbb{F}$. In the end we will assume that $\mathbb{F}$ is a field of characteristic zero, however, almost all of the results in this section hold over an arbitrary field. We delay introducing the quantum group $U_q(\widehat{\mathfrak{sl}}_e)$ until we actually need it because we want to emphasize the role that the quantum group is playing in the representation theory of $\mathscr{H}_n^\Lambda$.

For the time being fix an integer $n \geq 0$. Let $\mathcal{A} = \mathbb{Z}[q, q^{-1}]$ be the ring of Laurent polynomials in q over $\mathbb{Z}$.

Let $[\operatorname{Rep}(\mathscr{H}_n^\Lambda)]$ and $[\operatorname{Proj}(\mathscr{H}_n^\Lambda)]$ be the Grothendieck groups of the categories $\operatorname{Rep}(\mathscr{H}_n^\Lambda)$ and $\operatorname{Proj}(\mathscr{H}_n^\Lambda)$, respectively. If M is a finitely generated $\mathscr{H}_n^\Lambda$-module let $[M]$ be its image in $[\operatorname{Rep}(\mathscr{H}_n^\Lambda)]$. Abusing notation slightly, if M is projective we also let $[M]$ be its image in $[\operatorname{Proj}(\mathscr{H}_n^\Lambda)]$. Consider $[\operatorname{Rep}(\mathscr{H}_n^\Lambda)]$ and $[\operatorname{Proj}(\mathscr{H}_n^\Lambda)]$ as $\mathcal{A}$-modules by letting q act as the grading shift functor: $[M\langle d\rangle] = q^d[M]$, for $d \in \mathbb{Z}$.

Definition 3.5.1. Suppose that $\boldsymbol{\mu} \in \mathcal{K}_n^\Lambda$. Let $Y^{\boldsymbol{\mu}}$ be the projective cover of $D^{\boldsymbol{\mu}}$ in $\operatorname{Rep}(\mathscr{H}_n^\Lambda)$.

Importantly, the module $Y^{\boldsymbol{\mu}}$ is graded. Since $Y^{\boldsymbol{\mu}}$ is indecomposable, the grading on $Y^{\boldsymbol{\mu}}$ is uniquely determined by the surjection $Y^{\boldsymbol{\mu}} \twoheadrightarrow D^{\boldsymbol{\mu}}$, for $\boldsymbol{\mu} \in \mathcal{K}_n^\Lambda$. We use the notation $Y^{\boldsymbol{\mu}}$ because these modules are special cases of the graded lifts of the *Young modules* constructed in [105]; see [55, §5.1] and [99, §2.6]. (The symbol $P^{\boldsymbol{\mu}}$ is usually reserved for the projective indecomposable modules of the cyclotomic Schur algebras [20, 31, 55, 128].)

By definition, the Grothendieck groups $[\mathrm{Rep}(\mathscr{H}_n^\Lambda)]$ and $[\mathrm{Proj}(\mathscr{H}_n^\Lambda)]$ are free $\mathcal{A}$-modules that come equipped with distinguished bases:

$$[\mathrm{Rep}(\mathscr{H}_n^\Lambda)] = \bigoplus_{\boldsymbol{\mu}\in\mathcal{K}_n^\Lambda} \mathcal{A}[D^{\boldsymbol{\mu}}] \quad \text{and} \quad [\mathrm{Proj}(\mathscr{H}_n^\Lambda)] = \bigoplus_{\boldsymbol{\mu}\in\mathcal{K}_n^\Lambda} \mathcal{A}[Y^{\boldsymbol{\mu}}],$$

respectively. Recall from (3.2.8) that $\mathbf{d}_q = \big(\mathrm{d}_{\boldsymbol{\lambda}\boldsymbol{\mu}}(q)\big)$ is the graded decomposition matrix of $\mathscr{H}_n^\Lambda$. If $\boldsymbol{\lambda} \in \mathcal{P}_n^\Lambda$ and $\boldsymbol{\mu} \in \mathcal{K}_n^\Lambda$ then in $[\mathrm{Rep}(\mathscr{H}_n^\Lambda)]$,

$$[S^{\boldsymbol{\lambda}}] = \sum_{\substack{\boldsymbol{\tau}\in\mathcal{K}_n^\Lambda\\ \boldsymbol{\lambda}\trianglerighteq\boldsymbol{\tau}}} d_{\boldsymbol{\lambda}\boldsymbol{\tau}}(q)[D^{\boldsymbol{\tau}}] \quad \text{and} \quad [Y^{\boldsymbol{\mu}}] = \sum_{\substack{\boldsymbol{\sigma}\in\mathcal{P}_n^\Lambda\\ \boldsymbol{\sigma}\trianglerighteq\boldsymbol{\mu}}} d_{\boldsymbol{\sigma}\boldsymbol{\mu}}(q)[S^{\boldsymbol{\sigma}}],$$

where the second formula comes from Corollary 2.1.6. By Theorem 2.1.5(c), the submatrix $\mathbf{d}_q^{\mathcal{K}} = \big(d_{\boldsymbol{\lambda}\boldsymbol{\mu}}(q)\big)_{\boldsymbol{\lambda},\boldsymbol{\mu}\in\mathcal{K}_n^\Lambda}$ of the graded decomposition matrix $\mathbf{d}_q$ is invertible over $\mathcal{A}$ with inverse

$$\mathbf{e}_q^{\mathcal{K}} = (\mathbf{d}_q^{\mathcal{K}})^{-1} = \big(e_{\boldsymbol{\lambda}\boldsymbol{\mu}}(-q)\big)_{\boldsymbol{\lambda},\boldsymbol{\mu}\in\mathcal{K}_n^\Lambda}.$$

(The reason why we consider $e_{\boldsymbol{\lambda}\boldsymbol{\mu}}(-q)$ as a Laurent polynomial in $-q$ is explained after Corollary 3.5.27 below.) Hence, if $\boldsymbol{\lambda} \in \mathcal{K}_n^\Lambda$ then

$$[D^{\boldsymbol{\lambda}}] = \sum_{\boldsymbol{\mu}\in\mathcal{K}_n^\Lambda} e_{\boldsymbol{\lambda}\boldsymbol{\mu}}(-q)[S^{\boldsymbol{\mu}}].$$

Consequently, $\{\,[S^{\boldsymbol{\mu}}] \mid \boldsymbol{\mu} \in \mathcal{K}_n^\Lambda\,\}$ is a second $\mathcal{A}$-basis of $[\mathrm{Rep}(\mathscr{H}_n^\Lambda)]$.

The set of projective indecomposable $\mathscr{H}_n^\Lambda$-modules $\{[Y^{\boldsymbol{\mu}}]\}$ is the only natural basis of the split Grothendieck group $[\mathrm{Proj}(\mathscr{H}_n^\Lambda)]$. Somewhat artificially, but motivated by the formulas above, for $\boldsymbol{\mu} \in \mathcal{K}^\Lambda$ define

$$X_{\boldsymbol{\mu}} = \sum_{\boldsymbol{\lambda}\in\mathcal{K}_n^\Lambda} e_{\boldsymbol{\lambda}\boldsymbol{\mu}}(-q)[Y^{\boldsymbol{\lambda}}] \in [\mathrm{Proj}(\mathscr{H}_n^\Lambda)].$$

Then $\{\,X_{\boldsymbol{\mu}} \mid \boldsymbol{\mu} \in \mathcal{K}^\Lambda\,\}$ is an $\mathcal{A}$-basis of $[\mathrm{Proj}(\mathscr{H}_n^\Lambda)]$. We will use the bases $\{\,[S^{\boldsymbol{\mu}}] \mid \boldsymbol{\mu} \in \mathcal{K}^\Lambda\,\}$ and $\{\,X_{\boldsymbol{\mu}} \mid \mathcal{K}^\Lambda\,\}$ of $[\mathrm{Rep}(\mathscr{H}_n^\Lambda)]$ and $[\mathrm{Proj}(\mathscr{H}_n^\Lambda)]$, respectively, to construct new distinguished bases of these Grothendieck groups.

The **bar involution** on $\mathcal{A} = \mathbb{Z}[q, q^{-1}]$ is the unique $\mathbb{Z}$-linear map such that $\overline{q^d} = q^{-d}$, for $d \in \mathbb{Z}$. In particular, $\mathit{dim}_q\, M^{\circledast} = \overline{\mathit{dim}_q\, M}$, for any $\mathscr{R}_n^\Lambda$-module M. A **semilinear** map of $\mathcal{A}$-modules is a $\mathbb{Z}$-linear map $\theta\colon M \longrightarrow N$ such that $\theta(f(q)m) = \overline{f(q)}\theta(m)$, for all $f(q) \in \mathcal{A}$ and $m \in M$.

A **sesquilinear** map $f\colon M \times N \longrightarrow \mathcal{A}$, where M and N are $\mathcal{A}$-modules, is a function that is semilinear in the first variable and linear in the second. Let $\langle\ ,\ \rangle\colon [\mathrm{Proj}(\mathscr{H}_n^\Lambda)] \times [\mathrm{Rep}(\mathscr{H}_n^\Lambda)] \longrightarrow \mathcal{A}$ be the sesquilinear pairing

(3.5.2) $$\langle [P], [M] \rangle = \mathit{dim}_q\, \mathscr{H}\!om_{\mathscr{H}_n^\Lambda}(P, M),$$

for $P \in \mathrm{Proj}(\mathscr{H}_n^\Lambda)$ and $M \in \mathrm{Rep}(\mathscr{H}_n^\Lambda)$. This pairing is naturally **sesquilinear** because $\mathcal{H}om_{\mathscr{H}_n^\Lambda}(P\langle k\rangle, M) \cong \mathcal{H}om_{\mathscr{H}_n^\Lambda}(P, M\langle -k\rangle)$, for any $k \in \mathbb{Z}$.

The functors $\circledast$ and $\#$, of (2.1.4) and (3.4.3), induce semilinear automorphisms of the Grothendieck groups $[\mathrm{Rep}(\mathscr{H}_n^\Lambda)]$ and $[\mathrm{Proj}(\mathscr{H}_n^\Lambda)]$:

$$[P]^\# = [P^\#], \quad \text{and} \quad [M]^\circledast = [M^\circledast]$$

for $M \in \mathrm{Rep}(\mathscr{H}_n^\Lambda)$ and $P \in \mathrm{Proj}(\mathscr{H}_n^\Lambda)$. The next result is fundamental.

Lemma 3.5.3. *Suppose that* $[P] \in [\mathrm{Proj}_{\mathcal{A}}^\Lambda]$ *and* $[M] \in [\mathrm{Rep}_{\mathcal{A}}^\Lambda]$. *Then*

$$\langle [P], [M]^\circledast\rangle = \overline{\langle [P]^\#, [M]\rangle}.$$

Proof. Applying the definitions, and tensor-hom adjointness,

$$\begin{aligned}
\langle [P], [M]^\circledast\rangle &= \dim_q \mathcal{H}om_{\mathscr{R}_n^\Lambda}(P, M^\circledast) = \dim_q \mathcal{H}om_{\mathscr{R}_n^\Lambda}\big(P, \mathcal{H}om_{\mathscr{R}_n^\Lambda}(M, \mathbb{F})\big)\\
&= \dim_q \mathcal{H}om_{\mathscr{R}_n^\Lambda}(P \otimes_{\mathscr{R}_n^\Lambda} M, \mathbb{F}) = \dim_q (P \otimes_{\mathscr{R}_n^\Lambda} M)^\circledast\\
&= \overline{\dim_q P \otimes_{\mathscr{R}_n^\Lambda} M} = \overline{\dim_q \mathcal{H}om_{\mathscr{R}_n^\Lambda}(P^\#, \mathscr{R}_n^\Lambda) \otimes_{\mathscr{R}_n^\Lambda} M}\\
&= \overline{\dim_q \mathcal{H}om_{\mathscr{R}_n^\Lambda}(P^\#, M)} = \overline{\langle [P]^\#, [M]\rangle}.
\end{aligned}$$

For the second last line, note that $\mathcal{H}om_{\mathscr{R}_n^\Lambda}(Q, M) \cong \mathcal{H}om_{\mathscr{R}_n^\Lambda}(Q, \mathscr{R}_n^\Lambda) \otimes_{\mathscr{R}_n^\Lambda} M$ whenever Q is projective. □

Lemma 3.5.4. *Suppose that* $\boldsymbol{\lambda}, \boldsymbol{\mu} \in \mathcal{K}_n^\Lambda$. *Then*

$$\langle [Y^{\boldsymbol{\lambda}}], [D^{\boldsymbol{\mu}}]\rangle = \delta_{\boldsymbol{\lambda}\boldsymbol{\mu}} = \langle X_{\boldsymbol{\lambda}}, [S^{\boldsymbol{\mu}}]^\circledast\rangle.$$

Proof. The first equality is immediate from the definition of the sesquilinear form $\langle\ ,\ \rangle$ because $Y^{\boldsymbol{\lambda}}$ is the projective cover of $D^{\boldsymbol{\lambda}}$, for $\boldsymbol{\lambda} \in \mathcal{K}_n^\Lambda$. For the second equality, using the fact that $\circledast$ is semilinear,

$$\begin{aligned}
\langle X_{\boldsymbol{\lambda}}, [S^{\boldsymbol{\mu}}]^\circledast\rangle &= \sum_{\boldsymbol{\sigma} \in \mathcal{K}_n^\Lambda} \overline{e_{\boldsymbol{\sigma}\boldsymbol{\lambda}}(-q)} \langle [Y^{\boldsymbol{\sigma}}], [S^{\boldsymbol{\mu}}]^\circledast\rangle\\
&= \sum_{\boldsymbol{\sigma}, \boldsymbol{\tau} \in \mathcal{K}_n^\Lambda} \overline{e_{\boldsymbol{\sigma}\boldsymbol{\lambda}}(-q)}\, \overline{d_{\boldsymbol{\mu}\boldsymbol{\tau}}(q)} \langle [Y^{\boldsymbol{\sigma}}], [D^{\boldsymbol{\tau}}]\rangle\\
&= \sum_{\boldsymbol{\sigma} \in \mathcal{K}_n^\Lambda} \overline{d_{\boldsymbol{\mu}\boldsymbol{\sigma}}(q)}\, \overline{e_{\boldsymbol{\sigma}\boldsymbol{\lambda}}(-q)} = \delta_{\boldsymbol{\lambda}\boldsymbol{\mu}},
\end{aligned}$$

where the last equality follows because $\mathbf{e}_q^{\mathcal{K}} = (\mathbf{d}_q^{\mathcal{K}})^{-1}$. □

Lemma 3.5.5. *Suppose that* $\boldsymbol{\mu} \in \mathcal{K}_n^\Lambda$. *Then* $[Y^{\boldsymbol{\mu}}]^\# = [Y^{\boldsymbol{\mu}}]$, $[D^{\boldsymbol{\mu}}]^\circledast = [D^{\boldsymbol{\mu}}]$,

$$(X_{\boldsymbol{\mu}})^\# = X_{\boldsymbol{\mu}} + \sum_{\substack{\boldsymbol{\sigma} \in \mathcal{K}_n^\Lambda\\ \boldsymbol{\sigma} \rhd \boldsymbol{\mu}}} a_{\boldsymbol{\sigma}\boldsymbol{\mu}}(q) X_{\boldsymbol{\sigma}} \quad \text{and} \quad [S^{\boldsymbol{\mu}}]^\circledast = [S^{\boldsymbol{\mu}}] + \sum_{\substack{\boldsymbol{\tau} \in \mathcal{K}_n^\Lambda\\ \boldsymbol{\mu} \rhd \boldsymbol{\tau}}} a^{\boldsymbol{\mu}\boldsymbol{\tau}}(q) [S^{\boldsymbol{\tau}}],$$

for some Laurent polynomials $a_{\boldsymbol{\sigma}\boldsymbol{\mu}}(q), a^{\boldsymbol{\tau}\boldsymbol{\mu}}(q) \in \mathcal{A}$.

Proof. That $[D^{\boldsymbol{\mu}}]^{\circledast} = [D^{\boldsymbol{\mu}}]$ is immediate by Corollary 3.2.7, whereas $[Y^{\boldsymbol{\mu}}]^{\#} = Y^{\boldsymbol{\mu}}$ because $Y^{\boldsymbol{\mu}}$ is a direct summand of $\mathscr{H}_n^\Lambda$ — alternatively, use Lemma 3.5.4 and Lemma 3.5.3. If $\boldsymbol{\mu} \in \mathcal{K}_n^\Lambda$ then, by Theorem 2.1.5,

$$[S^{\boldsymbol{\mu}}]^{\circledast} = \Big(\sum_{\substack{\boldsymbol{\nu}\in\mathcal{K}_n^\Lambda \\ \boldsymbol{\mu}\trianglerighteq\boldsymbol{\nu}}} d_{\boldsymbol{\mu}\boldsymbol{\nu}}(q)[D^{\boldsymbol{\nu}}]\Big)^{\circledast} = \sum_{\substack{\boldsymbol{\nu}\in\mathcal{K}_n^\Lambda \\ \boldsymbol{\mu}\trianglerighteq\boldsymbol{\nu}}} \overline{d_{\boldsymbol{\mu}\boldsymbol{\nu}}(q)}\,[D^{\boldsymbol{\nu}}]$$

$$= [S^{\boldsymbol{\mu}}] + \sum_{\substack{\boldsymbol{\tau}\in\mathcal{K}_n^\Lambda \\ \boldsymbol{\mu}\triangleright\boldsymbol{\tau}}} \Big(\sum_{\substack{\boldsymbol{\nu}\in\mathcal{K}_n^\Lambda \\ \boldsymbol{\mu}\trianglerighteq\boldsymbol{\nu}\trianglerighteq\boldsymbol{\tau}}} \overline{d_{\boldsymbol{\mu}\boldsymbol{\nu}}(q)}\, e_{\boldsymbol{\nu}\boldsymbol{\tau}}(-q)\Big)[S^{\boldsymbol{\tau}}]$$

as claimed. Note that $d_{\boldsymbol{\mu}\boldsymbol{\mu}}(q) = 1 = e_{\boldsymbol{\mu}\boldsymbol{\mu}}(-q)$.

Finally, we can compute $(X_{\boldsymbol{\mu}})^{\#}$ by writing $X_{\boldsymbol{\mu}}=\sum_{\boldsymbol{\mu}} e_{\boldsymbol{\mu}\boldsymbol{\lambda}}(-q)[Y^{\boldsymbol{\lambda}}]$ and then using essentially the same argument to show that $(X_{\boldsymbol{\mu}})^{\#}$ can be written in the required form. Alternatively, use Lemma 3.5.4 and Lemma 3.5.3. □

The triangularity of the action of $\circledast$ and $\#$ on $[\mathrm{Rep}_{\mathcal{A}}^\Lambda]$ and $[\mathrm{Proj}_{\mathcal{A}}^\Lambda]$, respectively, has the following easy but important consequence.

Proposition 3.5.6. *Suppose that $\mathbb{F}$ is a field. Then there exist unique bases $\{ B_{\boldsymbol{\mu}} \mid \boldsymbol{\mu} \in \mathcal{K}_n^\Lambda \}$ and $\{ B^{\boldsymbol{\mu}} \mid \boldsymbol{\mu} \in \mathcal{K}_n^\Lambda \}$ of $[\mathrm{Proj}(\mathscr{H}_n^\Lambda)]$ and $[\mathrm{Rep}(\mathscr{H}_n^\Lambda)]$, respectively, such that $(B_{\boldsymbol{\mu}})^{\#} = B_{\boldsymbol{\mu}}$, $(B^{\boldsymbol{\mu}})^{\circledast} = B^{\boldsymbol{\mu}}$*

$$B_{\boldsymbol{\mu}} = X_{\boldsymbol{\mu}} + \sum_{\substack{\boldsymbol{\sigma}\in\mathcal{K}_n^\Lambda \\ \boldsymbol{\sigma}\triangleright\boldsymbol{\mu}}} b_{\boldsymbol{\sigma}\boldsymbol{\mu}}(q)X_{\boldsymbol{\sigma}} \quad \textit{and} \quad B^{\boldsymbol{\mu}} = [S^{\boldsymbol{\mu}}] + \sum_{\substack{\boldsymbol{\tau}\in\mathcal{K}_n^\Lambda \\ \boldsymbol{\mu}\triangleright\boldsymbol{\tau}}} b^{\boldsymbol{\mu}\boldsymbol{\tau}}(q)[S^{\boldsymbol{\tau}}]$$

for polynomials $b^{\boldsymbol{\mu}\boldsymbol{\sigma}}(q), b_{\boldsymbol{\sigma}\boldsymbol{\mu}}(q) \in \delta_{\boldsymbol{\sigma}\boldsymbol{\mu}} + q\mathbb{Z}[q]$.

Proof. The existence and uniqueness of these two bases follows immediately from Lemma 3.5.5 by *Lusztig's Lemma* [95, Lemma 24.2.1]. We give a variation of Lusztig's argument for the basis $\{B^{\boldsymbol{\mu}}\}$.

Fix a multipartition $\boldsymbol{\mu} \in \mathcal{K}_n^\Lambda$, for some $n \geq 0$, and suppose that $B^{\boldsymbol{\mu}}$ and $\dot{B}^{\boldsymbol{\mu}}$ are two elements of $[\mathrm{Rep}(\mathscr{H}_n^\Lambda)]$ with the required properties. By assumption the element $B^{\boldsymbol{\mu}} - \dot{B}^{\boldsymbol{\mu}}$ is $\circledast$-invariant and we can write

$$B^{\boldsymbol{\mu}} - \dot{B}^{\boldsymbol{\mu}} = \sum_{\boldsymbol{\mu}\triangleright\boldsymbol{\tau}} \dot{b}^{\boldsymbol{\mu}\boldsymbol{\tau}}(q)[S^{\boldsymbol{\tau}}],$$

for some polynomials $\dot{b}^{\boldsymbol{\mu}\boldsymbol{\tau}}(q) \in q\mathbb{Z}[q]$. As the left-hand side is $\circledast$-invariant and $\dot{b}^{\boldsymbol{\mu}\boldsymbol{\tau}}(q) \in q^{-1}\mathbb{Z}[q^{-1}]$, Lemma 3.5.5 forces $B^{\boldsymbol{\mu}} = \dot{B}^{\boldsymbol{\mu}}$.

To prove existence, we argue by induction on dominance. If $\boldsymbol{\mu}$ is minimal in $\mathcal{K}_n^\Lambda$ then we can take $B^{\boldsymbol{\mu}} = [S^{\boldsymbol{\mu}}] = [D^{\boldsymbol{\mu}}]$ by Lemma 3.5.5. If $\boldsymbol{\mu} \in \mathcal{K}_n^\Lambda$ is not minimal with respect to dominance then set $\dot{B}^{\mu} = [D^{\mu}]$. Then

$$(\dot{B}^{\mu})^{\circledast} = \dot{B}^{\mu} \quad \text{and} \quad \dot{B}^{\mu} = [S^{\mu}] + \sum_{\substack{\tau \in \mathcal{K}_n^{\Lambda} \\ \mu \rhd \tau}} \dot{b}^{\mu\tau}(q)[S^{\tau}],$$

for some Laurent polynomials $\dot{b}^{\mu\tau}(q) \in \mathbb{Z}[q, q^{-1}]$. If $\dot{b}^{\mu\tau}(q) \in q\mathbb{Z}[q]$, for all $\mu \rhd \tau$, then $B^{\mu} = \dot{B}^{\mu}$ has all of the required properties. Otherwise, pick any multipartition $\mu \rhd \nu$ that is maximal with respect to dominance such that $\dot{b}^{\mu\nu}(q) \notin q\mathbb{Z}[q]$. By induction, there exists an element B^{ν} with all of the required properties. Replace $\dot{B}^{\mu}$ with the element $\dot{B}^{\mu} - p^{\mu\nu}(q)B^{\nu}$, where $p^{\mu\nu}(q)$ is the unique Laurent polynomial such that $\overline{p^{\mu\nu}(q)} = p^{\mu\nu}(q)$ and $\dot{b}^{\mu\nu}(q) - p^{\mu\nu}(q) \in q\mathbb{Z}[q]$. Then $(\dot{B}^{\mu})^{\circledast} = \dot{B}^{\mu}$ and the coefficient of $[S^{\nu}]$ in $\dot{B}^{\mu}$ belongs to $q\mathbb{Z}[q]$. Continuing in this way, after finitely many steps we will construct an element B^{μ} with the required properties. □

Corollary 3.5.7. *Suppose that* $\lambda, \mu \in \mathcal{K}^{\Lambda}$. *Then*

$$\langle B_{\mu}, B^{\lambda} \rangle = \sum_{\substack{\sigma \in \mathcal{K}^{\Lambda} \\ \lambda \trianglerighteq \sigma \trianglerighteq \mu}} b^{\lambda\sigma}(q) b_{\sigma\mu}(q) = \delta_{\lambda\mu}.$$

Proof. If $\sigma, \tau \in \mathcal{K}_n^{\Lambda}$ then $\langle X_{\sigma}, [S^{\tau}]^{\circledast} \rangle = \delta_{\sigma\tau}$ by Lemma 3.5.4. Therefore, since the form $\langle\ ,\ \rangle$ is sesquilinear and $B^{\lambda\circledast} = B^{\lambda}$,

$$\begin{aligned} \langle B_{\mu}, B^{\lambda} \rangle = \langle B_{\mu}, B^{\lambda\circledast} \rangle &= \sum_{\sigma \trianglerighteq \mu} \sum_{\lambda \trianglerighteq \tau} \overline{b_{\sigma\mu}(q)}\, \overline{b^{\lambda\tau}(q)} \langle X_{\sigma}, [S^{\tau}]^{\circledast} \rangle \\ &= \sum_{\lambda \trianglerighteq \sigma \trianglerighteq \mu} \overline{b^{\lambda\sigma}(q)}\, \overline{b_{\sigma\mu}(q)}. \end{aligned}$$

In particular, $\langle B_{\mu}, B^{\lambda} \rangle \in \delta_{\lambda\mu} + q^{-1}\mathbb{Z}[q^{-1}]$. On the other hand,

$$\langle B_{\mu}, B^{\lambda} \rangle = \langle B_{\mu}^{\#}, B^{\lambda} \rangle = \overline{\langle B_{\mu}, B^{\lambda\circledast} \rangle} = \overline{\langle B_{\mu}, B^{\lambda} \rangle}$$

by Lemma 3.5.3. Therefore, $\langle B_{\mu}, B^{\lambda} \rangle = \delta_{\lambda\mu}$ as this is the only bar invariant polynomial in $\delta_{\lambda\mu} + q^{-1}\mathbb{Z}[q^{-1}]$. □

Applying Lemma 3.4.4 to Proposition 3.5.6 we obtain.

Corollary 3.5.8. *Suppose that* $\mu \in \mathcal{K}^{\Lambda}$. *Then* $(q^{-\operatorname{def}\mu} B_{\mu})^{\circledast} = q^{-\operatorname{def}\mu} B_{\mu}$ *and* $(q^{\operatorname{def}\mu} B^{\mu})^{\#} = q^{\operatorname{def}\mu} B^{\mu}$.

In order to link the bases $\{B_{\mu}\}$ and $\{B^{\mu}\}$ with the representation theory of $\mathscr{H}_n^{\Lambda}$ we need to introduce the quantum group $U_q(\widehat{\mathfrak{sl}}_e)$. A self-contained account of much of what we need can be found in Ariki's book [5]. See also [95, §3.1] and [22].

The **quantum group** $U_q(\widehat{\mathfrak{sl}}_e)$ associated with the quiver Γ_e is the $\mathbb{Q}(q)$-algebra generated by $\{\, E_i, F_i, K_i^{\pm} \mid i \in I \,\}$, subject to the relations:

$$K_iK_j = K_jK_i, \qquad K_iK_i^{-1} = 1, \qquad [E_i, F_j] = \delta_{ij}\frac{K_i - K_i^{-1}}{q - q^{-1}},$$

$$K_iE_jK_i^{-1} = q^{c_{ij}}E_j, \qquad K_iF_jK_i^{-1} = q^{-c_{ij}}F_j,$$

$$\sum_{0\le c\le 1-c_{ij}} (-1)^c \begin{bmatrix} 1-c_{ij} \\ c \end{bmatrix}_q E_i^{1-c_{ij}-c}E_jE_i^c = 0,$$

$$\sum_{0\le c\le 1-c_{ij}} (-1)^c \begin{bmatrix} 1-c_{ij} \\ c \end{bmatrix}_q F_i^{1-c_{ij}-c}F_jF_i^c = 0,$$

where $\left[\!\left[{d \atop c} \right]\!\right]_q = [\![d]\!]!/[\![c]\!]![\![d-c]\!]!$ and $[\![m]\!]! = \prod_{k=1}^m (q^k - q^{-k})/(q - q^{-1})$, for integers $c < d, m \in \mathbb{N}$. Then $U_q(\widehat{\mathfrak{sl}}_e)$ is a Hopf algebra with coproduct determined by $\Delta(K_i) = K_i \otimes K_i$, $\Delta(E_i) = E_i \otimes K_i + 1 \otimes E_i$ and $\Delta(K_i) = F_i \otimes 1 + K_i^{-1} \otimes F_i$, for $i \in I$.

The **combinatorial Fock space** $\mathscr{F}_{\mathcal{A}}^{\Lambda}$ is the free $\mathcal{A}$-module with basis the set of symbols $\{\, |\boldsymbol{\lambda}\rangle \mid \boldsymbol{\lambda} \in \mathcal{P}^{\Lambda} \,\}$, where $\mathcal{P}^{\Lambda} = \bigcup_{n\ge 0} \mathcal{P}_n^{\Lambda}$. For future use, let $\mathcal{K}^{\Lambda} = \bigcup_{n\ge 0} \mathcal{K}_n^{\Lambda}$. Set $\mathscr{F}_{\mathbb{Q}(q)}^{\Lambda} = \mathscr{F}_{\mathcal{A}}^{\Lambda} \otimes_{\mathcal{A}} \mathbb{Q}(q)$. Then, $\mathscr{F}_{\mathbb{Q}(q)}^{\Lambda}$ is an infinite dimensional $\mathbb{Q}(q)$-vector space. We consider $\{\, |\boldsymbol{\lambda}\rangle \mid \boldsymbol{\lambda} \in \mathcal{P}^{\Lambda} \,\}$ as a basis of $\mathscr{F}_{\mathbb{Q}(q)}^{\Lambda}$ by identifying $|\boldsymbol{\lambda}\rangle$ and $|\boldsymbol{\lambda}\rangle \otimes 1_{\mathbb{Q}(q)}$.

Recall the integers $d^A(\boldsymbol{\lambda})$, $d_B(\boldsymbol{\lambda})$ and $d_i(\boldsymbol{\lambda})$ from Definition 3.2.1.

Theorem 3.5.9 (Hayashi [52, 111]). *Suppose that $\Lambda \in P^+$. Then $\mathscr{F}_{\mathbb{Q}(q)}^{\Lambda}$ is an integrable $U_q(\widehat{\mathfrak{sl}}_e)$-module with $U_q(\widehat{\mathfrak{sl}}_e)$-action determined by*

$$E_i|\boldsymbol{\lambda}\rangle = \sum_{B\in \mathrm{Rem}_i(\boldsymbol{\lambda})} q^{d_B(\boldsymbol{\lambda})}|\boldsymbol{\lambda}-B\rangle \quad \text{and} \quad F_i|\boldsymbol{\lambda}\rangle = \sum_{A\in \mathrm{Add}_i(\boldsymbol{\lambda})} q^{-d^A(\boldsymbol{\lambda})}|\boldsymbol{\lambda}+A\rangle,$$

and $K_i|\boldsymbol{\lambda}\rangle = q^{d_i(\boldsymbol{\lambda})}|\boldsymbol{\lambda}\rangle$, for all $i \in I$ and $\boldsymbol{\lambda} \in \mathcal{P}_n^{\Lambda}$.

Remark 3.5.10. A slightly different action of $U_q(\widehat{\mathfrak{sl}}_e)$ on the Fock space is used in many places in the literature, such as [5, 89, 104]. As is already evident, and will be made precise in Proposition 3.5.16 below, the $U_q(\widehat{\mathfrak{sl}}_e)$-action on the Fock space is closely related to induction and restriction for the graded Specht modules. The $U_q(\widehat{\mathfrak{sl}}_e)$-action on the Fock space used in [5, 89, 104] corresponds to the action of the induction and restriction functors on the *dual* graded Specht modules. Equivalently, this difference in the $U_q(\widehat{\mathfrak{sl}}_e)$-action arises because, ultimately, we will work with an action of $U_q(\widehat{\mathfrak{sl}}_e)$ on the Grothendieck groups of the finitely generated $\mathscr{R}_n^{\Lambda}$-modules, whereas these other sources consider the corresponding adjoint action on the projective Grothendieck groups.

Hayashi [52] considered only the special case when $\Lambda = \Lambda_0$, however, this implies the general case using the coproduct of $U_q(\widehat{\mathfrak{sl}}_e)$ because

$$\mathscr{F}^{\Lambda}_{\mathbb{Q}(q)} \cong \mathscr{F}^{\Lambda_{\overline{\kappa_1}}}_{\mathbb{Q}(q)} \otimes \cdots \otimes \mathscr{F}^{\Lambda_{\overline{\kappa_\ell}}}_{\mathbb{Q}(q)}$$

as $U_q(\widehat{\mathfrak{sl}}_e)$-modules. The crystal and canonical bases of $\mathscr{F}^{\Lambda}_{\mathbb{Q}(q)}$, which were first studied in [65, 111, 131], play an important role in what follows. A self-contained proof of Theorem 3.5.9, stated with similar language, can be found in Ariki's book [6, Theorem 10.10].

An element $x \in \mathscr{F}^{\Lambda}_{\mathbb{Q}(q)}$ has **weight** $\mathrm{wt}(x) = \Gamma$ if $K_i x = q^{(\Gamma,\alpha_i)}x$, for $i \in I$. In particular, if $\mathbf{0}_\ell = (0|0|\dots|0) \in \mathcal{P}^{\Lambda}$ is the empty multipartition of level ℓ then $K_i|\mathbf{0}_\ell\rangle = q^{(\Lambda,\alpha_i)}|\mathbf{0}_\ell\rangle$, for $i \in I$, so that $|\mathbf{0}_\ell\rangle$ has weight Λ. More generally, if $\beta \in Q^+$ then writing $\boldsymbol{\lambda} = \boldsymbol{\mu} + A$ it follows by induction that

$$\text{if } \boldsymbol{\lambda} \in \mathcal{P}^{\Lambda}_{\beta} \text{ then } d_i(\boldsymbol{\lambda}) = (\Lambda - \beta, \alpha_i), \text{ for all } i \in I. \tag{3.5.11}$$

Therefore, $\mathrm{wt}(|\boldsymbol{\lambda}\rangle) = \Lambda - \beta$ by Theorem 3.5.9. Set $d_i(\beta) = (\Lambda - \beta, \alpha_i)$.

For each dominant weight $\Lambda \in P^+$ let $L(\Lambda) = U_q(\widehat{\mathfrak{sl}}_e)v_\Lambda$ be the irreducible integrable highest weight module of highest weight Λ, where v_Λ is a highest weight vector of weight Λ. By Theorem 3.5.9, $|\mathbf{0}_\ell\rangle$ is a highest vector of weight Λ in $\mathscr{F}^{\Lambda}_{\mathbb{Q}(q)}$. In fact, it follows from Theorem 3.5.9 that

$$L(\Lambda) \cong U_q(\widehat{\mathfrak{sl}}_e)|\mathbf{0}_\ell\rangle.$$

For example, see [5, Theorem 10.10]. Henceforth, we set $v_\Lambda = |\mathbf{0}_\ell\rangle$.

To compare the Grothendieck groups $[\mathrm{Rep}(\mathscr{H}^{\Lambda}_n)]$ and $[\mathrm{Proj}(\mathscr{H}^{\Lambda}_n)]$ with the Fock space we need to consider all $n \geq 0$ simultaneously. Define

$$[\mathrm{Rep}^{\Lambda}_{\mathcal{A}}] = \bigoplus_{n\geq 0}[\mathrm{Rep}(\mathscr{H}^{\Lambda}_n)] \quad \text{and} \quad [\mathrm{Proj}^{\Lambda}_{\mathcal{A}}] = \bigoplus_{n\geq 0}[\mathrm{Proj}(\mathscr{H}^{\Lambda}_n)].$$

Set $[\mathrm{Rep}^{\Lambda}_{\mathbb{Q}(q)}] = [\mathrm{Rep}^{\Lambda}_{\mathcal{A}}] \otimes_{\mathcal{A}} \mathbb{Q}(q)$ and $[\mathrm{Proj}^{\Lambda}_{\mathbb{Q}(q)}] = [\mathrm{Proj}^{\Lambda}_{\mathcal{A}}] \otimes_{\mathcal{A}} \mathbb{Q}(q)$.

Proposition 3.5.12. *Suppose that $\Lambda \in P^+$. Then the i-induction and i-restriction functors of $[\mathrm{Rep}^{\Lambda}_{\mathbb{Q}(q)}]$ induce a $U_q(\widehat{\mathfrak{sl}}_e)$-module structure on $[\mathrm{Proj}^{\Lambda}_{\mathbb{Q}(q)}]$ and $[\mathrm{Rep}^{\Lambda}_{\mathbb{Q}(q)}]$ such that, as $U_q(\widehat{\mathfrak{sl}}_e)$-modules,*

$$[\mathrm{Proj}^{\Lambda}_{\mathbb{Q}(q)}] \cong L(\Lambda) \cong [\mathrm{Rep}^{\Lambda}_{\mathbb{Q}(q)}].$$

Proof. Recall that $\mathbf{d}_q$ is the graded decomposition matrix of $\mathscr{H}^{\Lambda}_n$ and $\mathbf{d}_q^T$ is its transpose. Abusing notation slightly, by simultaneously using these matrices for all $n \geq 0$, define linear maps

$$\begin{array}{ccc} [\mathrm{Proj}^{\Lambda}_{\mathbb{Q}(q)}] & \xhookrightarrow{\ \mathbf{d}_q^T\ } & \mathscr{F}^{\Lambda}_{\mathbb{Q}(q)} \\ & \searrow^{\mathbf{c}_q} & \Big\downarrow \mathbf{d}_q \\ & & [\mathrm{Rep}^{\Lambda}_{\mathbb{Q}(q)}] \end{array}$$

where $\mathbf{d}_q^T([Y^{\boldsymbol{\mu}}]) = \sum_{\boldsymbol{\lambda}} d_{\boldsymbol{\lambda}\boldsymbol{\mu}}(q)|\boldsymbol{\lambda}\rangle$, $\mathbf{d}_q(|\boldsymbol{\lambda}\rangle) = \sum_{\boldsymbol{\mu}} d_{\boldsymbol{\lambda}\boldsymbol{\mu}}(q)[D^{\boldsymbol{\mu}}]$ and where $\mathbf{c}_q = \mathbf{d}_q \circ \mathbf{d}_q^T$ is the Cartan map. As vector space homomorphisms, $\mathbf{d}_q^T$ is injective and $\mathbf{d}_q$ is surjective. As defined these maps are only vector space homomorphisms, however, we claim that both maps can be made into $U_q(\widehat{\mathfrak{sl}}_e)$-module homomorphisms.

The i-induction and i-restriction functors are exact, for $i \in I$, because they are exact when we forget the grading [49, Corollary 8.9]. Therefore, they send projective modules to projectives and they induce endomorphisms of the Grothendieck groups $[\mathrm{Rep}^{\Lambda}_{\mathbb{Q}(q)}]$ and $[\mathrm{Proj}^{\Lambda}_{\mathbb{Q}(q)}]$. By Theorem 3.4.6,

$$\begin{aligned} [i\text{-Res}\, S^{\boldsymbol{\lambda}}] &= \sum_{B \in \mathrm{Rem}_i(\boldsymbol{\lambda})} q^{d_B(\boldsymbol{\lambda})}[S^{\boldsymbol{\lambda}-B}], \\ [i\text{-Ind}\, S^{\boldsymbol{\lambda}}\langle 1 - d_i(\boldsymbol{\lambda})\rangle] &= \sum_{A \in \mathrm{Add}_i(\boldsymbol{\lambda})} q^{d_A(\boldsymbol{\lambda})+1-d_i(\boldsymbol{\lambda})}[S^{\boldsymbol{\lambda}+A}] \\ &= \sum_{A \in \mathrm{Add}_i(\boldsymbol{\lambda})} q^{-d^A(\boldsymbol{\lambda})}[S^{\boldsymbol{\lambda}+A}], \end{aligned}$$

where the last equality uses Lemma 3.3.2(a). Identifying E_i with i-Res, and F_i with $q\,i$-Ind K_i^{-1}, the linear maps $\mathbf{d}_q$ and $\mathbf{d}_q^T$ become well-defined $U_q(\widehat{\mathfrak{sl}}_e)$-module homomorphisms by Theorem 3.5.9. As $U_q(\widehat{\mathfrak{sl}}_e)$-modules, $[\mathrm{Rep}^{\Lambda}_{\mathbb{Q}(q)}]$ and $[\mathrm{Proj}^{\Lambda}_{\mathbb{Q}(q)}]$ are both cyclic because they are both generated by $[Y^{\mathbf{0}}_{\ell}] = [S^{\mathbf{0}}_{\ell}] = [D^{\mathbf{0}}_{\ell}]$. By definition, $\mathbf{d}_q^T([Y^{\mathbf{0}}_{\ell}]) = v_\Lambda$ and $\mathbf{d}_q(v_\Lambda) = [S^{\mathbf{0}}_{\ell}]$, so the proposition follows because $L(\Lambda) \cong U_q(\widehat{\mathfrak{sl}}_e)v_\Lambda$ is irreducible. □

Let $U_{\mathcal{A}}(\widehat{\mathfrak{sl}}_e)$ be Lusztig's $\mathcal{A}$-form of $U_q(\widehat{\mathfrak{sl}}_e)$, which is the $\mathcal{A}$-subalgebra of $U_q(\widehat{\mathfrak{sl}}_e)$ generated by the quantised divided powers $E_i^{(k)} = E_i^k/[\![k]\!]!$ and $F_i^{(k)} = F_i^k/[\![k]\!]!$, for $i \in I$ and $k \geq 0$. Theorem 3.5.9 implies that $U_{\mathcal{A}}(\widehat{\mathfrak{sl}}_e)$ acts on the $\mathcal{A}$-submodule $\mathscr{F}^{\Lambda}_{\mathcal{A}}$ of $\mathscr{F}^{\Lambda}_{\mathbb{Q}(q)}$; compare with [89, Lemma 6.2] and with [104, Lemma 6.16]. Therefore, by Proposition 3.5.12, $[\mathrm{Rep}^{\Lambda}_{\mathcal{A}}]$ and $[\mathrm{Proj}^{\Lambda}_{\mathcal{A}}]$ are $U_{\mathcal{A}}(\widehat{\mathfrak{sl}}_e)$-modules. Moreover, if we set $L_{\mathcal{A}}(\Lambda) = U_{\mathcal{A}}(\widehat{\mathfrak{sl}}_e)v_\Lambda$

then there are $U_{\mathcal{A}}(\widehat{\mathfrak{sl}}_e)$-module homomorphisms $[\mathrm{Proj}_{\mathcal{A}}^{\Lambda}] \hookrightarrow L_{\mathcal{A}}(\Lambda) \twoheadrightarrow [\mathrm{Rep}_{\mathcal{A}}^{\Lambda}]$. In particular, $L_{\mathcal{A}}(\Lambda) \cong [\mathrm{Proj}_{\mathcal{A}}^{\Lambda}]$ as $U_{\mathcal{A}}(\widehat{\mathfrak{sl}}_e)$-modules by Proposition 3.5.12.

Lemma 3.5.13. *Suppose that $i \in I$. The involution $\circledast$ commutes with the actions of E_i and F_i on $[\mathrm{Rep}_{\mathcal{A}}^{\Lambda}]$ and on $[\mathrm{Proj}_{\mathcal{A}}^{\Lambda}]$.*

Proof. By Proposition 3.4.5, there are isomorphisms $i\text{-Res} \circ \circledast \cong \circledast \circ i\text{-Res}$ and $i\text{-Ind} \circ \# \cong \# \circ i\text{-Ind}$. In particular, the actions of $E_i = i\text{-Res}$ and $\circledast$ commute. Fix $\beta \in Q^+$. Recall from after (3.5.11) that $d_i(\beta) = (\Lambda - \beta, \alpha_i)$. Identifying F_i with the functor $q \circ i\text{-Ind} \circ K_i^{-1} = q^{1-d_i(\beta)}\, i\text{-Ind}$ on $\mathrm{Rep}(\mathscr{H}_\beta^\Lambda)$, there are isomorphisms

$$\begin{aligned}
F_i \circ \circledast &\cong q\, i\text{-Ind}\, K_i^{-1} \circ q^{-2\operatorname{def}\beta} \# && \text{by Lemma 3.4.4,}\\
&\cong q^{1-d_i(\beta)-2\operatorname{def}\beta}\, i\text{-Ind} \circ \# \\
&\cong q^{1-d_i(\beta)-2\operatorname{def}\beta} \# \circ i\text{-Ind} && \text{by Proposition 3.4.5,}\\
&\cong q^{-2\operatorname{def}(\beta+\alpha_i)} \# \circ q^{d_i(\beta)-1} \circ i\text{-Ind} && \text{by Lemma 3.3.2(a),}\\
&\cong \circledast \circ q^{-1}\, i\text{-Ind}\, K_i \cong \circledast \circ F_i, && \text{by Lemma 3.4.4.}
\end{aligned}$$

Hence, E_i and F_i commute with $\circledast$ on $[\mathrm{Rep}_{\mathcal{A}}^{\Lambda}]$ and $[\mathrm{Proj}_{\mathcal{A}}^{\Lambda}]$ as claimed. □

In contrast, E_i and F_i on $[\mathrm{Rep}_{\mathcal{A}}^{\Lambda}]$ and $[\mathrm{Proj}_{\mathcal{A}}^{\Lambda}]$ do not commute with $\#$.

We want to relate the Cartan pairing $\langle\ ,\ \rangle$ on $[\mathrm{Proj}_{\mathcal{A}}^{\Lambda}] \times [\mathrm{Rep}_{\mathcal{A}}^{\Lambda}]$ with the representation theory of $L_{\mathcal{A}}(\Lambda)$. Define a non-degenerate symmetric bilinear form $(\ ,\ )$ on the Fock space $\mathscr{F}_{\mathcal{A}}^{\Lambda}$ by

$$(|\boldsymbol{\lambda}\rangle, |\boldsymbol{\mu}\rangle) = \delta_{\boldsymbol{\lambda}\boldsymbol{\mu}} q^{\operatorname{def}\boldsymbol{\lambda}}, \qquad \text{for } \boldsymbol{\lambda}, \boldsymbol{\mu} \in \mathcal{P}^{\Lambda}. \tag{3.5.14}$$

By Theorem 3.5.9, $(K_i x, y) = q^{\mathrm{wt}(y)}(x, y) = (x, K_i y)$, for weight vectors $x, y \in \mathscr{F}_{\mathcal{A}}^{\Lambda}$ and $i \in I$. By restriction, we also consider $(\ ,\ )$ as a bilinear form on $L_{\mathcal{A}}(\Lambda)$.

Lemma 3.5.15. *The bilinear form $(\ ,\ )$ on $L_{\mathcal{A}}(\Lambda)$ is characterised by the three properties: $(v_\Lambda, v_\Lambda) = 1$, $(E_i x, y) = (x, F_i y)$ and $(F_i x, y) = (x, E_i y)$, for all $i \in I$ and $x, y \in L_{\mathcal{A}}(\Lambda)$.*

Proof. By definition, $(v_\Lambda, v_\Lambda) = 1$. If $i \in I$ then in order to check that E_i and F_i are biadjoint with respect to $(\ ,\ )$ it is enough to consider the cases when $x = |\boldsymbol{\lambda}\rangle$ and $y = |\boldsymbol{\mu}\rangle$, for $\boldsymbol{\lambda}, \boldsymbol{\mu} \in \mathcal{P}^{\Lambda}$. By Theorem 3.5.9, $(F_i|\boldsymbol{\lambda}\rangle, |\boldsymbol{\mu}\rangle) = 0 = (|\boldsymbol{\lambda}\rangle, E_i|\boldsymbol{\mu}\rangle)$ unless $\boldsymbol{\mu} = \boldsymbol{\lambda} + A$ for some $A \in \mathrm{Add}_i(\boldsymbol{\lambda})$. On the other hand, if $A \in \mathrm{Add}_i(\boldsymbol{\lambda})$ and $\boldsymbol{\mu} = \boldsymbol{\lambda} + A$ then, using Lemma 3.3.2(a) for the second equality,

$$\begin{aligned}
(F_i|\boldsymbol{\lambda}\rangle, |\boldsymbol{\mu}\rangle) &= q^{\operatorname{def}\boldsymbol{\mu} - d^A(\boldsymbol{\lambda})} = q^{\operatorname{def}\boldsymbol{\lambda} + d_i(\boldsymbol{\lambda}) - 1 - d^A(\boldsymbol{\lambda})}\\
&= q^{\operatorname{def}\boldsymbol{\lambda} + d_A(\boldsymbol{\mu})} = (|\boldsymbol{\lambda}\rangle, E_i|\boldsymbol{\mu}\rangle).
\end{aligned}$$

A similar calcuation shows that $(E_i|\boldsymbol{\lambda}\rangle, |\boldsymbol{\mu}\rangle) = (|\boldsymbol{\lambda}\rangle, F_i|\boldsymbol{\mu}\rangle)$, for all $\boldsymbol{\lambda}, \boldsymbol{\mu} \in \mathcal{P}_n^\Lambda$. As v_Λ is the highest weight vector in the irreducible module $L_\mathcal{A}(\Lambda)$, these three properties uniquely determine the bilinear form $(\ ,\ )$ on $L_\mathcal{A}(\Lambda)$ by induction on weight. □

By restriction, the next result categorifies the pairing $(\ ,\ )$ on $L_\mathcal{A}(\Lambda)$.

Proposition 3.5.16. *Let $x \in [\mathrm{Proj}_\mathcal{A}^\Lambda]$ and $y \in \mathscr{F}_\mathcal{A}^\Lambda$ with $\mathrm{wt}(\boldsymbol{\lambda}) = \beta$. Then*

$$\langle x, \mathbf{d}_q(y)\rangle = q^{-\operatorname{def}\beta}(\mathbf{d}_q^T(x^\#), y).$$

Proof. As $(\ ,\ )$ is bilinear and $\langle\ ,\ \rangle$ is sesquilinear it is enough to verify this identity when $x = X_{\boldsymbol{\mu}}$ and $y = |\boldsymbol{\lambda}\rangle$, for $\boldsymbol{\mu} \in \mathcal{K}^\Lambda$ and $\boldsymbol{\lambda} \in \mathcal{P}_\beta^\Lambda$. Then

$$\begin{aligned}
\langle x^\#, \mathbf{d}_q(y)\rangle = \langle X_{\boldsymbol{\mu}}, [S^{\boldsymbol{\lambda}}]\rangle &= \sum_{\boldsymbol{\sigma}\in\mathcal{K}_\beta^\Lambda} d_{\boldsymbol{\lambda}\boldsymbol{\sigma}}(q)\langle X_{\boldsymbol{\mu}}, [D^{\boldsymbol{\sigma}}]\rangle \\
&= \sum_{\boldsymbol{\tau}\in\mathcal{K}_\beta^\Lambda}\sum_{\boldsymbol{\sigma}\in\mathcal{K}_\beta^\Lambda} d_{\boldsymbol{\lambda}\boldsymbol{\sigma}}(q)\overline{e_{\boldsymbol{\sigma}\boldsymbol{\tau}}(-q)}\langle X_{\boldsymbol{\mu}}, [S^{\boldsymbol{\tau}}]^{\circledast}\rangle \\
&= \sum_{\boldsymbol{\sigma}\in\mathcal{K}_\beta^\Lambda} d_{\boldsymbol{\lambda}\boldsymbol{\sigma}}(q)\overline{e_{\boldsymbol{\sigma}\boldsymbol{\mu}}(-q)},
\end{aligned}$$

where the last equality uses Lemma 3.5.4. For the right hand side,

$$\begin{aligned}
(\mathbf{d}_q^T(x^\#), y) = (\mathbf{d}_q^T(X_{\boldsymbol{\mu}}^\#), |\boldsymbol{\lambda}\rangle) &= \sum_{\boldsymbol{\sigma}\in\mathcal{K}_\beta^\Lambda} \overline{e_{\boldsymbol{\sigma}\boldsymbol{\mu}}(-q)}(\mathbf{d}_q^T([Y^{\boldsymbol{\sigma}}]), |\boldsymbol{\lambda}\rangle) \\
&= \sum_{\substack{\boldsymbol{\nu}\in\mathcal{P}_n^\Lambda \\ \boldsymbol{\nu}\trianglerighteq\boldsymbol{\mu}}} \Big(\sum_{\substack{\boldsymbol{\sigma}\in\mathcal{K}_n^\Lambda \\ \boldsymbol{\nu}\trianglerighteq\boldsymbol{\sigma}\trianglerighteq\boldsymbol{\mu}}} d_{\boldsymbol{\tau}\boldsymbol{\sigma}}(q)\overline{e_{\boldsymbol{\sigma}\boldsymbol{\mu}}(-q)}\Big)(|\boldsymbol{\tau}\rangle, |\boldsymbol{\lambda}\rangle) \\
&= q^{\operatorname{def}\beta}\langle x, \mathbf{d}_q(y)\rangle,
\end{aligned}$$

by (3.5.14) and calculation above. The proof is complete. □

Now we can prove the results that we are really interested in.

Corollary 3.5.17. *Let $P \in \mathrm{Proj}(\mathscr{H}_n^\Lambda)$, $y \in \mathrm{Rep}(\mathscr{H}_{n+1}^\Lambda)$, and $i \in I$. Then*

$$\langle i\text{-Ind}\, x, y\rangle = \langle x, i\text{-Res}\, y\rangle \ \textit{and}\ \langle i\text{-Res}\, x, y\rangle = \langle x, i\text{-Ind}\, y\rangle.$$

Proof. By Theorem 3.4.2, (i-Res, i-Ind) is a biadjoint pair so the corollary follows directly from the definition of the Cartan pairing in (3.5.2). As it is non-trivial to show that i-Res is left adjoint to i-Ind we prove this at the level of Grothendieck groups. Write $\dot{x} = \mathbf{d}_q^T(x)$ and $y = \mathbf{d}_q(\dot{y})$ where $\dot{x}, \dot{y} \in L_\mathcal{A}(\Lambda)$

and $\operatorname{wt}(\dot{y}) = \Lambda - \beta$. Then $\langle i\text{-Res}\, x, y\rangle = 0$ unless $\operatorname{wt}(x) = \Lambda - (\beta + \alpha_i)$. To improve readability, identify x and $\dot{x} = \mathbf{d}_q^T(x)$ below. Then,

$$\begin{aligned}
\langle i\text{-Res}\, x, y\rangle &= q^{-\operatorname{def}\beta}\big((E_i x)^{\#}, \dot{y}\big) && \text{by Proposition 3.5.16,}\\
&= q^{\operatorname{def}\beta}\big(E_i x^{\circledast}, \dot{y}\big), && \text{by 3.4.4 and 3.5.13,}\\
&= q^{\operatorname{def}\beta}\big(x^{\circledast}, F_i \dot{y}\big), && \text{by Lemma 3.5.15,}\\
&= q^{\operatorname{def}\beta - 2\operatorname{def}(\beta+\alpha_i)}\big(x^{\#}, F_i \dot{y}\big), && \text{by Lemma 3.4.4,}\\
&= q^{\operatorname{def}\beta - \operatorname{def}(\beta+\alpha_i)}\langle x^{\#}, F_i y\rangle, && \text{by Proposition 3.5.16,}\\
&= \langle x, i\text{-Ind}\, y\rangle,
\end{aligned}$$

where the last equality uses Lemma 3.3.2 and the identification of F_i and $q\, i\text{-Ind} \circ K_i^{-1}$ on $[\operatorname{Rep}_{\mathcal{A}}^{\Lambda}]$, via Proposition 3.5.12. □

Let τ be the unique semilinear anti-isomorphism of $U_q(\widehat{\mathfrak{sl}}_e)$ such that $\tau(K_i) = K_i^{-1}$, $\tau(E_i) = qF_iK_i^{-1}$ and $\tau(F_i) = q^{-1}K_iE_i$, for all $i \in I$. Then the biadjointness of induction and restriction with respect to the Cartan pairing translates into the following more Lie theoretic statement.

Corollary 3.5.18. *Suppose that* $x \in [\operatorname{Proj}_{\mathcal{A}}^{\Lambda}]$ *and* $y \in [\operatorname{Rep}_{\mathcal{A}}^{\Lambda}]$. *Then*

$$\langle ux, y\rangle = \langle x, \tau(u)y\rangle, \qquad \textit{for all } u \in U_{\mathcal{A}}(\widehat{\mathfrak{sl}}_e).$$

The bar involution of $\mathcal{A}$ extends to a semilinear involution of $U_{\mathcal{A}}(\widehat{\mathfrak{sl}}_e)$ determined by $\overline{K_i} = K_i^{-1}$, $\overline{E_i} = E_i$ and $\overline{F_i} = F_i$, for all $i \in I$. Similarly, define a bar involution on $L_{\mathcal{A}}(\Lambda)$ by

$$\overline{v}_\Lambda = v_\Lambda \quad \text{and} \quad \overline{ux} = \overline{u}\,\overline{x}, \qquad \text{for } u \in U_{\mathcal{A}}(\widehat{\mathfrak{sl}}_e) \text{ and } x \in L_{\mathcal{A}}(\Lambda).$$

As noted in [22, §3.1], it follows from the relations that $\tau \circ \overline{} = \overline{} \circ \tau^{-1}$.

As in [22, §3.3], the **Shapovalov form** on $L(\Lambda)$ is the sesquilinear map

$$\langle x, y\rangle = q^{\operatorname{def}\beta}(\overline{x}, y),$$

for $x, y \in L(\Lambda)$ with $\operatorname{wt}(y) = \Lambda - \beta$, for $\beta \in Q^+$. As our notation suggests, the Shapovalov form is categorified by the Cartan pairing.

Corollary 3.5.19. *Suppose that* $x \in [\operatorname{Proj}_{\mathcal{A}}^{\Lambda}]$ *and* $y \in L_{\mathcal{A}}(\Lambda)$. *Then*

$$\langle \mathbf{d}_q^T(x), y\rangle = \langle x, \mathbf{d}_q(y)\rangle.$$

Proof. By Lemma 3.5.15, the pairing $(\ ,\ )$ on $L_{\mathcal{A}}(\Lambda)$ is unique symmetric bilinear map on $L_{\mathcal{A}}(\Lambda)$ that is biadjoint with respect to E_i and F_i and such that $(v_\Lambda, v_\Lambda) = 1$. This implies that the Shapovalov form is the unique sesquilinear form on $L_{\mathcal{A}}(\Lambda)$ such that $\langle v_\Lambda, v_\Lambda\rangle = 1$ and $\langle ux, y\rangle = \langle x, \tau(u)y\rangle$, for $x, y \in L_{\mathcal{A}}(\Lambda)$ and $u \in U_{\mathcal{A}}(\widehat{\mathfrak{sl}}_e)$. Hence, the result follows from Corollary 3.5.18. □

The module $L_{\mathcal{A}}(\Lambda) = U_{\mathcal{A}}(\widehat{\mathfrak{sl}}_e)v_\Lambda$ is the *standard* $\mathcal{A}$*-form* of the irreducible $U_q(\widehat{\mathfrak{sl}}_e)$-module $L(\Lambda)$. The *costandard* $\mathcal{A}$*-form* of $L(\Lambda)$ is the dual lattice

$$\begin{aligned} L_{\mathcal{A}}(\Lambda)^* &= \{\, y \in L(\Lambda) \mid (x,y) \in \mathcal{A} \text{ for all } x \in L_{\mathcal{A}}(\Lambda) \,\} \\ &= \{\, y \in L(\Lambda) \mid \langle x,y \rangle \in \mathcal{A} \text{ for all } x \in L_{\mathcal{A}}(\Lambda) \,\}. \end{aligned}$$

We can now identify both $[\mathrm{Proj}_{\mathcal{A}}^{\Lambda}]$ and $[\mathrm{Rep}_{\mathcal{A}}^{\Lambda}]$ as $U_{\mathcal{A}}(\widehat{\mathfrak{sl}}_e)$-modules.

Corollary 3.5.20. *Suppose that* $\Lambda \in Q^+$. *Then, as* $U_{\mathcal{A}}(\widehat{\mathfrak{sl}}_e)$*-modules,*

$$[\mathrm{Proj}_{\mathcal{A}}^{\Lambda}] \cong L_{\mathcal{A}}(\Lambda) \qquad \textit{and} \qquad [\mathrm{Rep}_{\mathcal{A}}^{\Lambda}] \cong L_{\mathcal{A}}(\Lambda)^*.$$

Proof. The first isomorphism we noted already after Proposition 3.5.12. The second isomorphism follows from Corollary 3.5.19 and Lemma 3.5.4. □

By Lemma 3.5.13, the action of F_i on $[\mathrm{Rep}_{\mathcal{A}}^{\Lambda}]$ and $[\mathrm{Proj}_{\mathcal{A}}^{\Lambda}]$, for $i \in I$, commutes with ⊛. In the language of [22, §3.1], ⊛ is a *compatible bar-involution.* As is easily proved by induction on weight, every integrable $U_{\mathcal{A}}(\widehat{\mathfrak{sl}}_e)$-module has a unique bar-compatible involution, so

$$\mathbf{d}_q(\overline{y}) = \mathbf{d}_q(y)^{\circledast} \qquad \text{for all } y \in \mathscr{F}_{\mathbb{Q}(q)}^{\Lambda}. \tag{3.5.21}$$

Proposition 3.5.6 implies that $\{\, B^{\boldsymbol{\mu}} \mid \boldsymbol{\mu} \in \mathcal{K}^{\Lambda} \,\}$ is Kashiwara's **upper global basis** at $q = 0$ [69], or Lusztig's **dual canonical basis** [94, §14.4], of $L(\Lambda)$. By Corollary 3.5.8, $q^{-\operatorname{def}\boldsymbol{\mu}} B_{\boldsymbol{\mu}}$ is bar invariant and, thinking of (,) as a pairing from $[\mathrm{Proj}_{\mathcal{A}}^{\Lambda}] \times [\mathrm{Rep}_{\mathcal{A}}^{\Lambda}]$ to $\mathcal{A}$, we have

$$(q^{-\operatorname{def}\boldsymbol{\mu}} B_{\boldsymbol{\mu}}, B^{\boldsymbol{\lambda}}) = \langle B_{\boldsymbol{\mu}}^{\#}, B^{\boldsymbol{\lambda}} \rangle = \langle B_{\boldsymbol{\mu}}, B^{\boldsymbol{\lambda}} \rangle = \delta_{\boldsymbol{\lambda}\boldsymbol{\mu}},$$

by Proposition 3.5.16 and Corollary 3.5.19. Hence, $\{\, q^{-\operatorname{def}\boldsymbol{\mu}} B_{\boldsymbol{\mu}} \mid \boldsymbol{\mu} \in \mathcal{K}^{\Lambda} \,\}$ is the **canonical basis**, or the **lower global basis**, of $L(\Lambda)$.

The equivalence of parts (a)–(d) of the next result could have been proved without introducing $U_q(\widehat{\mathfrak{sl}}_e)$. For (e), however, we need the work of Misra and Miwa [111], and Kashiwara's theory of crystal bases [69, 70], to connect the crystal bases of the Fock space with those of $L(\Lambda)$.

Proposition 3.5.22. *Suppose that* $\mathbb{F}$ *is an arbitrary field and that* $n \geq 0$. *Then the following are equivalent:*

a) *For all* $\boldsymbol{\mu} \in \mathcal{K}_n^{\Lambda}$, $B^{\boldsymbol{\mu}} = [D^{\boldsymbol{\mu}}]$.
b) *For all* $\boldsymbol{\lambda}, \boldsymbol{\mu} \in \mathcal{K}_n^{\Lambda}$, $e_{\boldsymbol{\lambda}\boldsymbol{\mu}}(-q) \in \delta_{\boldsymbol{\lambda}\boldsymbol{\mu}} + q\mathbb{N}[q]$.
c) *For all* $\boldsymbol{\mu} \in \mathcal{K}_n^{\Lambda}$, $B_{\boldsymbol{\mu}} = [Y^{\boldsymbol{\mu}}]$.
d) *For all* $\boldsymbol{\lambda}, \boldsymbol{\mu} \in \mathcal{K}_n^{\Lambda}$, $d_{\boldsymbol{\lambda}\boldsymbol{\mu}}(q) \in \delta_{\boldsymbol{\lambda}\boldsymbol{\mu}} + q\mathbb{N}[q]$.
e) *For all* $\boldsymbol{\lambda} \in \mathcal{P}_n^{\Lambda}$ *and* $\boldsymbol{\mu} \in \mathcal{K}_n^{\Lambda}$, $d_{\boldsymbol{\lambda}\boldsymbol{\mu}}(q) \in \delta_{\boldsymbol{\lambda}\boldsymbol{\mu}} + q\mathbb{N}[q]$.

Proof. In the Grothendieck groups, $[D^{\boldsymbol{\mu}}] = [S^{\boldsymbol{\mu}}] + \sum_{\boldsymbol{\mu} \rhd \boldsymbol{\tau}} e_{\boldsymbol{\mu}\boldsymbol{\tau}}(-q)[S^{\boldsymbol{\tau}}]$ and $[Y^{\boldsymbol{\mu}}] = X_{\boldsymbol{\mu}} + \sum_{\boldsymbol{\mu} \rhd \boldsymbol{\sigma}} d_{\boldsymbol{\sigma}\boldsymbol{\mu}}(q) X_{\boldsymbol{\sigma}}$, where in the sums $\boldsymbol{\sigma}, \boldsymbol{\tau} \in \mathcal{K}_n^{\Lambda}$. Moreover, by Lemma 3.5.5, $[Y^{\boldsymbol{\mu}}]^{\#} = [Y^{\boldsymbol{\mu}}]$ and $[D^{\boldsymbol{\mu}}]^{\circledast} = [D^{\boldsymbol{\mu}}]$, for all $\boldsymbol{\mu} \in \mathcal{K}_n^\Lambda$. By definition, $d_{\boldsymbol{\lambda\mu}}(q) \in \mathbb{N}[q, q^{-1}]$ and $e_{\boldsymbol{\lambda\mu}}(-q) \in \mathbb{Z}[q, q^{-1}]$. Hence, parts (a) and (b), and parts (c) and (d), are equivalent by Proposition 3.5.6. Moreover, $\mathbf{e}_q^{\mathcal{K}} = (\mathbf{d}_q^{\mathcal{K}})^{-1}$, $d_{\boldsymbol{\mu\mu}}(1) = 1 = e_{\boldsymbol{\mu\mu}}(-q)$ and the Laurent polynomials $d_{\boldsymbol{\lambda\mu}}(q)$ and $e_{\boldsymbol{\lambda\mu}}(-q)$ are non-zero only if $\boldsymbol{\lambda} \trianglerighteq \boldsymbol{\mu}$ by Theorem 1.3.4, so parts (b) and (d) are also equivalent. Certainly, (e) implies (d) so to complete the proof it is enough to show that (a) implies (e).

Suppose that (a) holds so that $B_{\boldsymbol{\mu}} = D^{\boldsymbol{\mu}}$, for all $\boldsymbol{\mu} \in \mathcal{K}^\Lambda$. To prove that (e) holds we need the machinery of *crystal bases* [69, 70] in the special case of the Fock space $\mathscr{F}^\Lambda_{\mathbb{Q}(q)}$. We will refer the reader to the literature for the definitions and results that we need.

Following [70, §2] define rings $\mathbb{A} = \mathbb{Q}[q, q^{-1}] = \mathbb{Q} \otimes_{\mathbb{Z}} \mathcal{A}$, $\mathbb{A}_0 = \mathbb{A}_{(q)}$ and $\mathbb{A}_\infty = \mathbb{A}_{(q^{-1})}$, so that $\mathbb{A}_0$ and $\mathbb{A}_\infty$ are the rational functions in $\mathbb{Q}(q)$ that are regular at 0 and ∞, respectively. Set

$$L_0(\Lambda) = \bigoplus_{\boldsymbol{\mu} \in \mathcal{K}^\Lambda} \mathbb{A}_0 B^{\boldsymbol{\mu}} = \bigoplus_{\boldsymbol{\mu} \in \mathcal{K}^\Lambda} \mathbb{A}_0 [S^{\boldsymbol{\mu}}]$$

and $B_0(\Lambda) = \{\, [S^{\boldsymbol{\mu}}] + qL_0(\Lambda) \mid \boldsymbol{\mu} \in \mathcal{K}^\Lambda \,\}$. As $\{B^{\boldsymbol{\mu}}\}$ is the upper crystal basis, the pair $\big(L_0(\Lambda), B_0(\Lambda)\big)$ is an upper crystal base at $q = 0$ for $L(\Lambda)$ as defined by Kashiwara [69, §2]. Similarly, in the Fock space define

$$\mathscr{F}_0^\Lambda = \bigoplus_{\boldsymbol{\lambda} \in \mathcal{P}^\Lambda} \mathbb{A}_0 |\boldsymbol{\lambda}\rangle \quad \text{and} \quad C_0^\Lambda = \{\, |\boldsymbol{\lambda}\rangle + q\mathscr{F}_0^\Lambda \mid \boldsymbol{\lambda} \in \mathcal{P}^\Lambda \,\}.$$

Misra and Miwa [111] showed that $(\mathscr{F}_0^\Lambda, C_0^\Lambda)$ is an upper crystal basis for $\mathscr{F}^\Lambda_{\mathbb{Q}(q)}$. (As we discuss below they also explicitly describe the crystal graph of $\mathscr{F}^\Lambda_{\mathbb{Q}(q)}$.) By [69, Theorem 7], the Fock space $\mathscr{F}^\Lambda_{\mathbb{Q}(q)}$ has a unique basis $\{\, C^{\boldsymbol{\lambda}} \mid \boldsymbol{\lambda} \in \mathcal{P}^\Lambda \,\}$, Kashiwara's upper global basis, such that

$$\text{(3.5.23)} \qquad \overline{C^{\boldsymbol{\lambda}}} = C^{\boldsymbol{\lambda}} \quad \text{and} \quad C^{\boldsymbol{\lambda}} \equiv |\boldsymbol{\lambda}\rangle \pmod{q\mathscr{F}_0^\Lambda}, \qquad \text{for } \boldsymbol{\lambda} \in \mathcal{P}^\Lambda.$$

Let $\mathscr{F}_\bullet^\Lambda = \big(\mathscr{F}_{\mathbb{A}}^\Lambda, \mathscr{F}_0^\Lambda, \overline{\mathscr{F}_0^\Lambda}\big)$ and $L_\bullet(\Lambda) = \big(L_{\mathbb{A}}(\Lambda), L_0(\Lambda), L_0(\Lambda)^{\circledast}\big)$, where $\mathscr{F}_{\mathbb{A}}^\Lambda = \mathbb{A} \otimes_{\mathbb{A}_0} \mathscr{F}_0^\Lambda$ and $L_{\mathbb{A}}(\Lambda) = \mathbb{A} \otimes_{\mathbb{A}_0} L_0(\Lambda)$. Then $\mathscr{F}_\bullet^\Lambda$ and $L_\bullet(\Lambda)$ are balanced triples in the sense of [70, §2]. The Λ-weight space of $\mathscr{F}^\Lambda_{\mathbb{Q}(q)}$ is $\mathbb{Q}(q)v_\Lambda$ so, up to a scalar, the decomposition map $\mathbf{d}_q$ is the unique $U_q(\widehat{\mathfrak{sl}}_e)$-module homomorphism $\mathbf{d}_q : \mathscr{F}^\Lambda_{\mathbb{Q}(q)} \longrightarrow [\mathrm{Rep}^\Lambda_{\mathbb{Q}(q)}]$. By [69, Proposition 5.2.1], the image of $\mathscr{F}_\bullet^\Lambda$ under $\mathbf{d}_q$ is a balanced triple contained in $L(\Lambda)$. In fact, we have $\mathbf{d}_q(\mathscr{F}_\bullet^\Lambda) = L_\bullet(\Lambda)$ by [69, Proposition 5.2.2] because $\mathbf{d}_q$ sends $v_\Lambda = |\mathbf{0}_\ell\rangle$ to $[S^{\mathbf{0}_\ell}]$. Consequently, if $\boldsymbol{\lambda} \in \mathcal{P}^\Lambda$ then $\mathbf{d}_q(|\boldsymbol{\lambda}\rangle) \in L_0(\Lambda)$. That is,

$$[S^{\boldsymbol{\lambda}}] = \mathbf{d}_q(|\boldsymbol{\lambda}\rangle) \in \bigoplus_{\boldsymbol{\mu} \in \mathcal{K}^\Lambda} \mathbb{A}_0 [S^{\boldsymbol{\mu}}] = \bigoplus_{\boldsymbol{\mu} \in \mathcal{K}^\Lambda} \mathbb{A}_0 [D^{\boldsymbol{\mu}}].$$

As $[S^{\boldsymbol\lambda}] = \sum_{\boldsymbol\mu} d_{\boldsymbol{\lambda\mu}}(q)[D^{\boldsymbol\mu}]$ it follows that $d_{\boldsymbol{\lambda\mu}}(q) \in \mathbb{N}[q,q^{-1}] \cap \mathbb{A}_0 = \mathbb{N}[q]$. Moreover, because of (3.5.21), $\mathbf{d}_q$ sends canonical basis elements in $\mathscr{F}_0^\Lambda$ to canonical basis elements in $L_0(\Lambda)$, or to zero. It follows that

$$\mathbf{d}_q(C^{\boldsymbol\lambda}) = \begin{cases} B^{\boldsymbol\lambda} = [D^{\boldsymbol\lambda}], & \text{if } \boldsymbol\lambda \in \mathcal{K}^\Lambda \\ 0, & \text{otherwise.} \end{cases}$$

By (3.5.23), $|\boldsymbol\lambda\rangle - C^{\boldsymbol\lambda} \in q\mathscr{F}_0^\Lambda$ for all $\boldsymbol\lambda \in \mathcal{P}^\Lambda$. Consequently, if $\boldsymbol\lambda \notin \mathcal{K}^\Lambda$ then

$$[S^{\boldsymbol\lambda}] = \mathbf{d}_q(|\boldsymbol\lambda\rangle) = \mathbf{d}_q(|\boldsymbol\lambda\rangle - C^{\boldsymbol\lambda}) \in \mathbf{d}_q(q\mathscr{F}_0^\Lambda) = qL_0(\Lambda).$$

Hence, $d_{\boldsymbol{\lambda\mu}}(q) \in \delta_{\boldsymbol{\lambda\mu}} + q\mathbb{N}[q]$, for all $\boldsymbol\lambda \in \mathcal{P}^\Lambda$ and all $\boldsymbol\mu \in \mathcal{K}^\Lambda$. Thus, (e) holds and the proposition is proved. □

Remark 3.5.24. The difference between the upper and lower crystal bases, or the dual canonical and canonical bases, can be interpreted as changing between the bases of Specht modules and dual Specht modules. The global bases and their crystal lattices are:

upper	$q=0$	$B^{\boldsymbol\mu} \equiv [S^{\boldsymbol\mu}]$	$(\text{mod } \sum_{\boldsymbol\lambda\in\mathcal{K}^\Lambda} \mathbb{A}_0[S^{\boldsymbol\lambda}])$
lower	$q=\infty$	$\mathbf{c}_q(q^{-\operatorname{def}\boldsymbol\mu} B_{\boldsymbol\mu}) \equiv [S_{\mathbf{m}(\boldsymbol\mu)'}]$	$(\text{mod } \sum_{\boldsymbol\lambda\in\mathcal{K}^\Lambda} \mathbb{A}_\infty[S_{\boldsymbol\lambda}])$

where $\mathbf{m}$ is an involution on $\mathcal{K}_n^\Lambda$ that generalises the well-known Mullineux map for the symmetric groups. See Theorem 3.6.6 below.

Remark 3.5.25. As mentioned in Remark 3.5.10, a different action on the Fock space is commonly used in the literature. With respect to the Cartan pairing, as in Corollary 3.5.18, this action is the adjoint of the action in Theorem 3.5.9. As a consequence, the papers that use a different $U_q(\widehat{\mathfrak{sl}}_e)$-action also use a different coproduct for $U_q(\widehat{\mathfrak{sl}}_e)$, as they have to if they want Kashiwara's tensor product rule to connect the crystal bases at different levels for a fixed $\Lambda \in P^+$. In the dual set up, # categorifies the bar involution on $L(\Lambda)$, $\{\, B_{\boldsymbol\mu} \mid \boldsymbol\mu \in \mathcal{K}^\Lambda \,\}$ is the canonical basis, or lower global crystal basis at $q=0$ for $L(\Lambda)$ and $\{\, q^{\operatorname{def}\boldsymbol\mu} B^{\boldsymbol\mu} \mid \boldsymbol\mu \in \mathcal{K}^\Lambda \,\}$ is the dual canonical basis.

It is natural to ask when the conditions of Proposition 3.5.22 are satisfied. This is a difficult open problem. The next result implies that the conditions of Proposition 3.5.22 hold whenever $\mathbb{F}$ is a field of characteristic zero.

We can now state Ariki's celebrated Categorification Theorem. By specializing $q=1$ the quantum group $U_{\mathcal{A}}(\widehat{\mathfrak{sl}}_e) \otimes \mathbb{Q}$ becomes the Kac-Moody algebra $U(\widehat{\mathfrak{sl}}_e)$. Let $L_1(\Lambda)$ be the irreducible integrable highest weight $U(\widehat{\mathfrak{sl}}_e)$-module of high weight Λ. The canonical bases of $L_1(\Lambda)$ are obtained by specializing $q = 1$ in the canonical bases of $L_{\mathcal{A}}(\Lambda)$. Forgetting the grading in the results above, $\underline{\mathrm{Rep}_{\mathbb{Q}}^{\Lambda}} \cong L_1(\Lambda) \cong \underline{\mathrm{Proj}_{\mathbb{Q}}^{\Lambda}}$, where $\underline{\mathrm{Rep}_{\mathbb{Q}}^{\Lambda}} = \bigoplus_n \mathrm{Rep}(\underline{\mathscr{H}_n^{\Lambda}}) \otimes_{\mathbb{Z}} \mathbb{Q}$ and $\underline{\mathrm{Proj}_{\mathbb{Q}}^{\Lambda}} = \bigoplus_n \mathrm{Proj}(\underline{\mathscr{H}_n^{\Lambda}}) \otimes_{\mathbb{Z}} \mathbb{Q}$.

Theorem 3.5.26 (Ariki's Categorification Theorem [3,23]). *Suppose that $\mathbb{F}$ is a field of characteristic zero. Then the canonical basis of $L_1(\Lambda)$ coincides with the basis of (ungraded) projective indecomposable $\mathscr{H}_n^{\Lambda}$-modules $\{ [\underline{Y}^{\boldsymbol{\mu}}] \mid \boldsymbol{\mu} \in \mathcal{K}^{\Lambda} \}$ of $\underline{\mathrm{Proj}_{\mathbb{Q}}^{\Lambda}}$.*

This theorem was proved by Ariki [3, Theorem 4.4] when $v^2 \neq 1$ and by Brundan and Kleshchev when $v^2 = 1$ [23, Theorem 3.10]. For a detailed proof of this important result when $v^2 \neq 1$ see [5, Theorem 12.5]. For an overview and historical account of Ariki's theorem see [44].

Combining Theorem 3.5.26 with Proposition 3.5.22 we obtain the main result of this section.

Corollary 3.5.27 (Brundan and Kleshchev [22, Theorem 5.14]). *Suppose that $\mathbb{F}$ is a field of characteristic zero. Then the canonical basis of $L_{\mathcal{A}}(\Lambda)$ coincides with the basis $\{ q^{-\operatorname{def}\boldsymbol{\mu}}[Y^{\boldsymbol{\mu}}] \mid \boldsymbol{\mu} \in \mathcal{K}^{\Lambda} \}$ of $[\mathrm{Proj}_{\mathbb{Q}(q)}^{\Lambda}]$. In particular, $d_{\boldsymbol{\lambda\mu}}(q) \in \delta_{\boldsymbol{\lambda\mu}} + q\mathbb{N}[q]$, for all $\boldsymbol{\lambda} \in \mathcal{P}^{\Lambda}$ and $\boldsymbol{\mu} \in \mathcal{K}^{\Lambda}$.*

When Λ is a weight of level 2 and $e = \infty$ this was first proved by Brundan and Stroppel [26, Theorem 9.2]. For extensions of this result to cyclotomic quiver Hecke algebras of arbitrary type see [67,91,122,134].

Corollary 3.5.27 implies that the graded decomposition numbers $d_{\boldsymbol{\lambda\mu}}(q) = [S^{\boldsymbol{\lambda}} : D^{\boldsymbol{\mu}}]_q = b_{\boldsymbol{\lambda\mu}}(q)$ are *parabolic Kazhdan-Lusztig polynomials*. Explicit formulas are given in [55, Appendix A] and [99, Lemma 2.46].

For the canonical basis $\{B_{\boldsymbol{\mu}}\}$ it is immediate that the Laurent polynomials $b_{\boldsymbol{\lambda\mu}}(q) \in \mathbb{Z}[q]$ are polynomials, for $\boldsymbol{\lambda} \in \mathcal{P}_n^{\Lambda}$ and $\boldsymbol{\mu} \in \mathcal{K}_n^{\Lambda}$, however, it is a deep fact that their coefficients are *non-negative* integers. In contrast, it is immediate that $d_{\boldsymbol{\lambda\mu}}(q) \in \mathbb{N}[q, q^{-1}]$ but it is a deep fact that the graded decomposition numbers are polynomials rather than *Laurent polynomials*. Thus, the difficult result changes from positivity of coefficients in the ungraded setting, to positivity of exponents in the graded setting. In fact, it is also true when $\mathbb{F} = \mathbb{C}$ that the inverse graded decomposition numbers $e_{\boldsymbol{\lambda\mu}}(q) = b_{\boldsymbol{\lambda\mu}}(-q)$ are polynomials in q with non-negative integer coefficients. This is perhaps best explained by passing to the Koszul dual of the corresponding graded cyclotomic Schur algebras [7,55,128] using [55,99]; see [55, Lemma 2.15] where this is stated explicitly.

Brundan and Kleshchev's proof of Corollary 3.5.27 is quite different to the one given here. They have to work quite hard to define triangular bar involutions on $L_{\mathcal{A}}(\Lambda)$ whereas we have done this by exploiting the representation theory of $\mathscr{H}_n^\Lambda$. One benefit of Brundan and Kleshchev's approach is that they have an explicit description of the bar involution on $\mathscr{F}_{\mathcal{A}}^\Lambda$. In contrast, we have no hope of working with our bar involution unless we already know the graded decomposition matrices. On the other hand, the approach here works for an arbitrary multicharge $\boldsymbol{\kappa}$.

To complete the proof of Corollary 3.5.27, Brundan and Kleshchev lift Grojnowski's approach [49] to the representation theory of $\mathscr{H}_n^\Lambda$ to the graded setting. As a result they obtain graded analogues of Kleshchev's modular branching rules [18, 77, 78]. Under categorification, these branching rules correspond to the action of the crystal operators on the crystal graph of $L(\Lambda)$; see [22, Theorem 4.12]. By invoking Ariki's theorem they deduce an analogue of Corollary 3.5.27, although with a possibly different labelling of the irreducible modules. Finally, they prove that the labelling of the irreducible $\mathscr{H}_n^\Lambda$-modules coming from the branching rules agrees with the labelling of Corollary 1.5.2; compare with [6, 9].

We have not yet given an explicit description of the labelling of the irreducible $\mathscr{H}_n^\Lambda$-modules because we defined $\mathcal{K}_n^\Lambda = \{\,\boldsymbol{\mu} \in \mathcal{P}_n^\Lambda \mid \underline{D}^{\boldsymbol{\mu}} \neq 0\,\}$. Extending Definition 3.2.1, if $\boldsymbol{\mu} \in \mathcal{P}_n^\Lambda$ and given nodes $A < C$ define

$$d_A^C(\boldsymbol{\mu}) = \#\{\, B \in \operatorname{Add}_i(\boldsymbol{\mu}) \mid A{<}B{<}C\,\} - \#\{\, B \in \operatorname{Rem}_i(\boldsymbol{\mu}) \mid A{<}B{<}C\,\}.$$

Following Misra and Miwa [111] (but using Kleshchev's terminology [76]), a removable i-node A is **normal** if $d_A(\boldsymbol{\mu}) \le 0$ and $d_A^C(\boldsymbol{\mu}) < 0$ whenever $C \in \operatorname{Rem}_i(\boldsymbol{\mu})$ and $A < C$. A normal i-node A is **good** if $A \le B$ whenever B is a normal i-node. Write $\boldsymbol{\lambda} \xrightarrow{\text{good}} \boldsymbol{\mu}$ if $\mu = \lambda{+}A$ for some good node A. Misra and Miwa [111, Theorem 3.2] show that the crystal graph of $L_{\mathcal{A}}(\Lambda)$, considered as a submodule $\mathscr{F}_{\mathcal{A}}^\Lambda$, is the graph with vertex set

$$\mathscr{L}_0^\Lambda = \{\,\boldsymbol{\mu} \in \mathcal{P}^\Lambda \mid \boldsymbol{\mu} = v_\Lambda \text{ or } \boldsymbol{\lambda} \xrightarrow{\text{good}} \boldsymbol{\mu} \text{ for some } \boldsymbol{\lambda} \in \mathscr{L}_0^\Lambda\,\},$$

and with labelled edges $\boldsymbol{\lambda} \xrightarrow{i} \boldsymbol{\mu}$ whenever $\boldsymbol{\mu}$ is obtained from $\boldsymbol{\lambda}$ by adding a good i-node, for some $i \in I$. See [5, Theorem 11.11] for a self-contained proof of this result, couched in similar language.

Corollary 3.5.28 (Ariki [4, 8, 22]). *Suppose that $\mathbb{F}$ is an arbitrary field and that $\boldsymbol{\mu} \in \mathcal{P}_n^\Lambda$. Then $\mathcal{K}^\Lambda = \mathscr{L}_0^\Lambda$. That is, if $\boldsymbol{\mu} \in \mathcal{P}_n^\Lambda$ then $\underline{D}_{\mathbb{F}}^{\boldsymbol{\mu}} \neq 0$ if and only if $\boldsymbol{\mu} \in \mathscr{L}_0^\Lambda$.*

Proof. If $\mathbb{F}$ is a field of characteristic zero then $\{\,\boldsymbol{\mu} + qL_{\mathcal{A}}(\Lambda) \mid \boldsymbol{\mu} \in \mathcal{K}^\Lambda\,\}$ is a basis of $L_{\mathcal{A}}(\Lambda)/qL_{\mathcal{A}}(\Lambda)$ by Proposition 3.5.22 and Corollary 3.5.27. This basis of $L_{\mathcal{A}}(\Lambda)/qL_{\mathcal{A}}(\Lambda)$ is exactly the crystal basis of $L(\Lambda)$ by Corollary 3.5.27, so $\mathcal{K}_n^\Lambda = \mathscr{L}_0^\Lambda$ in characteristic zero. If $\mathbb{F}$ is a field of positive characteristic then a straightforward modular reduction argument shows that $\underline{D}^{\boldsymbol{\mu}}_{\mathbb{F}} \neq 0$ only if $\underline{D}^{\boldsymbol{\mu}}_{\mathbb{C}} \neq 0$, for $\boldsymbol{\mu} \in \mathcal{P}^\Lambda_n$ (for example, see §3.7 below). So, $\mathcal{K}^\Lambda \subseteq \mathscr{L}^\Lambda_0$. By Proposition 3.5.12, the number of irreducible $\mathscr{H}^\Lambda_n$-modules depends only on e, and not on the field $\mathbb{F}$, so $\mathcal{K}^\Lambda = \mathscr{L}^\Lambda_0$ as required. □

The idea in Proposition 3.5.12 that over any field the natural bases $\{[D^{\boldsymbol{\mu}}]\}$ and $\{[Y^{\boldsymbol{\mu}}]\}$ of $[\mathrm{Rep}^\Lambda_{\mathcal{A}}]$ and $[\mathrm{Proj}^\Lambda_{\mathcal{A}}]$, respectively, are distinguished bases of $L(\Lambda)$ goes back to at least Lascoux, Leclerc and Thibon [89]. These bases are often called *p-canonical bases*. This idea was generalized to higher levels by Ariki [3] and it played an important role in the classification of the irreducible $\mathscr{H}^\Lambda_n$-modules [4, 12] and in Grojnowski and Vazirani's work [49, 51, 79, 133]. The role of the crystal graphs in the representation theory of $\mathscr{R}^\Lambda_n$ is explored further in [68, 91].

3.6 *Homogeneous Garnir relations*

By Theorem 3.2.9, $\mathscr{R}^\Lambda_n$ is a graded cellular algebra and, as a consequence, there exist graded lifts of the Specht modules for arbitrary $\Lambda \in P^+$. At this point we cannot compute inside the graded Specht modules because we do not know how to write basis elements indexed by non-standard tableaux in terms of standard ones. This section shows how to do this. First, some combinatorics.

Fix a multipartition $\boldsymbol{\lambda}$ and a node $A = (l, r, c) \in \boldsymbol{\lambda}$. A (row) **Garnir node** of $\boldsymbol{\lambda}$ is any node $A = (l, r, c)$ such that $(l, r+1, c) \in \boldsymbol{\lambda}$. The (e, A)-**Garnir belt** is the set of nodes

$$\begin{aligned}\mathbf{B}_A = &\{\, (l,r,k) \in \boldsymbol{\lambda} \mid k \geq c \text{ and } e\lceil \tfrac{k-c+1}{e} \rceil \leq \lambda^{(l)}_r - c + 1 \,\}\\ \cup\ &\{\, (l,r+1,k) \in \boldsymbol{\lambda} \mid k \leq c \text{ and } c \geq e\lceil \tfrac{c-k+1}{e} \rceil \,\}.\end{aligned}$$

Let $b_A = \#\mathbf{B}_A/e$ and write $b_A = a_A + c_A$ where ea_A is the number of nodes in $\mathbf{B}_A$ in row (l, r). Let $\mathscr{D}_A$ be the set of minimal length right coset representatives of $\mathfrak{S}_{a_A} \times \mathfrak{S}_{c_A}$ in $\mathfrak{S}_{b_A}$; see, for example, [104, Proposition 3.3]. When $e = \infty$ these definitions should be interpreted as $\mathbf{B}_A = \emptyset$, $b_A = 0 = a_A = c_A$ and $\mathscr{D}_A = 1$.

Suppose A is a Garnir node of $\boldsymbol{\lambda}$. The rows of $\boldsymbol{\lambda}$ are indexed by pairs (l, r), corresponding to row r in $\lambda^{(l)}$ where $1 \leq l \leq \ell$ and $r \geq 1$. Order the row indices lexicographically. Let t_A be the $\boldsymbol{\lambda}$-tableau that agrees with $\mathsf{t}^{\boldsymbol{\lambda}}$ for all numbers $k < \mathsf{t}^{\boldsymbol{\lambda}}(A) = \mathsf{t}^{\boldsymbol{\lambda}}(l, r, c)$ and $k > \mathsf{t}^{\boldsymbol{\lambda}}(l, r+1, c)$ and where the remaining entries in rows (l, r) and $(l, r+1)$ are filled in increasing order from left to right first along the nodes in row $(l, r+1)$ that are in the first c columns but not in $\mathbf{B}_A$, then along the nodes in row (l, r) of $\mathbf{B}_A$ followed by the nodes in row $(l, r+1)$ of $\mathbf{B}_A$, and then along the remaining nodes in row (l, r).

3.6.1. **Example** As Garnir belts are contained in consecutive rows of the same component, the general case can be understood by looking at a two-rowed partition (of level one), so we consider the case $e = 3$, $\lambda = (14, 6)$ and $A = (1, 1, 4)$. Then

$\mathfrak{t}_A =$

1	2	3	5	6	7	8	9	10	11	12	13	17	18
4	14	15	16	19	20								

The lines in $\mathfrak{t}_A$ show how the $(3, A)$-Garnir belt decomposes into a disjoint union of "e-bricks". In general, b_A is equal to the number of e-bricks in the Garnir belt and a_A is the number of e-bricks in its first row. In this case, $b_A = 4$ and $a_A = 3$. Therefore, $\mathscr{D}_A = \{1, s_3, s_3s_2, s_3s_2s_1\}$. ◊

Let $k_A = \mathfrak{t}_A(A)$ be the number occupying A in $\mathfrak{t}_A$ and define

$$w_r^A = \prod_{a=k_A+e(r-1)}^{k_A+re-1} (a, a+e),$$

for $1 \le r < b_A$. The subgroup $\langle w_r^A \mid 1 \le r < b_A \rangle$ of $\mathfrak{S}_n$ is isomorphic to $\mathfrak{S}_{b_A}$ via the map $w_r^A \mapsto s_r$, for $1 \le r < b_A$. Set $\mathbf{i}^A = \mathbf{i}^{\mathfrak{t}_A}$ and $\tau_r^A = e(\mathbf{i}^A)(\psi_{w_r^A} + 1)$, for $1 \le r < b_A$. If $d \in \mathscr{D}_A$ choose a reduced expression $d = s_{r_1} \dots s_{r_k}$ for d and define

$$\tau_d^A = \tau_{r_1}^A \dots \tau_{r_k}^A \in e(\mathbf{i}^A)\mathscr{R}_n^\Lambda.$$

The elements τ_d^A of $\mathscr{R}_n^\Lambda$ seem to be very special and deserving of further study. They are homogeneous elements in $\mathscr{R}_n^\Lambda$ of degree zero that are independent of all choices of reduced expressions. Moreover, by [81, Theorem 4.13], the elements $\{\, \tau_r^A \mid 1 \le r < b_A \,\}$ satisfy the braid relations when they act on $S_{\mathcal{Z}}^{\boldsymbol{\lambda}}$ and they generate a copy of $\mathfrak{S}_{b_A}$ inside $\operatorname{End}_{\mathcal{Z}}(S_{\mathcal{Z}}^{\boldsymbol{\lambda}})$!

Theorem 3.6.2 (Kleshchev, Mathas and Ram [81, Theorem 6.23]**).** *Suppose that $\boldsymbol{\lambda} \in \mathcal{P}_n^\Lambda$ and that $\mathcal{Z}$ is an integral domain. The graded Specht module $S_{\mathcal{Z}}^{\boldsymbol{\lambda}}$ of $\mathscr{R}_n^\Lambda(\mathcal{Z})$ is isomorphic to the graded $\mathscr{R}_n^\Lambda$-module generated by a homogeneous element $v_{\mathfrak{t}^{\boldsymbol{\lambda}}}$ of degree $\deg \mathfrak{t}^{\boldsymbol{\lambda}}$ subject to the relations:*

a) $v_{\mathfrak{t}^{\boldsymbol{\lambda}}} e(\mathbf{i}) = \delta_{\mathbf{i}\mathbf{i}^{\boldsymbol{\lambda}}} v_{\mathfrak{t}^{\boldsymbol{\lambda}}}$.

b) $v_{\mathfrak{t}^{\boldsymbol{\lambda}}} y_s = 0$, *for* $1 \le s \le n$.

c) $v_{\mathfrak{t}^{\boldsymbol{\lambda}}} \psi_r = 0$ *whenever* $s_r \in \mathfrak{S}_{\boldsymbol{\lambda}}$, *for* $1 \le r < n$.

d) $\sum_{d \in \mathscr{D}_A} v_{\mathfrak{t}^{\boldsymbol{\lambda}}} \psi_{\mathfrak{t}_A} \tau_d^A = 0$, *for all Garnir nodes* $A \in \boldsymbol{\lambda}$.

Relations (a)–(c) already appear in [25] and, in terms of the cellular basis machinery, they are a consequence of Proposition 3.2.10.

The relations in part (d) are the **homogeneous Garnir relations**. These relations are a homogeneous form of the well-known Garnir relations of the symmetric group [59, Theorem 7.2]. There is an analogous description of the dual Specht modules $S_{\boldsymbol{\lambda}}$ in terms of *column Garnir relations* [81, §7]. Using Dyck tilings, Fayers [41] has shown how to write the homogeneous Garnir relations in terms of ψ-basis of the Specht module.

The most difficult part of the proof of Theorem 3.6.2 is showing that the τ_d^A satisfy the braid relations. This is proved using the Khovanov-Lauda diagram calculus that was briefly mentioned in §2.2. Like Theorem 3.2.9 this result holds over an arbitrary ring. To prove that the graded module defined by Theorem 3.6.2 has the correct rank the construction of the graded Specht module $S^{\boldsymbol{\lambda}}$ over a field in Theorem 3.2.4, from [25, 54], is used.

One of the main points of Theorem 3.6.2 is that it makes it possible to calculate in the graded Specht module over any ring. Prior to Theorem 3.6.2 the only way to compute inside the graded Specht modules was, in effect, to use the isomorphism $\mathscr{R}_n^\Lambda \xrightarrow{\sim} \mathscr{H}_n^\Lambda$ of Theorem 3.1.1 to work in the ungraded setting then use the inverse isomorphism $\mathscr{H}_n^\Lambda \xrightarrow{\sim} \mathscr{R}_n^\Lambda$ to get back to the graded setting. This made it difficult to keep track of, and to exploit, the grading on $S^{\boldsymbol{\lambda}}$ — and it was only possible to work with Specht modules defined over a field.

Theorem 3.6.2 also gives the relations for $S^{\boldsymbol{\lambda}}$ as a $\mathscr{R}_n$-module. From this perspective Theorem 3.6.2 can be used to give another construction of the graded Specht modules. For $\alpha, \beta \in Q^+$ let $\mathscr{R}_{\alpha,\beta} = \mathscr{R}_\alpha \otimes \mathscr{R}_\beta$. Definition 2.2.1 implies that there is a non-unital embedding $\mathscr{R}_{\alpha,\beta} \hookrightarrow \mathscr{R}_{\alpha+\beta}$ that maps $e(\mathbf{i}) \otimes e(\mathbf{j})$ to $e(\mathbf{i} \vee \mathbf{j})$, where $\mathbf{i} \vee \mathbf{j}$ is the concatenation of $\mathbf{i}$ and $\mathbf{j}$. Under this embedding the identity element of $\mathscr{R}_{\alpha,\beta}$ maps to

$$e_{\alpha,\beta} = \sum_{\mathbf{i} \in I^\alpha, \mathbf{j} \in I^\beta} e(\mathbf{i} \vee \mathbf{j}).$$

Definition 2.2.1 implies that $\mathscr{R}_{\alpha+\beta}$ is free as an $\mathscr{R}_{\alpha,\beta}$-module, so the functor

$$\mathrm{Ind}_{\alpha,\beta}^{\alpha+\beta}(M \boxtimes N) = (M \boxtimes N)e_{\alpha,\beta} \otimes_{\mathscr{R}_{\alpha,\beta}} \mathscr{R}_{\alpha+\beta}$$

is a left adjoint to the natural restriction map. Iterating this construction, given $\beta_1, \ldots, \beta_\ell \in Q^+$ and $\mathscr{R}_{\beta_k}$ modules M_k, for $1 \le k \le \ell$, define

$$M_1 \circ \cdots \circ M_\ell = \mathrm{Ind}_{\beta_1,\ldots,\beta_\ell}^{\beta_1+\cdots+\beta_\ell}(M_1 \boxtimes \cdots \boxtimes M_\ell).$$

The definition of the graded Specht modules by generators and relations in Theorem 3.6.2 makes the following result almost obvious. This description of the Specht modules is part of the folklore of these algebras with several authors [23, 133] using it as the definition of Specht modules.

Corollary 3.6.3 (Kleshchev, Mathas and Ram [81, Theorem 8.2]**).** *Suppose that $\lambda^{(k)} \in \mathcal{P}_{1,\beta_k}$, for $\beta_k \in Q^+$ and $1 \le k \le \ell$, so that $\boldsymbol{\lambda} \in \mathcal{P}^\Lambda_\beta$, where $\beta = \beta_1 + \cdots + \beta_\ell$. Then there is an isomorphism of graded $\mathscr{R}^\Lambda_n$-modules (and graded $\mathscr{R}_n$-modules),*

$$S^{\boldsymbol{\lambda}}\langle \deg \mathsf{t}^{\lambda^{(1)}} + \cdots + \deg \mathsf{t}^{\lambda^{(\ell)}} \rangle \cong (S^{\lambda^{(1)}} \circ \cdots \circ S^{\lambda^{(\ell)}})\langle \deg \mathsf{t}^{\boldsymbol{\lambda}} \rangle,$$

where on the right hand side $S^{\lambda^{(k)}}$ is considered as an $\mathscr{R}_{\beta_k}$-module, for $1 \le k \le \ell$.

A second application of Theorem 3.6.2 is a generalization of James' famous result [59, Theorem 8.15] for symmetric groups that describes what happens to the Specht modules when they are tensored with the sign representation. First some notation.

Following [81, §3.3], for $\mathbf{i} \in I^n$ let $-\mathbf{i} = (-i_1, \cdots - i_n) \in I^n$. Recalling the multicharge $\boldsymbol{\kappa}$ from §1.2, set $\boldsymbol{\kappa}' = (-\kappa_\ell, \ldots, -\kappa_1)$ and let $\Lambda' = \Lambda(\boldsymbol{\kappa}') \in P^+$. Similarly, if $\beta = \sum_i a_i \alpha_i \in Q^+$ let $\beta' = \sum_{i \in I} a_i \alpha_{-i}$. Inspecting Definition 2.2.9, there is a unique isomorphism of graded algebras

$$(3.6.4)\quad \mathbf{sgn}\colon \mathscr{R}^\Lambda_\beta \longrightarrow \mathscr{R}^{\Lambda'}_{\beta'}; \quad e(\mathbf{i}) \mapsto e(-\mathbf{i}), \quad y_r \mapsto -y_r, \quad \text{and} \quad \psi_s \mapsto -\psi_s,$$

for all admissible r and s and $\mathbf{i} \in I^\beta$. The involution $\mathbf{sgn}$ induces an equivalence of categories $\operatorname{Rep}(\mathscr{R}^{\Lambda'}_{\beta'}) \longrightarrow \operatorname{Rep}(\mathscr{R}^\Lambda_\beta)$ that sends an $\mathscr{R}^{\Lambda'}_{\beta'}$-module M to the $\mathscr{R}^\Lambda_\beta$-module $M^{\mathbf{sgn}}$, where the $\mathscr{R}^\Lambda_\beta$-action is twisted by $\mathbf{sgn}$.

Corollary 3.6.5 (Kleshchev, Mathas and Ram [81, Theorem 8.5]**).** *Suppose that $\boldsymbol{\mu} \in \mathcal{P}^\Lambda_\beta$, for $\beta \in Q^+$. Then*

$$S^{\boldsymbol{\mu}} \cong (S_{\boldsymbol{\mu}'})^{\mathbf{sgn}} \quad \text{and} \quad S_{\boldsymbol{\mu}} \cong (S^{\boldsymbol{\mu}'})^{\mathbf{sgn}}$$

as $\mathscr{R}^\Lambda_\beta$-modules.

In [81] this is proved by checking the relations in Theorem 3.6.2. As noted in [55, Proposition 3.26], this can be proved more transparently by noting that, up to sign, the involution $\mathbf{sgn}$ maps the ψ-basis of $\mathscr{R}^\Lambda_n$ to the ψ'-basis of $\mathscr{R}^{\Lambda'}_{\beta'}$. Some care must be taken with the notation here. For example, if $\boldsymbol{\mu} \in \mathcal{P}^\Lambda_\beta$ then $\boldsymbol{\mu}' \in \mathcal{P}^\Lambda_{\beta'}$. See [55, §3.7] for more details.

We give an application of these results to the graded decomposition numbers. First, by Corollary 3.5.28 if $\boldsymbol{\mu} \in \mathcal{K}^\Lambda_n$ there exists $\mathbf{i} \in I^n$ and a sequence of multipartitions $\boldsymbol{\mu}_0 = \upsilon_\Lambda, \boldsymbol{\mu}_1, \ldots, \boldsymbol{\mu}_n = \boldsymbol{\mu}$ in $\mathcal{K}^\Lambda$ such that $\boldsymbol{\mu}_{k+1}$ is obtained from $\boldsymbol{\mu}_k$ by adding a good i_k-node, for $0 \le k < n$. It follows from the modular branching rules [22, Theorem 4.12], and properties of crystal graphs, that there exists a unique sequence of multipartitions $\mathbf{m}(\boldsymbol{\mu}_0) = v_\Lambda, \mathbf{m}(\boldsymbol{\mu}_1), \dots, \mathbf{m}(\boldsymbol{\mu}_n) = \mathbf{m}(\boldsymbol{\mu})$ such that $\mathbf{m}(\boldsymbol{\mu}_{k+1})$ is obtained from $\mathbf{m}(\boldsymbol{\mu}_k)$ by adding a good $-i_k$-node and $\mathbf{m}(\boldsymbol{\mu}_{k+1}) \in \mathcal{K}^{\Lambda'}_{k+1}$, for $1 \le k \le n$. The **Mullineux conjugate** of $\boldsymbol{\mu}$ is the multipartition $\mathbf{m}(\boldsymbol{\mu})$. Thus, $D^{\mathbf{m}(\boldsymbol{\mu})}$ is a non-zero irreducible $\mathscr{R}^{\Lambda'}_{\beta'}$-module. We emphasize that the $\mathscr{R}^{\Lambda'}_{\beta'}$-module $D^{\mathbf{m}(\boldsymbol{\mu})}$ is defined using the ψ-basis of $\mathscr{R}^{\Lambda'}_{\beta'}$ and hence the crystal theory used in §3.5, with respect to the multicharge $\boldsymbol{\kappa}'$.

Theorem 3.6.6. *Suppose that $\boldsymbol{\mu} \in \mathcal{K}^\Lambda_\beta$, for $\beta \in Q^+$. Then*

$$(D^{\mathbf{m}(\boldsymbol{\mu})})^{\mathtt{sgn}} \cong D^{\boldsymbol{\mu}}$$

as $\mathscr{R}^\Lambda_\beta$-modules.

Proof. As $\mathtt{sgn}$ is an equivalence of categories, $(D^{\mathbf{m}(\boldsymbol{\mu})})^{\mathtt{sgn}} \cong D^{\boldsymbol{\nu}}\langle d\rangle$ for some $\boldsymbol{\nu} \in \mathcal{K}^\Lambda_\beta$ and $d \in \mathbb{Z}$ by Corollary 3.2.7. Since $\mathtt{sgn}$ is homogeneous, by Theorem 2.1.5(a),

$$\mathit{dim}_q (D^{\mathbf{m}(\boldsymbol{\mu})})^{\mathtt{sgn}} = \mathit{dim}_q\, D^{\mathbf{m}(\boldsymbol{\mu})} = \overline{\mathit{dim}_q\, D^{\mathbf{m}(\boldsymbol{\mu})}} = \overline{\mathit{dim}_q (D^{\mathbf{m}(\boldsymbol{\mu})})^{\mathtt{sgn}}},$$

so that $d = 0$ and $(D^{\mathbf{m}(\boldsymbol{\mu})})^{\mathtt{sgn}} \cong D^{\boldsymbol{\nu}}$. To show that $\boldsymbol{\nu} = \boldsymbol{\mu}$ it is now enough to work in the ungraded setting. Therefore, we can either use the modular branching rules of [6, 49], or their graded counterparts from [22, Theorem 4.12], together with what is by now a standard argument due to Kleshchev [77, Theorem 4.7], to show that $\boldsymbol{\nu} = \boldsymbol{\mu}$. □

The $\mathtt{sgn}$ map induces an equivalence $\operatorname{Rep}(\mathscr{R}^{\Lambda'}_{\beta'}) \longrightarrow \operatorname{Rep}(\mathscr{R}^\Lambda_\beta)$. As $\mathtt{sgn}$ is an involution, we also write $\mathtt{sgn}\colon \operatorname{Rep}(\mathscr{R}^\Lambda_\beta) \longrightarrow \operatorname{Rep}(\mathscr{R}^{\Lambda'}_{\beta'})$ for the inverse equivalence. The last two results can now be written as $(S^{\boldsymbol{\lambda}})^{\mathtt{sgn}} \cong S_{\boldsymbol{\lambda}'}$ and $(D^{\boldsymbol{\mu}})^{\mathtt{sgn}} \cong D^{\mathbf{m}(\boldsymbol{\mu})}$ as $\mathscr{R}^{\Lambda'}_{\beta'}$-modules, for $\boldsymbol{\lambda} \in \mathcal{P}^\Lambda_\beta$ and $\boldsymbol{\mu} \in \mathcal{K}^\Lambda_\beta$.

Corollary 3.6.7. *Suppose that F is a field and that $\boldsymbol{\lambda} \in \mathcal{P}^\Lambda_\beta$ and $\boldsymbol{\mu} \in \mathcal{K}^\Lambda_\beta$. Then $d_{\boldsymbol{\mu}\boldsymbol{\mu}}(q) = 1$, $d_{\mathbf{m}(\boldsymbol{\mu})'\boldsymbol{\mu}}(q) = q^{\operatorname{def}\boldsymbol{\mu}}$ and $d_{\boldsymbol{\lambda}\boldsymbol{\mu}}(q) \ne 0$ only if $\mathbf{m}(\boldsymbol{\mu})' \trianglerighteq \boldsymbol{\lambda} \trianglerighteq \boldsymbol{\mu}$. Moreover, if $F = \mathbb{C}$ then $0 < \deg d^{\mathbb{C}}_{\boldsymbol{\lambda}\boldsymbol{\mu}}(q) < \operatorname{def}\boldsymbol{\mu}$ whenever $\mathbf{m}(\boldsymbol{\mu})' \triangleright \boldsymbol{\lambda} \triangleright \boldsymbol{\mu}$.*

Proof. Suppose that $\boldsymbol{\lambda} \in \mathcal{P}^\Lambda_\beta$ and $\boldsymbol{\mu} \in \mathcal{K}^\Lambda_\beta$. Then

$$[S^{\boldsymbol{\lambda}} : D^{\boldsymbol{\mu}}]_q = [(S^{\boldsymbol{\lambda}})^{\mathtt{sgn}} : (D^{\boldsymbol{\mu}})^{\mathtt{sgn}}]_q = [S_{\boldsymbol{\lambda}'} : D^{m(\boldsymbol{\mu})}]_q,$$

by Corollary 3.6.5 and Theorem 3.6.6, respectively. Therefore, using Corollary 3.3.5 and Theorem 2.1.5(a),

$$[S^{\boldsymbol{\lambda}} : D^{\boldsymbol{\mu}}]_q = q^{\operatorname{def}\boldsymbol{\lambda}}[(S^{\boldsymbol{\lambda}'})^{\circledast} : D^{m(\boldsymbol{\mu})}]_q, = q^{\operatorname{def}\boldsymbol{\mu}}\overline{[S^{\boldsymbol{\lambda}'} : D^{m(\boldsymbol{\mu})}]_q}.$$

By Theorem 2.1.5(c), if $\boldsymbol{\tau} \in \mathcal{K}_\beta^\Lambda$ and $\boldsymbol{\sigma} \in \mathcal{P}_\beta^\Lambda$ then $d_{\boldsymbol{\tau}\boldsymbol{\tau}}(1) = 1$ and $d_{\boldsymbol{\sigma}\boldsymbol{\tau}}(q) \neq 0$ only if $\boldsymbol{\sigma} \trianglerighteq \boldsymbol{\tau}$. Therefore, $d_{\mathbf{m}(\boldsymbol{\mu})'\boldsymbol{\mu}}(q) = q^{\operatorname{def}\boldsymbol{\mu}}\overline{d_{\mathbf{m}(\boldsymbol{\mu})\mathbf{m}(\boldsymbol{\mu})}(q)} = q^{\operatorname{def}\boldsymbol{\mu}}$ and $d_{\boldsymbol{\lambda}\boldsymbol{\mu}}(q) \neq 0$ only if $\mathbf{m}(\boldsymbol{\mu})' \trianglerighteq \boldsymbol{\lambda} \trianglerighteq \boldsymbol{\mu}$. The argument so far is valid over any field. Now suppose that $F = \mathbb{C}$. Then $d_{\boldsymbol{\lambda}\boldsymbol{\mu}}(q) \in \delta_{\boldsymbol{\lambda}\boldsymbol{\mu}} + q\mathbb{N}[q]$, by Corollary 3.5.27, so the remaining statement about the degrees of the graded decomposition numbers follows. □

Corollary 3.6.7 is the easy half of a conjecture of Fayers [40], which he was interested in because it leads to a faster algorithm for computing the graded decomposition numbers of $\mathscr{H}_n^\Lambda$. At the level of canonical bases the last two results correspond to the fact shifting by the defect transforms an upper crystal base into a lower crystal base [69, Lemma 2.4.1]. See also [22, Remark 3.19].

3.7 *Graded adjustment matrices*

All of the results in this section have their origin in the work of James [60] and Geck [43] on *adjustment matrices.* Brundan and Kleshchev have given two different approaches to graded decomposition matrices in [21, §6] and [22, §5.6]. In this section we give third cellular algebra approach. Even though our definitions and proofs are different, it is easy to see that everything in this section is equivalent to definitions or theorems of Brundan and Kleshchev — or to graded analogues of results of James and Geck.

Before we introduce the adjustment matrices, let $\mathcal{A}[I^n]$ be the free $\mathcal{A}$-module generated by I^n. The **q-character** of a finite dimensional $\mathscr{R}_n$-module M is

$$\operatorname{Ch}_q M = \sum_{\mathbf{i}\in I^n} \operatorname{dim}_q M_{\mathbf{i}} \cdot \mathbf{i} \in \mathcal{A}[I^n],$$

where $M_{\mathbf{i}} = Me(\mathbf{i})$, for $\mathbf{i} \in I^n$. For example, $\operatorname{Ch}_q S^{\boldsymbol{\lambda}} = \sum_{\mathfrak{t}\in\operatorname{Std}(\boldsymbol{\lambda})} q^{\deg(\mathfrak{t})} \cdot \mathbf{i}^{\mathfrak{t}}$.

Theorem 3.7.1 ([74, Theorem 3.17]). *Let $\mathcal{Z}$ be a field. Then the map*

$$\operatorname{Ch}_q : [\operatorname{Rep}(\mathscr{R}_n)] \longrightarrow \mathcal{A}[I^n]; [M] \mapsto \operatorname{Ch}_q M$$

is injective.

As every $\mathscr{R}_n^\Lambda$-module can be considered as an $\mathscr{R}_n$-module by inflation, it follows that the restriction of Ch_q to $[\operatorname{Rep}(\mathscr{R}_n^\Lambda)]$ is still injective. Extend the map ⊛ to $\mathcal{A}[I^n]$ by defining $\big(\sum_{\mathbf{i}} f_{\mathbf{i}}(q)\cdot\mathbf{i}\big)^{\circledast} = \sum_{\mathbf{i}} \overline{f_{\mathbf{i}}(q)}\cdot\mathbf{i}$. Then $(\operatorname{Ch}_q[M])^{\circledast} = \operatorname{Ch}_q[M^{\circledast}]$, for all $M \in \operatorname{Rep}(\mathscr{R}_n^\Lambda)$.

This section compares representations of cyclotomic KLR algebras over different fields. Write $S^{\boldsymbol{\lambda}}_{\mathcal{Z}}$ and $D^{\boldsymbol{\mu}}_{\mathcal{Z}}$ to emphasize that these modules are $\mathscr{R}^{\Lambda}_n(\mathcal{Z})$-modules, for $\boldsymbol{\lambda} \in \mathcal{P}^{\Lambda}_n$ and $\boldsymbol{\mu} \in \mathcal{K}^{\Lambda}_n$. If $\mathcal{Z} = F$ is a field, and K is an extension of F, then $D^{\boldsymbol{\mu}}_K \cong D^{\boldsymbol{\mu}}_F \otimes_F K$ since $D^{\boldsymbol{\mu}}_F$ is absolutely irreducible by Theorem 2.1.5. Therefore $\operatorname{Ch}_q D^{\boldsymbol{\mu}}_F$ depends only on $\boldsymbol{\mu}$ and the characteristic of F.

By Theorem 3.2.9, or by Theorem 3.6.2, the graded Specht module $S^{\boldsymbol{\mu}}_{\mathbb{Z}}$ is defined over $\mathbb{Z}$ and $S^{\boldsymbol{\mu}}_{\mathcal{Z}} \cong S^{\boldsymbol{\mu}}_{\mathbb{Z}} \otimes_{\mathbb{Z}} \mathcal{Z}$ for any commutative ring $\mathcal{Z}$. The graded Specht module $S^{\boldsymbol{\mu}}_{\mathbb{Z}}$ has basis $\{\, \psi_{\mathfrak{t}} \mid \mathfrak{t} \in \operatorname{Std}(\boldsymbol{\mu}) \,\}$ and it comes equipped with a $\mathbb{Z}$-valued bilinear form $\langle\ ,\ \rangle$ that is determined by

$$(3.7.2) \qquad \langle \psi_{\mathfrak{s}}, \psi_{\mathfrak{t}} \rangle \psi_{\mathfrak{t}^{\boldsymbol{\lambda}}} = \psi_{\mathfrak{s}} \psi_{\mathfrak{t}\mathfrak{t}^{\boldsymbol{\lambda}}} = \psi_{\mathfrak{s}} \psi^{\star}_{d(\mathfrak{t})} y^{\boldsymbol{\mu}} e(\mathbf{i}^{\boldsymbol{\lambda}}).$$

Following (1.3.3), define the (integral) radical of $S^{\boldsymbol{\mu}}_{\mathbb{Z}}$ to be

$$\operatorname{rad} S^{\boldsymbol{\mu}}_{\mathbb{Z}} = \{\, x \in S^{\boldsymbol{\mu}}_{\mathbb{Z}} \mid \langle x, y \rangle = 0 \text{ for all } y \in S^{\boldsymbol{\mu}}_{\mathbb{Z}} \,\}.$$

In fact, by (3.7.2), $\operatorname{rad} S^{\boldsymbol{\mu}}_{\mathbb{Z}} = \{\, x \in S^{\boldsymbol{\mu}}_{\mathbb{Z}} \mid xa = 0 \text{ for all } a \in (\mathscr{R}^{\Lambda}_n)^{\trianglerighteq \boldsymbol{\mu}} \,\}$.

Definition 3.7.3. Suppose that $\boldsymbol{\mu} \in \mathcal{P}^{\Lambda}_n$. Let $D^{\boldsymbol{\mu}}_{\mathbb{Z}} = S^{\boldsymbol{\mu}}_{\mathbb{Z}} / \operatorname{rad} S^{\boldsymbol{\mu}}_{\mathbb{Z}}$.

By definition, $\operatorname{rad} S^{\boldsymbol{\mu}}_{\mathbb{Z}}$ is a graded submodule of $S^{\boldsymbol{\mu}}_{\mathbb{Z}}$, so $D^{\boldsymbol{\mu}}_{\mathbb{Z}}$ is a graded $\mathscr{R}^{\Lambda}_n(\mathbb{Z})$-module. Hence, $D^{\boldsymbol{\mu}}_{\mathbb{Z}} \otimes_{\mathbb{Z}} \mathcal{Z}$ is a graded $\mathscr{R}^{\Lambda}_n(\mathcal{Z})$-module for any ring $\mathcal{Z}$.

The following result should be compared with [21, Theorem 6.5].

Theorem 3.7.4. *Suppose that* $\boldsymbol{\mu} \in \mathcal{P}^{\Lambda}_n$*. Then* $\operatorname{rad} S^{\boldsymbol{\mu}}_{\mathbb{Z}}$ *is a* $\mathbb{Z}$*-lattice in* $\operatorname{rad} S^{\boldsymbol{\mu}}_{\mathbb{Q}}$ *and* $D^{\boldsymbol{\mu}}_{\mathbb{Z}}$ *is a* $\mathbb{Z}$*-lattice in* $D^{\boldsymbol{\mu}}_{\mathbb{Q}}$*. Consequently,* $D^{\boldsymbol{\mu}}_{\mathbb{Q}} = D^{\boldsymbol{\mu}}_{\mathbb{Z}} \otimes_{\mathbb{Z}} \mathbb{Q}$ *and* $\operatorname{Ch}_q D^{\boldsymbol{\mu}}_{\mathbb{Z}} = \operatorname{Ch}_q D^{\boldsymbol{\mu}}_{\mathbb{Q}}$.

Proof. Let $G^{\boldsymbol{\mu}}_{\mathbb{Z}} = (\langle \psi_{\mathfrak{s}}, \psi_{\mathfrak{t}} \rangle)$ be the (integral) Gram matrix of $S^{\boldsymbol{\mu}}_{\mathbb{Z}}$. As $\mathbb{Z}$ is a principal ideal domain, by the Smith normal form there exists a pair of bases $\{a_r\}$ and $\{b_s\}$ of $S^{\boldsymbol{\mu}}_{\mathbb{Z}}$ such that $(\langle a_r, b_s \rangle) = \operatorname{diag}(d_1, d_2, \dots, d_z)$ for some non-negative integers such that $d_1 \mid d_2 \mid \cdots \mid d_z$, where $d_r = 0$ only if $d_s = 0$ for all $s \geq r$. That is, $d_1, \dots, d_z$ are the elementary divisors of $G^{\boldsymbol{\mu}}_{\mathbb{Z}}$. As the form is homogeneous, we may assume that the bases $\{a_r\}$ and $\{b_s\}$ are homogeneous with $\deg a_r = \deg \mathfrak{t}_r = -\deg b_r$, for some ordering $\operatorname{Std}(\boldsymbol{\mu}) = \{\mathfrak{t}_1, \dots, \mathfrak{t}_z\}$. Moreover, in view of Proposition 3.2.10(a), we can also assume that $a_r e(\mathbf{i}) = \delta_{\mathbf{i}^{\mathfrak{t}_r}, \mathbf{i}} a_r$ and $b_s e(\mathbf{i}) = \delta_{\mathbf{i}^{\mathfrak{t}_s}, \mathbf{i}} b_s$, for $1 \leq r, s \leq z$ and $\mathbf{i} \in I^n$. Comparing with the definitions above, it follows that $\{\, a_r \mid d_r = 0 \,\}$ is a basis of $\operatorname{rad} S^{\boldsymbol{\mu}}_{\mathbb{Z}}$ and that $\{\, a_r + \operatorname{rad} S^{\boldsymbol{\mu}}_{\mathbb{Z}} \mid d_r \neq 0 \,\}$ is a basis of $D^{\boldsymbol{\mu}}_{\mathbb{Z}}$. All of our claims now follow. □

For an arbitrary field F, it is usually not the case that $D^{\boldsymbol{\mu}}_F$ is isomorphic to $D^{\boldsymbol{\mu}}_{\mathbb{Z}} \otimes_{\mathbb{Z}} F$ as an $\mathscr{R}^\Lambda_n(F)$-module. Indeed, if F is a field of characteristic $p > 0$ then the argument of Theorem 3.7.4 shows that

$$\dim_F D^{\boldsymbol{\mu}}_F = \{\, 1 \le r \le z \mid d_r \not\equiv 0 \pmod p \,\} \le \operatorname{rank}_{\mathbb{Z}} D^{\boldsymbol{\mu}}_{\mathbb{Z}} = \dim_{\mathbb{Q}} D^{\boldsymbol{\mu}}_{\mathbb{Q}},$$

with equality if and only if all of the non-zero elementary divisors of $G^{\boldsymbol{\mu}}_{\mathbb{Z}}$ are coprime to p.

Definition 3.7.5 (cf. Brundan and Kleshchev [22, §5.6]**).** *Suppose that F is a field. For $\boldsymbol{\lambda}, \boldsymbol{\mu} \in \mathcal{K}^\Lambda_n$ define Laurent polynomials $a^F_{\boldsymbol{\lambda\mu}}(q) \in \mathbb{N}[q, q^{-1}]$ by*

$$a^F_{\boldsymbol{\lambda\mu}}(q) = \sum_{d\in\mathbb{Z}} [D^{\boldsymbol{\lambda}}_{\mathbb{Z}} \otimes_{\mathbb{Z}} F : D^{\boldsymbol{\mu}}_F\langle d\rangle]\, q^d.$$

*The matrix $\mathbf{a}^F_q = \big(a^F_{\boldsymbol{\lambda\mu}}(q)\big)$ is the **graded adjustment matrix** of $\mathscr{R}^\Lambda_n(F)$.*

Recall that $d_{\boldsymbol{\lambda\mu}}(q)$ is a graded decomposition number of $\mathscr{R}^\Lambda_n$. If we want to emphasize the field F then we write $d^F_{\boldsymbol{\lambda\mu}}(q) = [S^{\boldsymbol{\lambda}}_F : D^{\boldsymbol{\mu}}_F]_q$ and $\mathbf{d}^F_q = \big(d^F_{\boldsymbol{\lambda\mu}}(q)\big)$. Note that e is always fixed.

Theorem 3.7.6 (cf. Brundan and Kleshchev [22, Theorem 5.17]**).** *Suppose that F is a field. Then:*

a) *If $\boldsymbol{\lambda}, \boldsymbol{\mu} \in \mathcal{K}^\Lambda_n$ then $a^F_{\boldsymbol{\lambda\lambda}}(1) = 1$ and $a^F_{\boldsymbol{\lambda\mu}}(q) \ne 0$ only if $\boldsymbol{\lambda} \trianglerighteq \boldsymbol{\mu}$. Moreover, $\overline{a^F_{\boldsymbol{\lambda\mu}}(q)} = a^F_{\boldsymbol{\lambda\mu}}(q)$.*
b) *We have, $\mathbf{d}^F_q = \mathbf{d}^{\mathbb{Q}}_q \circ \mathbf{a}^F_q$. That is, if $\boldsymbol{\lambda} \in \mathcal{P}^\Lambda_n$ and $\boldsymbol{\mu} \in \mathcal{K}^\Lambda_n$ then*

$$[S^{\boldsymbol{\lambda}}_F : D^{\boldsymbol{\mu}}_F]_q = d^F_{\boldsymbol{\lambda\mu}}(q) = \sum_{\boldsymbol{\nu}\in\mathcal{K}^\Lambda_n} d^{\mathbb{Q}}_{\boldsymbol{\lambda\nu}}(q) a^F_{\boldsymbol{\nu\mu}}(q).$$

Proof. By construction, every composition factor of $D^{\boldsymbol{\lambda}}_{\mathbb{Z}} \otimes F$ is a composition factor of $S^{\boldsymbol{\lambda}}_F$, so the first two properties of the Laurent polynomials $a^F_{\boldsymbol{\lambda\mu}}(q)$ follow from Theorem 2.1.5. By Theorem 3.7.4, the adjustment matrix induces a well-defined map of Grothendieck groups $\mathbf{a}^F_q : [\operatorname{Rep}(\mathscr{R}^\Lambda_n(\mathbb{Q}))] \longrightarrow [\operatorname{Rep}(\mathscr{R}^\Lambda_n(F))]$ given by

$$\mathbf{a}^F_q\big([D^{\boldsymbol{\lambda}}_{\mathbb{Q}}]\big) = [D^{\boldsymbol{\lambda}}_{\mathbb{Z}} \otimes F] = \sum_{\boldsymbol{\mu}\in\mathcal{K}^\Lambda_n} a^F_{\boldsymbol{\lambda\mu}}(q)[D^{\boldsymbol{\mu}}_F].$$

Taking q-characters, $\operatorname{Ch}_q D^{\boldsymbol{\lambda}}_{\mathbb{Q}} = \sum_{\boldsymbol{\mu}} a^F_{\boldsymbol{\lambda\mu}}(q) \operatorname{Ch}_q D^{\boldsymbol{\mu}}_F$. Applying ⊛ to both sides gives $\operatorname{Ch}_q D^{\boldsymbol{\lambda}}_{\mathbb{Q}} = \sum_{\boldsymbol{\mu}} \overline{a^F_{\boldsymbol{\lambda\mu}}(q)} \operatorname{Ch}_q D^{\boldsymbol{\mu}}_F$. Therefore, $\overline{a^F_{\boldsymbol{\lambda\mu}}(q)} = a^F_{\boldsymbol{\lambda\mu}}(q)$ by Theorem 3.7.1, completing the proof of part (a). For (b), since $S^{\boldsymbol{\lambda}}_F \cong S^{\boldsymbol{\lambda}}_{\mathbb{Z}} \otimes_{\mathbb{Z}} F$,

$$[S^{\boldsymbol{\lambda}}_F] = \mathbf{a}^F_q\big([S^{\boldsymbol{\lambda}}_{\mathbb{Q}}]\big) = \mathbf{a}^F_q\Big(\sum_{\boldsymbol{\nu}\in\mathcal{K}^\Lambda_n} d^{\mathbb{Q}}_{\boldsymbol{\lambda\nu}}(q)[D^{\boldsymbol{\nu}}_{\mathbb{Q}}]\Big) = \sum_{\boldsymbol{\nu}\in\mathcal{K}^\Lambda_n} \sum_{\boldsymbol{\mu}\in\mathcal{K}^\Lambda_n} d^{\mathbb{Q}}_{\boldsymbol{\lambda\nu}}(q) a^F_{\boldsymbol{\nu\mu}}(q)[D^{\boldsymbol{\mu}}_F].$$

Comparing the coefficient of $[D^{\boldsymbol{\mu}}_F]$ on both sides completes the proof. □

Corollary 3.5.27 determines the graded decomposition numbers of the cyclotomic Hecke algebras in characteristic zero. There are several different algorithms for computing the graded decomposition numbers in characteristic zero [40, 46, 55, 82, 89, 131]. To determine the graded decomposition numbers in positive characteristic it is enough to compute the adjustment matrices of Theorem 3.7.6. The simplest case will be when $a^F_{\boldsymbol{\lambda}\boldsymbol{\mu}}(q) = \delta_{\boldsymbol{\lambda}\boldsymbol{\mu}}$, for all $\boldsymbol{\lambda}, \boldsymbol{\mu} \in \mathcal{K}^\Lambda_n$. Unfortunately, we currently have no idea when this happens. Two failed conjectures for when $\mathbf{a}^F_q$ is the identity matrix are discussed in Example 3.7.10 and Example 3.7.11 below.

We now compute the integral Gram matrices $G^{\boldsymbol{\lambda}}_{\mathbb{Z}} = \big(\langle\psi_{\mathsf{s}}, \psi_{\mathsf{t}}\rangle\big)$ and some adjustment matrix entries in several examples.

3.7.7. **Example** (Semisimple algebras) Suppose that $e > n$ and that $(\Lambda, \alpha_{i,n}) \le 1$, for all $i \in I$. Let $\boldsymbol{\lambda} \in \mathcal{P}^\Lambda_n$ and $\mathsf{s}, \mathsf{t} \in \mathrm{Std}(\boldsymbol{\lambda})$. Then $\langle\psi_{\mathsf{s}}, \psi_{\mathsf{t}}\rangle = \delta_{\mathsf{st}}$ because $\mathbf{i}^{\mathsf{s}} = \mathbf{i}^{\mathsf{t}}$ if and only if $\mathsf{s} = \mathsf{t}$ by Lemma 2.4.1. Hence, $G^{\boldsymbol{\lambda}}_{\mathbb{Z}}$ is the identity matrix for all $\boldsymbol{\lambda} \in \mathcal{P}^\Lambda_n$. $\diamondsuit$

3.7.8. **Example** (Nil-Hecke algebras) Suppose that $\Lambda = n\Lambda_i$ and $\beta = n\alpha_i$, for some $i \in I$. Let $\boldsymbol{\lambda} = (1|1|\dots|1) \in \mathcal{P}^\Lambda_n$, as in §2.5, and suppose $\mathsf{s}, \mathsf{t} \in \mathrm{Std}(\boldsymbol{\lambda})$ then $\langle\psi_{\mathsf{s}}, \psi_{\mathsf{t}}\rangle\psi_{\mathsf{t}^{\boldsymbol{\lambda}}} = \psi_{\mathsf{s}}\psi^\star_{d(\mathsf{t})} y_1^{n-1} y_2^{n-2} \dots y_{n-1}$, by (3.7.2) and Example 3.2.5. By Proposition 2.5.2, $\psi_{\mathsf{s}}\psi^\star_{d(\mathsf{t})} = \psi_{\mathsf{u}}$, where $\mathsf{u} = \mathsf{s}d(\mathsf{t})^{-1}$, if $\ell(d(\mathsf{u})) = \ell(d(\mathsf{s})) + \ell(d(\mathsf{t}))$ and otherwise $\psi_{\mathsf{s}}\psi^\star_{d(\mathsf{t})} = 0$. On the other hand, by the last paragraph of the proof of Proposition 2.5.2, or simply by counting degrees, $\psi_{\mathsf{u}} y_1^{n-1} y_2^{n-2} \dots y_{n-1} = 0$ if $\mathsf{u} \ne \mathsf{t}_{\boldsymbol{\lambda}}$ and $\psi_{\mathsf{t}_{\boldsymbol{\lambda}}} y_1^{n-1} y_2^{n-2} \dots y_{n-1} = (-1)^{n(n-2)/2}\psi_{\mathsf{t}^{\boldsymbol{\lambda}}}$. So $\langle\psi_{\mathsf{s}}, \psi_{\mathsf{t}}\rangle = \delta_{\mathsf{st}'}$, where $\mathsf{t}' = \mathsf{t}_{\boldsymbol{\lambda}} d'(\mathsf{t})$ is the tableau that is conjugate to t. Hence, $G^{\boldsymbol{\lambda}}_{\mathbb{Z}}$ is $(-1)^{n(n-2)/2}$ times the anti-diagonal identity matrix. Consequently, $D^{\boldsymbol{\lambda}}_{\mathbb{Z}} = S^{\boldsymbol{\lambda}}_{\mathbb{Z}}$ and $S^{\boldsymbol{\lambda}}_F$ is irreducible for any field F. $\diamondsuit$

3.7.9. **Example** Suppose $e = 2$, $\Lambda = \Lambda_0$ and $\lambda = (2, 2, 1)$. Then $\mathrm{Std}(\lambda)$ contains the five tableaux:

	$\mathsf{t}_1 = \mathsf{t}^\lambda$	t_2	t_3	t_4	t_5
t	$\begin{array}{\|c\|c\|}\hline 1&2\\\hline 3&4\\\hline 5\\\cline{1-1}\end{array}$	$\begin{array}{\|c\|c\|}\hline 1&3\\\hline 2&4\\\hline 5\\\cline{1-1}\end{array}$	$\begin{array}{\|c\|c\|}\hline 1&3\\\hline 2&5\\\hline 4\\\cline{1-1}\end{array}$	$\begin{array}{\|c\|c\|}\hline 1&2\\\hline 3&5\\\hline 4\\\cline{1-1}\end{array}$	$\begin{array}{\|c\|c\|}\hline 1&4\\\hline 2&5\\\hline 3\\\cline{1-1}\end{array}$
$d(\mathsf{t})$	1	s_2	s_2s_4	s_4	$s_2s_4s_3$
$\deg \mathsf{t}$	2	0	-2	0	0
$\mathbf{i}^{\mathsf{t}}$	01100	01100	01100	01100	01010

We want to compute the Gram matrix $G^{\boldsymbol{\lambda}}_{\mathbb{Z}} = \big(\langle\psi_{\mathsf{s}}, \psi_{\mathsf{t}}\rangle\big)$. Now $\langle\psi_{\mathsf{t}}, \psi_{\mathsf{t}}\rangle \ne 0$ only if $\mathbf{i}^{\mathsf{s}} = \mathbf{i}^{\mathsf{t}}$, by Proposition 3.2.10(a), and only if $\deg \mathsf{s} + \deg \mathsf{t} = 0$, since the bilinear form is homogeneous of degree zero. Hence, the only possible non-zero inner products are

$$\langle\psi_{\mathfrak{t}_1},\psi_{\mathfrak{t}_3}\rangle=\langle\psi_{\mathfrak{t}^\lambda},\psi_{\mathfrak{t}^\lambda}\psi_2\psi_4\rangle=\langle\psi_{\mathfrak{t}^\lambda}\psi_4,\psi_{\mathfrak{t}^\lambda}\psi_2\rangle=\langle\psi_{\mathfrak{t}_4},\psi_{\mathfrak{t}_2}\rangle,$$

together with $\langle\psi_{\mathfrak{t}_2},\psi_{\mathfrak{t}_2}\rangle$, $\langle\psi_{\mathfrak{t}_4},\psi_{\mathfrak{t}_4}\rangle$ and $\langle\psi_{\mathfrak{t}_5},\psi_{\mathfrak{t}_5}\rangle$. If $a\in\{2,4\}$ then

$$\langle\psi_{\mathfrak{t}^\lambda}\psi_a,\psi_{\mathfrak{t}^\lambda}\psi_a\rangle=\langle\psi_{\mathfrak{t}^\lambda}\psi_a^2,\psi_{\mathfrak{t}^\lambda}\rangle=\pm\langle\psi_{\mathfrak{t}^\lambda}(y_a-y_{a+1}),\psi_{\mathfrak{t}^\lambda}\rangle=0,$$

because $\psi_{\mathfrak{t}^\lambda}y_r=0$ by (2.2.3), for $1\le r\le 5$. To compute the remaining inner products we have to go back to the definition of the bilinear form (3.7.2). By Definition 3.2.2, $y^\lambda=y_2y_4$ so

$$\langle\psi_{\mathfrak{t}_1},\psi_{\mathfrak{t}_3}\rangle\psi_{\mathfrak{t}^\lambda}=\psi_{\mathfrak{t}^\lambda}\psi_2\psi_4y_2y_4=\psi_{\mathfrak{t}^\lambda}\psi_2y_2\psi_4y_4=\psi_{\mathfrak{t}^\lambda}(y_3\psi_2+1)(y_5\psi_4+1)=\psi_{\mathfrak{t}^\lambda},$$

by Proposition 3.2.10(c). Hence, $\langle\psi_{\mathfrak{t}_1},\psi_{\mathfrak{t}_3}\rangle=1=\langle\psi_{\mathfrak{t}_2},\psi_{\mathfrak{t}_4}\rangle$. Finally,

$$\langle\psi_{\mathfrak{t}_5},\psi_{\mathfrak{t}_5}\rangle\psi_{\mathfrak{t}^\lambda}=\psi_{\mathfrak{t}^\lambda}\psi_2\psi_4\psi_3^2\psi_2\psi_4y_2y_4=\psi_{\mathfrak{t}^\lambda}\psi_2\psi_4(2y_3y_4-y_3^2-y_4^2)\psi_2\psi_4y_2y_4.$$

where the second equality uses (2.2.3). Now $v_{\mathfrak{t}^\lambda}\psi_2y_3=v_{\mathfrak{t}^\lambda}(y_2\psi_1+1)=v_{\mathfrak{t}^\lambda}$ and, similarly, $v_{\mathfrak{t}^\lambda}y_4\psi_4=-v_{\mathfrak{t}^\lambda}$. Consequently $v_{\mathfrak{t}^\lambda}\psi_2\psi_4y_a^2=0$, for $a=3,4$, so it follows that $\psi_{\mathfrak{t}^\lambda}\psi_2\psi_4\psi_3^2=-2\psi_{\mathfrak{t}^\lambda}$ and hence that $\langle\psi_{\mathfrak{t}_5},\psi_{\mathfrak{t}_5}\rangle=-2$. Therefore, the Gram matrix of $S^{(2,2,1)}$ is

$$G^\lambda_{\mathbb{Z}}=\begin{pmatrix}0&1&0&0&0\\1&0&0&0&0\\0&0&0&1&0\\0&0&1&0&0\\0&0&0&0&-2\end{pmatrix}.$$

Consequently, the elementary divisors of $G^\lambda_{\mathbb{Z}}$ are $1,1,1,1,2$. Therefore, if $e=2$ and $\mathcal{Z}=\mathbb{Q}$ (so $v=-1$), then $S^\lambda_{\mathbb{Q}}=D^\lambda_{\mathbb{Q}}$ is irreducible, as is easily checked using Corollary 1.7.6. Now suppose that $\mathcal{Z}=\mathbb{F}_2$ (so $v=1$), so that $\mathscr{H}_n^\Lambda\cong\mathbb{F}_2\mathfrak{S}_5$. Then the calculation of $G^\lambda_{\mathbb{Z}}$ shows that the Specht module S^λ is reducible with $\dim_{\mathbb{F}_2}D^\lambda_{\mathbb{F}_2}=4<5=\dim_{\mathbb{Q}}D^\lambda_{\mathbb{Q}}$. It follows that if $e=p=2$ then $D^{(1^5)}$ is also a composition factor of $S^{\boldsymbol{\lambda}}$, so $a^{\mathbb{F}_2}_{(2,2,1),(1^5)}(q)=1$. ◊

3.7.10. **Example** Kleshchev and Ram [84, Conjecture 7.3] made a conjecture that, in type A, is equivalent to saying that the adjustment matrices $\mathbf{a}^F_q$ of the (cyclotomic) KLR algebras are trivial when $e=\infty$. Williamson [136] has given an example that shows that, in general, this is not true. Williamson's example comes from geometry [72], however, when it is translated into the language that we are using it corresponds to a statement about the simple module $D^{\boldsymbol{\mu}}$, for $\boldsymbol{\mu}=(2|2|1|1|3|3|2|2)$, for the cyclotomic quiver Hecke algebra $\mathscr{R}^\Lambda_{16}$ with $e=\infty$ and $\Lambda=2\Lambda_1+2\Lambda_2+2\Lambda_3+2\Lambda_4$. Fix the multicharge $\boldsymbol{\kappa}=(4,4,3,3,2,2,1,1)$ and set

$\mathbf{i} = (4,5,3,4,2,3,4,5,2,3,1,2,3,4,1,2)$. So $y^{\boldsymbol{\mu}} = y_1 y_9 y_{15} y_{19}$. There are 5 standard $\boldsymbol{\mu}$-tableaux of degree zero with residue sequence $\mathbf{i}$, namely:

t	$\ell(d(\mathsf{t}))$
(4 8 | 1 2 | 13 | 3 | 5 6 7 | 9 10 14 | 15 16 | 11 12)	23
(1 2 | 4 8 | 13 | 3 | 9 10 14 | 5 6 7 | 15 16 | 11 12)	28
(4 8 | 1 2 | 13 | 3 | 9 10 14 | 5 6 7 | 11 12 | 15 16)	28
(4 8 | 1 2 | 10 | 3 | 9 13 14 | 5 6 7 | 15 16 | 11 12)	31
(4 8 | 1 2 | 3 | 13 | 9 10 14 | 5 6 7 | 15 16 | 11 12)	31

The negative of the Gram matrix for $\mathbf{i}$-weight space $S^{\boldsymbol{\mu}}e(\mathbf{i})$ of $S^{\boldsymbol{\mu}}$ is

$$\begin{pmatrix} 0&0&1&1&0\\ 0&0&1&1&0\\ 1&1&0&1&1\\ 1&1&1&0&1\\ 0&0&1&1&0 \end{pmatrix}.$$

Calculating this matrix is non-trivial because the lengths of the permutations $d(\mathsf{t})$ are reasonably large. This matrix was computed using the author's implementation of the graded Specht modules in Sage [129]. Brundan, Kleshchev and McNamara [24, Example 2.16] obtain exactly the same matrix, up to a permutation of the rows and columns, as part of the Gram matrix for the homogeneous bilinear form of the corresponding proper standard module for $\mathscr{R}_n$.

The elementary divisors of this matrix are $1,1,2,0,0$, so the dimension of $D^{\boldsymbol{\mu}}e(\mathbf{i})$ is 2 in characteristic 2 and 3 in all other characteristics. Consequently, the dimension of $D^{\boldsymbol{\mu}}$, and hence the adjustment matrix $\mathbf{a}_q^F$ for $\mathscr{R}_{16}^{\Lambda}(F)$, depends on the characteristic of F. That $\dim D\boldsymbol{\mu}_F$ depends on F was first proved by Williamson who computed a one dimensional *intersection form* coming from geometry. ◊

3.7.11. **Example** Consider the case when $\Lambda = \Lambda_0$, so that $\mathscr{H}_n^{\Lambda}$ is the Iwahori-Hecke algebra of the symmetric group. The *James conjecture* [60, §4] says that if F is a field of characteristic $p > 0$ and $\lambda, \mu \in \mathcal{P}_n^{\Lambda}$ then $a_{\lambda\mu}(q) = \delta_{\lambda\mu}$ if $ep > n$. A natural strengthening of this conjecture is that the adjustment matrix of $\mathscr{R}_{\beta}^{\Lambda}$ is trivial whenever $\operatorname{def}\beta < p$. For the symmetric groups, the condition $\operatorname{def}\beta < p$ exactly corresponds to the case when the defect group of the block $\mathscr{R}_{\beta}^{\Lambda}$ is abelian.

The James conjecture is known to be true for blocks of weight at most 4 [38, 39, 60, 119]. Moreover, for every defect $w \geq 0$ there exists a *Rouquier block* of defect w for which the James conjecture holds [61]. Starting from the Rouquier blocks, there was some hope that the derived equivalences of Chuang and Rouquier [28] could be used to prove the James conjecture for all blocks.

Notwithstanding all of the evidence in favour of the James conjecture, it turns out that the conjecture is wrong! Again, Williamson [135, §6] has cruelly (but ultimately kindly) produced counter-examples to the James conjecture. At the same time he also found counter-examples to the *Lusztig conjecture* [93] for SL_n. These examples rely upon Williamson's recent work with Elias that gives generators and relations for the category of Soergel bimodules [34]. As of writing, the smallest known counter-example to the James conjecture occurs in a block of defect 561 in $\mathbb{F}_{839}\mathfrak{S}_{467874}$. Williamson has not revealed which Specht modules his counter-examples appear in, so the size of Gram matrix that needs to be computed in order to verify this example is not known. The Gram matrices of the Specht modules will be significantly larger, and harder to compute, than the one dimensional intersection form that Williamson reduces to (using a chain of deep results in geometric representation theory), and then calculates, using elementary techniques (and a computer). ◊

Williamson's counter-examples to the James and Lusztig conjectures suggest that there is no block theoretic criterion for the adjustment matrix of a block to be trivial, except asymptotically where the Lusztig conjecture is known to hold [1]. With hindsight, perhaps this is not so surprising because the condition given in Corollary 1.7.6 for a Specht module to be irreducible is rarely a block invariant. The failure of the James and Lusztig conjectures suggests that we should, instead, look for necessary and sufficient conditions for the $\mathscr{R}^\Lambda_n(F)$-modules $D^{\boldsymbol{\mu}}_{\mathbb{Z}} \otimes F$ to be irreducible, for $\boldsymbol{\mu} \in \mathcal{K}^\Lambda_n$. Some steps towards such a criterion are made in Conjecture 4.4.1 below.

Brundan and Kleshchev [22, §5.6] remarked that $a^F_{\boldsymbol{\lambda\mu}}(q) \in \mathbb{N}$ in all of the examples that they had computed. They asked whether this might always be the case. The next examples show that, in general, $a^F_{\lambda\mu}(q) \notin \mathbb{N}$.

3.7.12. **Example** (Evseev [35, Corollary 5]) Suppose that $e = 2$, $\Lambda = \Lambda_0$ and let $\lambda = (3, 2^2, 1^2)$ and $\mu = (1^9)$. Take $F = \mathbb{F}_2$ to be a field of characteristic 2 and let $\mathbf{a}^{\mathbb{F}_2}_q = (a^{\mathbb{F}_2}_{\boldsymbol{\lambda\mu}}(q))$ be the corresponding adjustment matrix.

As part of a general argument Evseev shows that $a^{\mathbb{F}_2}_{\lambda\mu}(q) \notin \mathbb{N}$. In fact, it is not hard to see directly that $a^{\mathbb{F}_2}_{\lambda\mu}(q) = q+q^{-1}$. Comparing the decomposition matrix for $\mathbb{F}_2\mathfrak{S}_9$ given by James [59] with the graded decomposition matrices when $e = 2$ given in [104], shows that $d^{\mathbb{Q}}_{\lambda\mu} = 0$, $d^{\mathbb{F}_2}_{\lambda\mu} = 2$, and that $a^{\mathbb{F}_2}_{\lambda\mu}(1) = 2$.

Now $D^{\mu}_{\mathbb{F}_2} = D^{\mu}_{\mathbb{F}_2} e(\mathbf{i}^{\mu})$ is one dimensional, so any composition factor of $S^{\lambda}_{\mathbb{F}_2}$ that is isomorphic to $D^{\mu}_{\mathbb{F}_2}\langle d\rangle$, for some $d \in \mathbb{Z}$, must be contained in $S^{\lambda}_{\mathbb{F}_2} e(\mathbf{i}^{\mu})$. There are exactly six standard λ-tableau with residue sequence $\mathbf{i}^{\mu}$, namely:

$$\begin{array}{ccccccc}
\deg \mathsf{t} & 1 & 1 & 1 & 1 & 1 & -1\\
\mathsf{t} &
\begin{array}{|c|c|c|}\hline 1&6&9\\\hline 2&7\\\cline{1-2} 3&8\\\cline{1-2} 4\\\cline{1-1} 5\\\cline{1-1}\end{array} &
\begin{array}{|c|c|c|}\hline 1&4&9\\\hline 2&5\\\cline{1-2} 3&8\\\cline{1-2} 6\\\cline{1-1} 7\\\cline{1-1}\end{array} &
\begin{array}{|c|c|c|}\hline 1&4&5\\\hline 2&7\\\cline{1-2} 3&8\\\cline{1-2} 6\\\cline{1-1} 9\\\cline{1-1}\end{array} &
\begin{array}{|c|c|c|}\hline 1&2&3\\\hline 4&7\\\cline{1-2} 5&8\\\cline{1-2} 6\\\cline{1-1} 9\\\cline{1-1}\end{array} &
\begin{array}{|c|c|c|}\hline 1&4&7\\\hline 2&5\\\cline{1-2} 3&6\\\cline{1-2} 8\\\cline{1-1} 9\\\cline{1-1}\end{array} &
\begin{array}{|c|c|c|}\hline 1&4&7\\\hline 2&5\\\cline{1-2} 3&8\\\cline{1-2} 6\\\cline{1-1} 9\\\cline{1-1}\end{array}
\end{array}$$

As D^{μ} is one dimensional, and concentrated in degree zero, it follows that $a^{\mathbb{F}_2}_{\lambda\mu} = d^{\mathbb{F}_2}_{\lambda\mu}(q) = q + q^{-1}$. We can see a shadow of the adjustment matrix entry in the Gram matrix of $S^{\lambda}_{\mathbb{Z}} e(\mathbf{i}^{\mu})$, that is equal to

$$\begin{pmatrix} 0&0&0&0&0&0\\ 0&0&0&0&0&0\\ 0&0&0&0&0&4\\ 0&0&0&0&0&-2\\ 0&0&0&0&0&2\\ 0&0&4&-2&2&0 \end{pmatrix}.$$

The elementary divisors of this matrix are $2, 2, 0, 0, 0, 0$, with the 2's in degrees ± 1. Therefore, the graded dimension of $D^{\lambda}_{\mathbb{F}_2} e(\mathbf{i}^{\mu})$ decreases by $q+q^{-1}$ in characteristic 2. ◊

3.7.13. **Example** Motivated by the runner removable theorems of [27, 63] and Example 3.7.12, take $e = 3$, $F = \mathbb{F}_2$, $\lambda = (3, 2^4, 1^3)$ and $\mu = (1^{14})$. (The partitions λ and μ are obtained from the corresponding partitions in Example 3.7.12 by conjugating, adding an empty runner, and then conjugating again.) Again, we work over $\mathbb{F}_2$ and consider the corresponding adjustment matrices.

Calculating with SPECHT [102] we find that $d^{\mathbb{Q}}_{\lambda\mu} = 0$ and that $d^{\mathbb{F}_2}_{\lambda\mu} = 2$. Once again, it turns out that there are six λ-tableaux with 3-residue sequence $\mathbf{i}^{\mu}$, with five of these having degree 1 and one having degree -1. (Moreover, the Gram matrix of $S^{\lambda} e(\mathbf{i}^{\mu})$ is the same as the Gram matrix given in Example 3.7.12.) Hence, as in Example 3.7.12, $a^{\mathbb{F}_2}_{\lambda\mu}(q) = q + q^{-1} = d^{\mathbb{F}_2}_{\lambda\mu}(q)$.

As the runner removable theorems compare blocks for different e over the same field we cannot expect to find an example of a non-polynomial adjustment matrix entry in odd characteristic in this way. Nonetheless, it seems fairly certain that non-polynomial adjustment matrix entries exist for all e and all $p > 0$.

Evseev [35, Corollary 5] gives three other examples of adjustment matrix entries that are equal to $q + q^{-1}$ when $e = p = 2$. All of them have similar analogues when $e = 3$ and $p = 2$. Finally, if we try adding further empty runners to the partitions λ and μ, so that $e \geq 4$, then the corresponding adjustment matrix entry is zero (all of these partitions have weight 4). ◊

4 Seminormal bases and the KLR grading

In this final section we link the KLR grading on $\mathscr{R}_n^\Lambda$ with the semisimple representation theory of $\mathscr{H}_n^\Lambda$ using the seminormal bases. We start by showing that by combining information from all of the KLR gradings for different cyclic quivers leads to an integral formula for the Gram determinants of the ungraded Specht modules.

4.1 *Gram determinants and graded dimensions*

In Theorem 1.7.3 we gave a "rational" formula for the Gram determinant of the ungraded Specht modules $\underline{S}^{\boldsymbol{\lambda}}$, for $\boldsymbol{\lambda} \in \mathcal{P}_n^\Lambda$. We now give an integral formula for these determinants and give both a combinatorial and a representation theoretic interpretation of this formula.

Suppose that the Hecke parameter v from Definition 1.1.1 is an indeterminate over $\mathbb{Q}$ and consider an integral cyclotomic Hecke algebra $\mathscr{H}_n^\Lambda$ over the field $\mathcal{Z} = \mathbb{Q}(v)$ where $\Lambda \in P^+$ such that $e > n$ and $(\Lambda, \alpha_{i,n}) \leq 1$, for all $i \in I$. Then $\mathscr{H}_n^\Lambda$ is semisimple by Corollary 1.6.11.

Definition 4.1.1. Suppose that $\boldsymbol{\lambda} \in \mathcal{P}_n^\Lambda$. For $e \geq 2$ and $\mathbf{i} \in I_e^n$ define

$$\deg_{e,\mathbf{i}}(\boldsymbol{\lambda}) = \sum_{\mathsf{t} \in \operatorname{Std}_{\mathbf{i}}(\boldsymbol{\lambda})} \deg_e \mathsf{t},$$

where $\operatorname{Std}_{\mathbf{i}}(\boldsymbol{\lambda}) = \{\,\mathsf{t} \in \operatorname{Std}(\boldsymbol{\lambda}) \mid \mathbf{i}^{\mathsf{t}} = \mathbf{i}\,\}$. Set $\deg_e(\boldsymbol{\lambda}) = \sum_{\mathbf{i} \in I_e^n} \deg_{e,\mathbf{i}}(\boldsymbol{\lambda})$. For a prime integer $p > 0$ set $\operatorname{Deg}_p(\boldsymbol{\lambda}) = \sum_{k \geq 1} \deg_{p^k}(\boldsymbol{\lambda})$.

By definition, $\deg_e(\boldsymbol{\lambda}), \operatorname{Deg}_p(\boldsymbol{\lambda}) \in \mathbb{Z}$. For $e > 0$ let $\Phi_e(x) \in \mathbb{Z}[x]$ be the eth cyclotomic polynomial in the indeterminate x.

Theorem 4.1.2 (Hu-Mathas [57, Theorem C]**).** *Suppose that $\Lambda \in P^+$, $e > n$ and that $(\Lambda, \alpha_{i,n}) \leq 1$, for all $i \in I$. Let $\boldsymbol{\lambda} \in \mathcal{P}_n^\Lambda$. Then*

$$\det \underline{G}^{\boldsymbol{\lambda}} = \prod_{e>1} \Phi_e(v^2)^{\deg_e(\boldsymbol{\lambda})}.$$

Consequently, if $v = 1$ then $\det \underline{G}^{\boldsymbol{\lambda}} = \prod_{p \text{ prime}} p^{\operatorname{Deg}_p(\boldsymbol{\lambda})}$.

Proving this result is not hard: it amounts to interpreting Definition 1.6.6 in light of the KLR degree functions on $\operatorname{Std}(\boldsymbol{\lambda})$. There is a power of v in the statement of this result in [57]. This is not needed here because we have renormalised the quadratic relations in the Hecke algebra given in Definition 1.1.1.

The Murphy basis is defined over $\mathbb{Z}[v,v^{-1}]$. Therefore, $\det \underline{G}^{\boldsymbol{\lambda}} \in \mathbb{Z}[v,v^{-1}]$ and Theorem 4.1.2 implies that $\deg_e(\boldsymbol{\lambda}) \geq 0$ for all $\boldsymbol{\lambda} \in \mathcal{P}_n^\Lambda$ and $e \geq 2$. In fact, [57, Theorem 3.24] gives an analogue of Theorem 4.1.2 for the determinant of the Gram matrix restricted to $\underline{S}^{\boldsymbol{\lambda}} e(\mathbf{i})$, suitably interpreted, and the following is true:

Corollary 4.1.3 ([57, Corollary 3.25]). *Suppose that* $e \geq 2$, $\boldsymbol{\lambda} \in \mathcal{P}_n^\Lambda$ *and* $\mathbf{i} \in I_e^n$. *Then* $\deg_{e,\mathbf{i}}(\boldsymbol{\lambda}) \geq 0$.

The definition of the integers $\deg_{e,\mathbf{i}}(\boldsymbol{\lambda})$ is purely combinatorial, so it should be possible to give a combinatorial proof of this result perhaps using Theorem 3.4.6. We think, however, that this is probably difficult.

Fix an integer $e \geq 2$ and a dominant weight $\Lambda \in P^+$ and consider the Hecke algebra $\mathscr{H}_n^\Lambda$ over a field F. If $\boldsymbol{\lambda} \in \mathcal{P}_n^\Lambda$ then, by definition,

$$\operatorname{Ch}_q S^{\boldsymbol{\lambda}} = \sum_{\boldsymbol{\mu} \in \mathcal{K}_n^\Lambda} d_{\boldsymbol{\lambda}\boldsymbol{\mu}}(q) \operatorname{Ch}_q D^{\boldsymbol{\mu}} \in \mathcal{A}[I^n].$$

Let $\partial \colon \mathcal{A}[I^n] \longrightarrow \mathbb{Z}[I^n]$ be the linear map given by $\partial(f(q) \cdot \mathbf{i}) = f'(1)\mathbf{i}$, where $f'(1)$ is the derivative of $f(q) \in \mathcal{A}$ evaluated at $q = 1$. Then $\partial \operatorname{Ch}_q S^{\boldsymbol{\lambda}} = \sum_{\mathbf{i}} \deg_{e,\mathbf{i}}(\boldsymbol{\lambda}) \cdot \mathbf{i}$. The KLR idempotents are orthogonal, so $\mathit{dim}_q D_{\mathbf{i}}^{\boldsymbol{\mu}} = \overline{\mathit{dim}_q D_{\mathbf{i}}^{\boldsymbol{\mu}}}$ since $(D^{\boldsymbol{\mu}})^{\circledast} \cong D^{\boldsymbol{\mu}}$. Therefore, $\partial \operatorname{Ch}_q D^{\boldsymbol{\mu}} = 0$. Hence, applying ∂ to the formula for $\operatorname{Ch}_q S^{\boldsymbol{\lambda}}$ shows that

$$\sum_{\mathbf{i} \in I^n} \deg_{e,\mathbf{i}}(\boldsymbol{\lambda}) \cdot \mathbf{i} = \partial \operatorname{Ch}_q S^{\boldsymbol{\lambda}} = \sum_{\mathbf{i} \in I^n} \sum_{\boldsymbol{\mu} \in \mathcal{K}_n^\Lambda} d'_{\boldsymbol{\lambda}\boldsymbol{\mu}}(1) \dim D_{\mathbf{i}}^{\boldsymbol{\mu}} \cdot \mathbf{i}. \tag{4.1.4}$$

Consequently, $\deg_{e,\mathbf{i}}(\boldsymbol{\lambda}) = \sum_{\boldsymbol{\mu}} d'_{\boldsymbol{\lambda}\boldsymbol{\mu}}(1) \dim D_{\mathbf{i}}^{\boldsymbol{\mu}}$. So far we have worked over an arbitrary field. If $F = \mathbb{C}$ then $d_{\boldsymbol{\lambda}\boldsymbol{\mu}}(q) \in \mathbb{N}[q]$, by Proposition 3.5.6, so that $d'_{\boldsymbol{\lambda}\boldsymbol{\mu}}(1) \geq 0$. Therefore, $\deg_{e,\mathbf{i}}(\boldsymbol{\lambda}) \geq 0$ as claimed. (In fact, by Theorem 3.7.6, the right-hand side of (4.1.4) is independent of F, as it must be.)

Theorem 1.7.4 shows that taking the $\mathfrak{p}$-adic valuation of the Gram determinant of $\underline{S}^{\boldsymbol{\lambda}}$ leads to the Jantzen sum formula for $\underline{S}^{\boldsymbol{\lambda}}$. Therefore, (4.1.4) suggests that

$$\sum_{k>0} [J_k(\underline{S}_{\mathbb{C}}^{\boldsymbol{\lambda}})] = \sum_{\boldsymbol{\mu} \rhd \boldsymbol{\lambda}} d'_{\boldsymbol{\lambda}\boldsymbol{\mu}}(1) [\underline{D}_{\mathbb{C}}^{\boldsymbol{\mu}}], \tag{4.1.5}$$

where we use the notation of Theorem 1.7.4. That is, Theorem 4.1.2 corresponds to writing the Jantzen sum formula as a non-negative linear combination of simple modules. In fact, we have not done enough to prove (4.1.5). (One way to do this would be to establish analogous statements for the Gram determinants of the Weyl modules of the cyclotomic Schur algebras [31].)

Nonetheless, (4.1.5) is true, being proved by Ryom-Hansen [124, Theorem 1] in level one and by Yvonne [138, Theorem 2.11] in general.

A better interpretation of (4.1.4) is in terms of *grading filtrations* [15, §2.4]. Let $\dot{\mathscr{R}}_n^\Lambda = \mathcal{H}om_{\mathscr{R}_n^\Lambda}(Y, Y)$, where $Y = \bigoplus_{\boldsymbol{\mu}\in\mathcal{K}_n^\Lambda} Y^{\boldsymbol{\mu}}$ is a progenerator for $\mathscr{R}_n^\Lambda$. Then $\dot{\mathscr{R}}_n^\Lambda$ is a graded basic algebra for $\mathscr{R}_n^\Lambda$ and the functor

$$\mathsf{F}_n : \operatorname{Rep}(\mathscr{R}_n^\Lambda) \longrightarrow \operatorname{Rep}(\dot{\mathscr{R}}_n^\Lambda); M \mapsto \mathcal{H}om_{\mathscr{R}_n^\Lambda}(Y, M), \qquad \text{for } M \in \operatorname{Rep}(\mathscr{R}_n^\Lambda),$$

is a graded Morita equivalence; see, for example, [55, §2.3-2.4]. Recall that the matrix $\mathbf{c}_q = \big(c_{\boldsymbol{\lambda\mu}}(q)\big) = \mathbf{d}_q^T \circ \mathbf{d}_q$ is the Cartan matrix of $\mathscr{R}_n^\Lambda$. By Corollary 2.1.6, $c_{\boldsymbol{\lambda\mu}}(q) = \mathit{dim}_q\, \mathcal{H}om_{\mathscr{R}_n^\Lambda}(Y^{\boldsymbol{\lambda}}, Y^{\boldsymbol{\mu}})$ so that

$$\mathit{dim}_q\, \dot{\mathscr{R}}_n^\Lambda = \sum_{\boldsymbol{\lambda},\boldsymbol{\mu}\in\mathcal{K}_n^\Lambda} c_{\boldsymbol{\lambda\mu}}(q) \in \mathbb{N}[q, q^{-1}].$$

For the rest of this section assume that $F = \mathbb{C}$. Then $c_{\boldsymbol{\lambda\mu}}(q) \in \mathbb{N}[q]$ by Corollary 3.5.27. Therefore, $\mathit{dim}_q\, \dot{\mathscr{R}}_n^\Lambda \in \mathbb{N}[q]$ so that $\dot{\mathscr{R}}_n^\Lambda$ is a positively graded algebra. Let $\dot{M} = \bigoplus_{d=a}^{z} \dot{M}_d$ be a $\dot{\mathscr{R}}_n^\Lambda$-module. The **grading filtration** of $\dot{M}$ is the filtration $\dot{M} = G_a(\dot{M}) \supseteq G_{a+1}(\dot{M}) \supseteq \cdots \supseteq G_z(\dot{M}) \supset 0$, where

$$G_d(\dot{M}) = \bigoplus_{k\geq d} \dot{M}_k.$$

Then $G_r(\dot{M})$ is a graded $\dot{\mathscr{R}}_n^\Lambda$-module precisely because $\dot{\mathscr{R}}_n^\Lambda$ is *positively* graded. The grading filtration of a $\mathscr{R}_n$-module M is the filtration given by $G_r(M) = \mathsf{F}_n^{-1}(G_r(\mathsf{F}_n(M)))$, for $r \in \mathbb{Z}$. By Corollary 3.6.7, $S^{\boldsymbol{\lambda}} = G_0(S^{\boldsymbol{\lambda}})$ and $G_r(S^{\boldsymbol{\lambda}}) = 0$ for $r > \operatorname{def}\boldsymbol{\lambda}$.

For $\boldsymbol{\lambda} \in \mathcal{P}_n^\Lambda$ and $\boldsymbol{\mu} \in \mathcal{K}_n^\Lambda$ write $d_{\boldsymbol{\lambda\mu}}(q) = \sum_{r\geq 0} d_{\boldsymbol{\lambda\mu}}^{(r)} q^r$, for $d_{\boldsymbol{\lambda\mu}}^{(r)} \in \mathbb{N}$.

Lemma 4.1.6. *Suppose that $F = \mathbb{C}$ and $\boldsymbol{\lambda} \in \mathcal{P}_n^\Lambda$. If $0 \leq r \leq \operatorname{def}\boldsymbol{\lambda}$ then*

$$G_r(S^{\boldsymbol{\lambda}})/G_{r+1}(S^{\boldsymbol{\lambda}}) \cong \bigoplus_{\boldsymbol{\mu}\in\mathcal{K}_n^\Lambda} \big(D^{\boldsymbol{\mu}}\langle r\rangle\big)^{\oplus d_{\boldsymbol{\lambda\mu}}^{(r)}}.$$

Proof. This is an immediate consequence of the definition of the grading filtration and Corollary 3.5.27. □

Comparing this with (4.1.5) suggests that $J_r(\underline{S}^{\boldsymbol{\lambda}}) = G_r(S^{\boldsymbol{\lambda}})$, for $r \geq 0$. Of course, there is no reason to expect that $J_r(\underline{S}^{\boldsymbol{\lambda}})$ is a graded submodule of $S^{\boldsymbol{\lambda}}$. Nonetheless, establishing a conjecture of Rouquier [89, (16)], Shan has proved the following when Λ is a weight of level 1.

Theorem 4.1.7 (Shan [126, Theorem 0.1]**).** *Suppose that F is a field of characteristic zero, $\Lambda = \Lambda_0$, and that $\boldsymbol{\lambda} \in \mathcal{P}_n^\Lambda$. Then $J_r(\underline{S}^{\boldsymbol{\lambda}}) = G_r(S^{\boldsymbol{\lambda}})$ is a graded submodule of $S^{\boldsymbol{\lambda}}$ and $[J_r(\underline{S}^{\boldsymbol{\lambda}})/J_{r+1}(\underline{S}^{\boldsymbol{\lambda}}) : D^{\boldsymbol{\mu}}\langle s\rangle] = \delta_{rs} d_{\boldsymbol{\lambda\mu}}^{(r)}$, for all $\boldsymbol{\mu} \in \mathcal{K}_n^\Lambda$ and $r \geq 0$.*

Shan actually proves that the Jantzen, radical and grading filtrations of graded Weyl modules coincide for the Dipper-James v-Schur algebras [30]. This implies the result above because the Schur functor maps Jantzen filtrations of Weyl modules to Jantzen filtrations of Specht modules. There is a catch, however, because Shan remarks that it is unclear how her geometrically defined grading relates to the grading on the v-Schur algebra given by Ariki [7] and hence to the KLR grading on $\mathscr{R}_n^\Lambda$. As we now sketch, Theorem 4.1.7 can be deduced from Shan's result using recent work.

Since Shan's paper cyclotomic quiver Schur algebras have been introduced for arbitrary dominant weights [7,55,128], thus giving a grading on all of the cyclotomic Schur algebras introduced by Dipper, James and the author [31]. The key point, which is non-trivial, is that the module categories of the cyclotomic quiver Schur algebras are Koszul. When $e = \infty$ this is proved in [55] by reducing to parabolic category $\mathcal{O}$ for the general linear groups, which is known to be Koszul by [14,15]. Using similar ideas, Maksimau [99] proves that Stroppel and Webster's cyclotomic quiver Schur algebras are Koszul for arbitrary e by using [123] to reduce to affine parabolic category $\mathcal{O}$.

As the module categories of the cyclotomic quiver Schur are Koszul, an elementary argument [15, Proposition 2.4.1] shows that the radical and grading filtrations of the graded Weyl modules of these algebras coincide. By definition, the analogue of Lemma 4.1.6 describes the graded composition factors of the grading (=radical) filtrations of the graded Weyl modules — compare with [55, Corollary 7.24] when $e = \infty$ and [99, Theorem 1.1] in general. The graded Schur functors of [55,99] send graded Weyl modules to graded Specht modules, graded simple modules to graded simple $\mathscr{R}_n^\Lambda$-modules (or zero), grading filtrations to grading filtrations and Jantzen filtrations to Jantzen filtrations. Combining these facts with Shan's work [126] implies Theorem 4.1.7 when $\Lambda = \Lambda_0$. We note that the v-Schur algebras were first shown to be Koszul by Shan, Varagnolo and Vasserot [127]. It is also possible to match up Shan's grading on the v-Schur algebras with the gradings of [7,128] using the uniqueness of Koszul gradings [15, Proposition 2.5.1]. As these papers use different conventions, it is necessary to work with the graded Ringel dual.

The obstacle to extending Theorem 4.1.7 to arbitrary weights $\Lambda \in P^+$ is in showing that the Jantzen and radical (=grading) filtrations of the graded Weyl modules of the cyclotomic quiver Schur algebras coincide. As the cyclotomic quiver Schur algebras are Koszul it is possible that this is straightforward. It seems to the author, however, that it is necessary to generalize Shan's arguments [126] to realize the Jantzen filtration geometrically using the language of [123].

4.2 A deformation of the KLR grading

Following [57], especially the appendix, we now sketch how to use the seminormal basis to prove that $\mathscr{R}_n^\Lambda \cong \mathscr{H}_n^\Lambda$ over a field (Theorem 3.1.1). The aim in doing this is not so much to give a new proof of the graded isomorphism theorem. Rather, we want to build a bridge between the KLR algebras and the well-understood semisimple representation theory of the cyclotomic Hecke algebras. In §4.3 we cross this bridge to construct a new graded cellular basis $\{B_{\mathsf{st}}\}$ of $\mathscr{H}_n^\Lambda$ that is independent of the choices of reduced expressions that are necessary in Theorem 3.2.4.

Throughout this section we consider a cyclotomic Hecke algebra $\mathscr{H}_n^\Lambda$ defined over a field F that has Hecke parameter $v \in F^\times$ of quantum characteristic $e \geq 2$. As in §1.2, $\Lambda \in P^+$ is determined by a multicharge $\kappa \in \mathbb{Z}^\ell$. We set up a modular system for studying $\mathscr{H}_n^\Lambda = \mathscr{H}_n^\Lambda(F)$.

Let x be an indeterminate over F and let $\mathcal{O} = F[x]_{(x)}$ be the localization of $F[x]$ at the principal ideal generated by x. Let $K = F(x)$ be the field of fractions of $\mathcal{O}$. Let $\mathscr{H}_n^{\mathcal{O}}$ be the cyclotomic Hecke algebra with Hecke parameter $t = x + v$, a unit in $\mathcal{O}$, and cyclotomic parameters $Q_l = x^l + [\kappa_l]_t$, for $1 \leq l \leq \ell$. Then $\mathscr{H}_n^K = \mathscr{H}_n^{\mathcal{O}} \otimes_{\mathcal{O}} K$ is a split semisimple algebra by Theorem 2.4.8. Moreover, by definition, $\mathscr{H}_n^\Lambda = \mathscr{H}_n^\Lambda(F) \cong \mathscr{H}_n^{\mathcal{O}} \otimes_{\mathcal{O}} F$, where we consider F as an $\mathcal{O}$-module by letting x act as multiplication by 0.

As the algebra $\mathscr{H}_n^K$ is semisimple, it has a seminormal basis $\{f_{\mathsf{st}}\}$ in the sense of Definition 1.6.4. With our choice of parameters, the content functions from (1.6.1) become

$$c_r^{\mathscr{Z}}(\mathsf{s}) = t^{2(c-b)}x^l + [\kappa_l + c - b]_t = t^{2(c-b)}x^l + [\mathsf{c}_r^{\mathbb{Z}}(\mathsf{s})]_t$$

if $\mathsf{s}(l,b,c) = r$, for $1 \leq k \leq n$. Then, $L_r f_{\mathsf{st}} = c_r^{\mathscr{Z}}(\mathsf{s}) f_{\mathsf{st}}$, for $(\mathsf{s},\mathsf{t}) \in \operatorname{Std}^2(\mathcal{P}_n^\Lambda)$. By Corollary 1.6.9, the basis $\{f_{\mathsf{st}}\}$ determines a seminormal coefficient system $\boldsymbol{\alpha} = \{\alpha_r(\mathsf{t}) \mid \mathsf{t} \in \operatorname{Std}(\mathcal{P}_n^\Lambda) \text{ and } 1 \leq r < n\}$ and a set of scalars $\{\gamma_{\mathsf{t}} \mid \mathsf{t} \in \operatorname{Std}(\mathcal{P}_n^\Lambda)\}$.

For $\mathbf{i} \in I^n$ let $\operatorname{Std}(\mathbf{i}) = \{\mathsf{s} \in \operatorname{Std}(\mathcal{P}_n^\Lambda) \mid \mathbf{i}^{\mathsf{s}} = \mathbf{i}\}$ be the set of standard tableaux with residue sequence $\mathbf{i}$. Define

(4.2.1) $$f_{\mathbf{i}}^{\mathcal{O}} = \sum_{\mathsf{t} \in \operatorname{Std}(\mathbf{i})} F_{\mathsf{t}}.$$

By definition, $f_{\mathbf{i}}^{\mathcal{O}} \in \mathscr{H}_n^K$ but, in fact, $f_{\mathbf{i}}^{\mathcal{O}} \in \mathscr{H}_n^{\mathcal{O}}$. This idempotent lifting result dates back to Murphy [112] for the symmetric groups. For higher levels it was first proved in [108]. In [57] it is proved for a more general class of rings $\mathcal{O}$.

Lemma 4.2.2 ([57, Lemma 4.4]). *Suppose that* $\mathbf{i} \in I^n$. *Then* $f_{\mathbf{i}}^{\mathcal{O}} \in \mathscr{H}_n^{\mathcal{O}}$.

We will see that $f_{\mathbf{i}}^{\mathcal{O}} \otimes_{\mathcal{O}} 1_F$ is the KLR idempotent $e(\mathbf{i})$, for $\mathbf{i} \in I^n$. Notice that $1 = \sum_{\mathbf{i}} f_{\mathbf{i}}^{\mathcal{O}}$ and, further, that $f_{\mathbf{i}}^{\mathcal{O}} f_{\mathbf{j}}^{\mathcal{O}} = \delta_{\mathbf{ij}} f_{\mathbf{i}}^{\mathcal{O}}$, for $\mathbf{i}, \mathbf{j} \in I^n$, by Theorem 1.6.7.

As detailed after Theorem 3.1.1, Brundan and Kleshchev construct their isomorphisms $\mathscr{R}_n^\Lambda \xrightarrow{\sim} \mathscr{H}_n^\Lambda$ using certain rational functions $P_r(\mathbf{i})$ and $Q_r(\mathbf{i})$ in $F[y_1, \dots, y_n]$. The advantage of working with seminormal forms is that, at least intuitively, these rational functions "converge" and can be replaced with "nicer" polynomials. The main tool for doing this is the following result, generalizing Lemma 4.2.2.

Let $M_r = 1 - t^{-1}L_r + tL_{r+1}$, for $1 \le r < n$. Then $M_r f_{\mathsf{st}} = M_r^{\mathcal{Z}}(\mathsf{s}) f_{\mathsf{st}}$, where $M_r^{\mathcal{Z}}(\mathsf{s}) = 1 - t^{-1}c_r^{\mathcal{Z}}(\mathsf{s}) + tc_{r+1}^{\mathcal{Z}}(\mathsf{s})$. The constant term of $M_r^{\mathcal{Z}}(\mathsf{s})$ is equal to $v^{2c_r^{\mathbb{Z}}(\mathsf{s})-1}[1 - c_r^{\mathbb{Z}}(\mathsf{s}) + c_{r+1}^{\mathbb{Z}}(\mathsf{s})]_v \neq 0$. Consequently, M_r acts invertibly on f_{st} whenever $\mathsf{s} \in \mathrm{Std}(\mathbf{i})$ and $1 - i_r + i_{r+1} \neq 0$ in $I = \mathbb{Z}/e\mathbb{Z}$. This observation is part of the proof of part (a) of the next result. Similarly, set $\rho_r^{\mathcal{Z}}(\mathsf{s}) = c_r^{\mathcal{Z}}(\mathsf{s}) - c_{r+1}^{\mathcal{Z}}(\mathsf{s})$. Then $\rho_r^{\mathcal{Z}}(\mathsf{s})$ is invertible in $\mathcal{O}$ if $i_r \neq i_{r+1}$.

Corollary 4.2.3 (Hu-Mathas [57, Corollary 4.6]**).** *Suppose that* $1 \le r < n$ *and* $\mathbf{i} \in I^n$.

a) *If* $i_r \neq i_{r+1} + 1$ *then* $\dfrac{1}{M_r} f_{\mathbf{i}}^{\mathcal{O}} = \displaystyle\sum_{\mathsf{s} \in \mathrm{Std}(\mathbf{i})} \frac{1}{M_r^{\mathcal{Z}}(\mathsf{s})} F_{\mathsf{s}} \in \mathscr{H}_n^{\mathcal{O}}$.

b) *If* $i_r \neq i_{r+1}$ *then* $\dfrac{1}{L_r - L_{r+1}} f_{\mathbf{i}}^{\mathcal{O}} = \displaystyle\sum_{\mathsf{s} \in \mathrm{Std}(\mathbf{i})} \frac{1}{\rho_r^{\mathcal{Z}}(\mathsf{s})} F_{\mathsf{s}} \in \mathscr{H}_n^{\mathcal{O}}$.

The invertibility of $M_r f_{\mathbf{i}}^{\mathcal{O}}$, when $i_r \neq i_{r+1} + 1$, allows us to define analogues of the KLR generators of $\mathscr{R}_n^\Lambda$ in $\mathscr{H}_n^{\mathcal{O}}$. The invertibility of $(L_r - L_{r+1}) f_{\mathbf{i}}^{\mathcal{O}}$ is needed to show that these new elements generate $\mathscr{H}_n^{\mathcal{O}}$.

Define an embedding $I \hookrightarrow \mathbb{Z}; i \mapsto \hat{\imath}$ by letting $\hat{\imath}$ be the smallest non-negative integer such that $i = \hat{\imath} + e\mathbb{Z}$, for $i \in I$.

Definition 4.2.4. Suppose that $1 \le r < n$. Define elements $\psi_r^{\mathcal{O}} = \sum_{\mathbf{i} \in I^n} \psi_r^{\mathcal{O}} f_{\mathbf{i}}^{\mathcal{O}}$ in $\mathscr{H}_n^{\mathcal{O}}$ by

$$\psi_r^{\mathcal{O}} f_{\mathbf{i}}^{\mathcal{O}} = \begin{cases} (T_r + t^{-1}) \frac{t^{2\hat{\imath}_r}}{M_r} f_{\mathbf{i}}^{\mathcal{O}}, & \text{if } i_r = i_{r+1}, \\ (T_r L_r - L_r T_r) t^{-2\hat{\imath}_r} f_{\mathbf{i}}^{\mathcal{O}}, & \text{if } i_r = i_{r+1} + 1, \\ (T_r L_r - L_r T_r) \frac{1}{M_r} f_{\mathbf{i}}^{\mathcal{O}}, & \text{otherwise.} \end{cases}$$

If $1 \le r \le n$ then define $y_r^{\mathcal{O}} = \sum_{\mathbf{i} \in I^n} t^{-2\hat{\imath}_r - 1}(L_r - [\hat{\imath}_r]) f_{\mathbf{i}}^{\mathcal{O}}$.

We now describe an $\mathcal{O}$-deformation of cyclotomic KLR algebra $\mathscr{R}_n^\Lambda$. This is a special case of one of the main results of [57], which allows greater flexibility in the choice of the ring $\mathcal{O}$.

Theorem 4.2.5 (Hu-Mathas [57, Theorem A]**).** *As an $\mathcal{O}$-algebra, the algebra $\mathscr{H}_n^\mathcal{O}$ is generated by the elements*

$$\{ f_{\mathbf{i}}^\mathcal{O} \mid \mathbf{i} \in I^n \} \cup \{ \psi_r^\mathcal{O} \mid 1 \le r < n \} \cup \{ y_r^\mathcal{O} \mid 1 \le r \le n \}$$

subject only to the following relations:

$$\prod_{\substack{1\le l\le \ell \\ \kappa_l \equiv i_1 (\mathrm{mod}\ e)}} (y_1^\mathcal{O} - x^l - [\kappa_l - i_1]) f_{\mathbf{i}}^\mathcal{O} = 0,$$

$$f_{\mathbf{i}}^\mathcal{O} f_{\mathbf{j}}^\mathcal{O} = \delta_{\mathbf{i}\mathbf{j}} f_{\mathbf{i}}^\mathcal{O}, \qquad \textstyle\sum_{\mathbf{i}\in I^n} f_{\mathbf{i}}^\mathcal{O} = 1, \qquad y_r^\mathcal{O} f_{\mathbf{i}}^\mathcal{O} = f_{\mathbf{i}}^\mathcal{O} y_r^\mathcal{O},$$

$$\psi_r^\mathcal{O} f_{\mathbf{i}}^\mathcal{O} = f_{s_r\cdot \mathbf{i}}^\mathcal{O} \psi_r^\mathcal{O}, \qquad y_r^\mathcal{O} y_s^\mathcal{O} = y_s^\mathcal{O} y_r^\mathcal{O},$$

$$\psi_r^\mathcal{O} y_{r+1}^\mathcal{O} f_{\mathbf{i}}^\mathcal{O} = (y_r^\mathcal{O} \psi_r^\mathcal{O} + \delta_{i_r i_{r+1}}) f_{\mathbf{i}}^\mathcal{O}, \quad y_{r+1}^\mathcal{O} \psi_r^\mathcal{O} f_{\mathbf{i}}^\mathcal{O} = (\psi_r^\mathcal{O} y_r^\mathcal{O} + \delta_{i_r i_{r+1}}) f_{\mathbf{i}}^\mathcal{O},$$

$$\psi_r^\mathcal{O} y_s^\mathcal{O} = y_s^\mathcal{O} \psi_r^\mathcal{O}, \qquad \text{if } s \ne r, r+1,$$

$$\psi_r^\mathcal{O} \psi_s^\mathcal{O} = \psi_s^\mathcal{O} \psi_r^\mathcal{O}, \qquad \text{if } |r-s| > 1,$$

$$(\psi_r^\mathcal{O})^2 f_{\mathbf{i}}^\mathcal{O} = \begin{cases} (y_r^{\langle 1+\rho_r(\mathbf{i})\rangle} - y_{r+1}^\mathcal{O})(y_{r+1}^{\langle 1-\rho_r(\mathbf{i})\rangle} - y_r^\mathcal{O}) f_{\mathbf{i}}^\mathcal{O}, & \text{if } i_r \leftrightarrows i_{r+1}, \\ (y_r^{\langle 1+\rho_r(\mathbf{i})\rangle} - y_{r+1}^\mathcal{O}) f_{\mathbf{i}}^\mathcal{O}, & \text{if } i_r \to i_{r+1}, \\ (y_{r+1}^{\langle 1-\rho_r(\mathbf{i})\rangle} - y_r^\mathcal{O}) f_{\mathbf{i}}^\mathcal{O}, & \text{if } i_r \leftarrow i_{r+1}, \\ 0, & \text{if } i_r = i_{r+1}, \\ f_{\mathbf{i}}^\mathcal{O}, & \text{otherwise,} \end{cases}$$

and $(\psi_r^\mathcal{O}\psi_{r+1}^\mathcal{O}\psi_r^\mathcal{O} - \psi_{r+1}^\mathcal{O}\psi_r^\mathcal{O}\psi_{r+1}^\mathcal{O}) f_{\mathbf{i}}^\mathcal{O}$ *is equal to*

$$\begin{cases} (y_r^{\langle 1+\rho_r(\mathbf{i})\rangle} + y_{r+2}^{\langle 1+\rho_r(\mathbf{i})\rangle} - y_{r+1}^{\langle 1+\rho_r(\mathbf{i})\rangle} - y_{r+1}^{\langle 1-\rho_r(\mathbf{i})\rangle}) f_{\mathbf{i}}^\mathcal{O}, & \text{if } i_{r+2} = i_r \rightleftarrows i_{r+1}, \\ -t^{1+\rho_r(\mathbf{i})} f_{\mathbf{i}}^\mathcal{O}, & \text{if } i_{r+2} = i_r \to i_{r+1}, \\ f_{\mathbf{i}}^\mathcal{O}, & \text{if } i_{r+2} = i_r \leftarrow i_{r+1}, \\ 0, & \text{otherwise,} \end{cases}$$

where $\rho_r(\mathbf{i}) = \hat{\imath}_r - \hat{\imath}_{r+1}$ *and* $y_r^{\langle d\rangle} = t^{2d} y_r^\mathcal{O} + t^{-1}[d]$, *for* $d \in \mathbb{Z}$.

The statement of Theorem 4.2.5 is slightly different to [57, Theorem A] because we are using a different choice of modular system $(K, \mathcal{O}, F)$ and because Definition 1.1.1 renormalises the quadratic relations for the generators T_r of $\mathscr{H}_n^\mathcal{O}$, for $1 \le r < n$.

The strategy behind the proof of Theorem 4.2.5 is quite simple: we compute the action of the elements defined in Definition 4.2.4 on the seminormal basis and use this to verify that they satisfy the relations in the theorem. To bound the rank of the algebra defined by the presentation in Theorem 4.2.5 we essentially count dimensions. By specializing $x = 0$, we obtain Theorem 3.1.1 as a corollary of Theorem 4.2.5.

To give a flavour of the type of calculations that were used to verify that the elements in Definition 4.2.4 satisfy the relations in Theorem 4.2.5, for $\mathsf{s} \in \operatorname{Std}(\mathbf{i})$ and $1 \le r < n$ define

$$\beta_r(\mathsf{s}) = \begin{cases} \dfrac{\alpha_r(\mathsf{s})t^{2\hat{\imath}_r}}{M_r^{\mathcal{Z}}(\mathsf{s})}, & \text{if } i_r = i_{r+1}, \\ \alpha_r(\mathsf{s})\rho_r^{\mathcal{Z}}(\mathsf{s})t^{-2\hat{\imath}_r}, & \text{if } i_r = i_{r+1}+1, \\ \dfrac{\alpha_r(\mathsf{s})\rho_r^{\mathcal{Z}}(\mathsf{s})}{M_r^{\mathcal{Z}}(\mathsf{s})}, & \text{otherwise.} \end{cases} \tag{4.2.6}$$

Then Theorem 1.6.7 easily yields the following.

Lemma 4.2.7. *Suppose that $1 \le r < n$ and that $(\mathsf{s}, \mathsf{t}) \in \operatorname{Std}^2(\mathcal{P}_n^\Lambda)$. Set $\mathbf{i} = \mathbf{i}^{\mathsf{s}}$, $\mathbf{j} = \mathbf{i}^{\mathsf{t}}$, $\mathsf{u} = \mathsf{s}(r, r+1)$ and $\mathsf{v} = \mathsf{t}(r, r+1)$. Then*

$$\psi_r^{\mathcal{O}} f_{\mathsf{st}} = \beta_r(\mathsf{s}) f_{\mathsf{ut}} - \delta_{i_r i_{r+1}} \frac{1}{\rho_r^{\mathcal{Z}}(\mathsf{s})} f_{\mathsf{st}}.$$

Moreover, if $\mathsf{s}(l, b, c) = r$ then

$$y_r^{\langle d\rangle} f_{\mathsf{st}} = t^{-1}\big(t^{2(c-b+d-i_r)}x^l + [c_k^{\mathbb{Z}}(\mathsf{s}) + d - \hat{\imath}_r]\big) f_{\mathsf{st}},$$

for $1 \le r \le n$ and $d \in \mathbb{Z}$.

Armed with Lemma 4.2.7, and Definition 1.6.6, it is an easy exercise to verify that all of the relations in Theorem 4.2.5 hold in $\mathscr{H}_n^{\mathcal{O}}$. For the quadratic relations, Lemma 4.2.7 implies that $(\psi_r^{\mathcal{O}})^2 f_{\mathsf{st}} = 0$ if $\mathsf{s} \in \operatorname{Std}(\mathbf{i})$ and $i_r = i_{r+1}$ whereas if $i_r \ne i_{r+1}$ then $(\psi_r^{\mathcal{O}})^2 f_{\mathsf{st}} = \beta_r(\mathsf{s})\beta_r(\mathsf{u}) f_{\mathsf{st}}$, where $\mathsf{u} = \mathsf{s}(r, r+1)$. The quadratic relations in Theorem 4.2.5 now follow using (4.2.6) and Lemma 4.2.7. For example, suppose that $i_r \to i_{r+1}$ and $\mathsf{s} \in \operatorname{Std}(\mathbf{i})$. Pick nodes (l, b, c) and (l', b', c') such that $\mathsf{s}(l, b, c) = r$ and $\mathsf{s}(l', b', c') = r + 1$. Then, using Lemma 4.2.7 and Definition 1.6.6,

$$(\psi_r^{\mathcal{O}})^2 f_{\mathsf{st}} = t^{-2\hat{\imath}_{r+1}} \beta_r(\mathsf{s})\beta_r(\mathsf{u}) f_{\mathsf{st}} = t^{-2\hat{\imath}_{r+1}} M_r^{\mathcal{Z}}(\mathsf{u}) f_{\mathsf{st}}.$$

On the other hand, by Lemma 4.2.7, $(y_r^{\langle 1+\rho_r(\mathbf{i})\rangle} - y_{r+1}^{\mathcal{O}})$ acts on f_{st} as multiplication by the same scalar. It follows that

$$\begin{aligned}(\psi_r^{\mathcal{O}})^2 f_{\mathbf{i}}^{\mathcal{O}} &= (\psi_r^{\mathcal{O}})^2 \sum_{\mathsf{s}\in\operatorname{Std}(\mathbf{i})} \frac{1}{\gamma_{\mathsf{s}}} f_{\mathsf{ss}} = (y_r^{\langle 1+\rho_r(\mathbf{i})\rangle} - y_{r+1}^{\mathcal{O}}) \sum_{\mathsf{s}\in\operatorname{Std}(\mathbf{i})} \frac{1}{\gamma_{\mathsf{s}}} f_{\mathsf{ss}} \\ &= (y_r^{\langle 1+\rho_r(\mathbf{i})\rangle} - y_{r+1}^{\mathcal{O}}) f_{\mathbf{i}}^{\mathcal{O}}\end{aligned}$$

when $i_r \to i_{r+1}$. These calculations are perhaps not very pretty, but nor are they are difficult. As indicated by Remark 2.2.6, the quadratic relations appear in, and simplify, the proof of the deformed braid relations.

4.3 *A distinguished homogeneous basis*

One of the advantages of Theorem 4.2.5 is that it allows us to transplant questions about the KLR algebra $\mathscr{R}_n^\Lambda$ into the language of seminormal bases. Definition 1.6.6 defines $*$-seminormal bases, which provide a good framework for studying the semisimple cyclotomic Hecke algebras. The algebra $\mathscr{H}_n^\Lambda$ comes with two cellular algebra automorphisms, $*$ and $\star$, where $\star$ is the unique anti-isomorphism fixing the homogeneous generators of Definition 2.2.9 and $*$ is the unique anti-isomorphism fixing the generators of Definition 1.1.1. In general, these automorphisms are different.

Definition 4.3.1 (Hu-Mathas [57, §5]**).** *A **$\star$-seminormal coefficient system** is a collection of scalars*

$$\boldsymbol{\beta} = \{\, \beta_r(\mathsf{t}) \mid \mathsf{t} \in \operatorname{Std}(\mathcal{P}_n^\Lambda) \text{ and } 1 \le r \le n \,\}$$

such that $\beta_r(\mathsf{t}) = 0$ *if* $\mathsf{v} = \mathsf{t}(r, r+1)$ *is not standard, if* $\mathsf{v} \in \operatorname{Std}(\mathcal{P}_n^\Lambda)$ *then* $\beta_r(\mathsf{v})\beta_r(\mathsf{t})$ *is given by the product of the particular* β*-coefficients in* (4.2.6), *and* $\beta_r(\mathsf{t})\beta_{r+1}(\mathsf{t}s_r)\beta_r(\mathsf{t}s_r s_{r+1}) = \beta_{r+1}(\mathsf{t})\beta_r(\mathsf{t}s_{r+1})\beta_{r+1}(\mathsf{t}s_{r+1}s_r)$, *and if* $|r - r'| > 1$ *then* $\beta_r(\mathsf{t})\beta_{r'}(\mathsf{t}s_r) = \beta_{r'}(\mathsf{t})\beta_r(\mathsf{t}s_{r'})$ *for* $1 \le r, r' < n$.

Exactly as in Corollary 1.6.9, a $\star$-seminormal coefficient system determines a $\star$-seminormal basis $\{f_{\mathsf{st}}\}$ that, like Definition 1.6.4, consists of elements $f_{\mathsf{st}} \in H_{\mathsf{st}}$ such that $f_{\mathsf{st}}^\star = f_{\mathsf{ts}}$, for $(\mathsf{s}, \mathsf{t}) \in \operatorname{Std}^2(\mathcal{P}_n^\Lambda)$. The left (and right) the action of $\psi_r^{\mathcal{O}}$ on f_{st} is exactly as in Lemma 4.2.7 but where the coefficients come from an arbitrary $\star$-seminormal coefficient system $\boldsymbol{\beta}$.

Definition 4.3.1 gives extra flexibility in choosing a $\star$-seminormal basis. By [57, (5.8)] there exists a $\star$-seminormal basis $\{f_{\mathsf{st}}\}$ such that the ψ-basis of Theorem 3.2.4 lifts to a $\psi^{\mathcal{O}}$-basis $\{\psi_{\mathsf{st}}^{\mathcal{O}}\}$ with the property that

$$\psi_{\mathsf{st}}^{\mathcal{O}} = f_{\mathsf{st}} + \sum_{(\mathsf{u},\mathsf{v}) \blacktriangleright (\mathsf{s},\mathsf{t})} r_{\mathsf{uv}} f_{\mathsf{uv}}, \tag{4.3.2}$$

for some $r_{\mathsf{uv}} \in K$. In this way we recover Theorem 3.2.4 and with quicker proof than the original arguments in [54]. Perhaps most significantly, by working with $\mathscr{H}_n^{\mathcal{O}}$ we can improve upon the ψ-basis.

Theorem 4.3.3 (Hu-Mathas [57, Theorem 6.2, Corollary 6.3]**).** *Suppose that* $(\mathsf{s}, \mathsf{t}) \in \operatorname{Std}^2(\mathcal{P}_n^\Lambda)$. *There exists a unique element* $B_{\mathsf{st}}^{\mathcal{O}} \in \mathscr{H}_n^{\mathcal{O}}$ *such that*

$$B_{\mathsf{st}}^{\mathcal{O}} = f_{\mathsf{st}} + \sum_{\substack{(\mathsf{u},\mathsf{v}) \in \operatorname{Std}^2(\mathcal{P}_n^\Lambda) \\ (\mathsf{u},\mathsf{v}) \blacktriangleright (\mathsf{s},\mathsf{t})}} p_{\mathsf{uv}}^{\mathsf{st}}(x^{-1}) f_{\mathsf{uv}},$$

where $p_{\mathsf{uv}}^{\mathsf{st}}(x) \in xK[x]$. *Moreover,* $\{\, B_{\mathsf{st}}^{\mathcal{O}} \mid (\mathsf{s}, \mathsf{t}) \in \operatorname{Std}^2(\mathcal{P}_n^\Lambda) \,\}$ *is a cellular basis of* $\mathscr{H}_n^{\mathcal{O}}$.

The existence and uniqueness of this basis essentially come down to Gaussian elimination, although for technical reasons it is necessary to work over the $x\mathcal{O}$-adic completion of $\mathcal{O}$. Proving that $\{B^{\mathcal{O}}_{\mathsf{st}}\}$ is a cellular basis is more involved and, ultimately, this relies on the uniqueness properties of the $B^{\mathcal{O}}$-basis elements.

As the $B^{\mathcal{O}}$-basis is determined by a $\star$-seminormal basis, the basis $\{B^{\mathcal{O}}_{\mathsf{st}}\}$ behaves well with respect to the KLR grading on $\mathscr{H}_n^\Lambda$. The main justification for using this seminormal basis as a proxy for choosing a "nice" basis for $\mathscr{H}_n^\Lambda$, apart from the fact that it works, is that Theorem 2.4.8 shows that the natural homogeneous basis of the semisimple cyclotomic quiver Hecke algebras is a $\star$-seminormal basis.

In characteristic zero the non-zero polynomials $p^{\mathsf{st}}_{\mathsf{uv}}(x)$ satisfy

$$0 < \deg p^{\mathsf{st}}_{\mathsf{uv}}(x) \le \tfrac{1}{2}(\deg \mathsf{u} - \deg \mathsf{s} + \deg \mathsf{v} - \deg \mathsf{t}), \tag{4.3.4}$$

whenever $(\mathsf{u},\mathsf{v}) \blacktriangleright (\mathsf{s},\mathsf{t})$ by [57, Proposition 6.4]. Moreover, if $\mathsf{s},\mathsf{t},\mathsf{u},\mathsf{v}$ are all standard tableaux of the same shape then $p^{\mathsf{st}}_{\mathsf{uv}}(x) = p^{\mathsf{s}}_{\mathsf{u}}(x)p^{\mathsf{t}}_{\mathsf{v}}(x)$, where $0 < \deg p^{\mathsf{s}}_{\mathsf{u}}(x) \le \frac{1}{2}(\deg \mathsf{u} - \deg \mathsf{s})$ and $0 < \deg p^{\mathsf{t}}_{\mathsf{v}} \le \frac{1}{2}(\deg \mathsf{v} - \deg \mathsf{t})$, whenever $\mathsf{u} \rhd \mathsf{s}$ and $\mathsf{v} \rhd \mathsf{t}$, respectively.

As the basis $\{B^{\mathcal{O}}_{\mathsf{st}}\}$ is defined over $\mathcal{O}$ we can reduce modulo the ideal $x\mathcal{O}$ to obtain a basis $\{B^{\mathcal{O}}_{\mathsf{st}} \otimes_{\mathcal{O}} 1_K\}$ of $\mathscr{H}_n^\Lambda = \mathscr{H}_n^\Lambda(K)$. This basis is hard to compute and we do not know if the elements of $\{B^{\mathcal{O}}_{\mathsf{st}} \otimes_{\mathcal{O}} 1_K\}$ are homogeneous in general. Nonetheless, it is possible to construct a homogeneous basis $\{B_{\mathsf{st}}\}$ of $\mathscr{H}_n^\Lambda$ from $\{B^{\mathcal{O}}_{\mathsf{st}}\}$. If $\boldsymbol{\lambda} \in \mathcal{P}_n^\Lambda$ then define $B_{\mathsf{t}^{\boldsymbol{\lambda}}\mathsf{t}^{\boldsymbol{\lambda}}}$ to be the homogeneous component of $B^{\mathcal{O}}_{\mathsf{t}^{\boldsymbol{\lambda}}\mathsf{t}^{\boldsymbol{\lambda}}} \otimes 1_K$ of degree $2\deg \mathsf{t}^{\boldsymbol{\lambda}}$. More generally, for $\mathsf{s},\mathsf{t} \in \operatorname{Std}(\boldsymbol{\lambda})$ we define $B_{\mathsf{st}} = D^\star_{\mathsf{s}} B_{\mathsf{t}^{\boldsymbol{\lambda}}\mathsf{t}^{\boldsymbol{\lambda}}} D_{\mathsf{t}}$, where $D_{\mathsf{s}}, D_{\mathsf{t}} \in \mathscr{H}_n^\Lambda$ are certain homogeneous elements in $\mathscr{H}_n^\Lambda$. In characteristic zero, B_{st} is essentially the homogeneous component of $B^{\mathcal{O}}_{\mathsf{st}} \otimes 1_K$ of degree $\deg \mathsf{s} + \deg \mathsf{t}$, and all other components are of larger degree. For any field, by (4.3.2) and Theorem 4.3.3,

$$B_{\mathsf{st}} = \psi_{\mathsf{st}} + \sum_{(\mathsf{u},\mathsf{v}) \blacktriangleright (\mathsf{s},\mathsf{t})} a_{\mathsf{uv}} \psi_{\mathsf{uv}}, \tag{4.3.5}$$

for some $a_{\mathsf{uv}} \in K$ that are non-zero only if $\mathbf{i}^{\mathsf{u}} = \mathbf{i}^{\mathsf{s}}$, $\mathbf{i}^{\mathsf{v}} = \mathbf{i}^{\mathsf{t}}$ and $\deg \mathsf{u} + \deg \mathsf{v} = \deg \mathsf{s} + \deg \mathsf{t}$. Therefore, the B-basis resolves the ambiguities of Proposition 3.2.10(b). More importantly, we have the following.

Theorem 4.3.6 (Hu-Mathas [57, Theorem 6.9]**).** *Suppose that K is a field. Then $\{\, B_{\mathsf{st}} \mid (\mathsf{s},\mathsf{t}) \in \operatorname{Std}^2(\mathcal{P}_n^\Lambda) \,\}$ is a graded cellular basis of $\mathscr{R}_n^\Lambda$ with weight poset $(\mathcal{P}_n^\Lambda, \unrhd)$, cellular algebra automorphism $\star$ and with $\deg B_{\mathsf{st}} = \deg \mathsf{s} + \deg \mathsf{t}$, for $(\mathsf{s},\mathsf{t}) \in \operatorname{Std}^2(\mathcal{P}_n^\Lambda)$. Moreover, if $(\mathsf{s},\mathsf{t}) \in \operatorname{Std}^2(\mathcal{P}_n^\Lambda)$ then $B_{\mathsf{st}} + \mathscr{H}_n^{\rhd\boldsymbol{\lambda}}$ depends only on s and t and not on the choice of reduced expressions for the permutations $d(\mathsf{s}), d(\mathsf{t}) \in \mathfrak{S}_n$.*

By construction, the basis $\{B_{st}\}$ depends on the field F. If F is a field of positive characteristic then B_{st} depends upon the choice of the elements D_s and D_t, which are uniquely determined modulo the ideal $\mathscr{H}_n^{\rhd\boldsymbol{\lambda}}$. If $e < \infty$ and $\ell > 1$ then $\{B_{st}\}$ can depend on the choice of multicharge.

4.4 *A simple conjecture*

The construction of the basis $\{B_{st}^{\mathcal{O}}\}$ of $\mathscr{H}_n^{\mathcal{O}}$ in Theorem 4.3.3, together with the degree constraints on the polynomials $p_{uv}^{st}(x)$ in (4.3.4), is reminiscent of the Kazhdan-Lusztig basis [73]. There is no known analogue of the Kazhdan-Lusztig bar involution in this setting. On the other hand, we do require that the basis elements B_{st} are homogeneous, which might be an appropriate substitute for being bar invariant in the graded setting. Partly motivated by this analogy with the Kazhdan-Lusztig basis, we now define analogues of *cell representations* for the B-basis.

The basis $\{B_{st}\}$ of Theorem 4.3.6 is a graded cellular basis so it defines a new homogeneous basis $\{\, B_t \mid t \in \mathrm{Std}(\boldsymbol{\lambda}) \,\}$ of the graded Specht module $S^{\boldsymbol{\lambda}}$. Let the pre-order $\succeq_B$ on $\mathrm{Std}(\boldsymbol{\lambda})$ be the transitive closure of the relation $\dot{\succeq}_B$ where $t \dot{\succeq}_B v$ if there exists $a \in \mathscr{R}_n^{\Lambda}$ such that $B_t a = \sum_s r_s B_s$ with $r_v \neq 0$. (So $\succeq_B$ is reflexive and transitive but not anti-symmetric.) Let $\sim_B$ be the equivalence relation on $\mathrm{Std}(\boldsymbol{\lambda})$ determined by $\succeq_B$ so that $t \sim_B v$ if and only if $t \succeq_B v \succeq_B t$. For example, $t^{\boldsymbol{\lambda}} \succeq_B t \succeq_B t_{\boldsymbol{\lambda}}$, for all $t \in \mathrm{Std}(\boldsymbol{\lambda})$.

Let $\mathrm{Std}[\boldsymbol{\lambda}]$ be the set of $\sim_B$-equivalence classes in $\mathrm{Std}(\boldsymbol{\lambda})$. The set $\mathrm{Std}[\boldsymbol{\lambda}]$ is partially ordered by $\succeq_B$, where $\mathsf{T} \succeq_B \mathsf{V}$ if $t \succeq_B v$ for some $t \in \mathsf{T}$ and $v \in \mathsf{V}$. Write $\mathsf{T} \succeq_B v$ if $t \succeq_B v$ for some $t \in \mathsf{T}$ and $\mathsf{T} \succ_B v$ if $\mathsf{T} \succeq_B v$ and $v \notin \mathsf{T}$. Define $S^{\boldsymbol{\lambda}}_{\mathsf{T}\succeq}$ to be the vector subspace of $S^{\boldsymbol{\lambda}}$ with basis $\{\, B_v \mid \mathsf{T} \succeq_B v \,\}$. Similarly, let $S^{\boldsymbol{\lambda}}_{\mathsf{T}\succ}$ be the vector space with basis $\{\, B_v \mid \mathsf{T} \succ_B v \,\}$. The definition of $\succeq_B$ ensures that $S^{\boldsymbol{\lambda}}_{\mathsf{T}\succeq}$ and $S^{\boldsymbol{\lambda}}_{\mathsf{T}\succ}$ are both graded $\mathscr{H}_n^{\Lambda}$-submodules of $S^{\boldsymbol{\lambda}}$ and that $S^{\boldsymbol{\lambda}}_{\mathsf{T}\succ} \subsetneq S^{\boldsymbol{\lambda}}_{\mathsf{T}\succeq}$. Therefore, $S^{\boldsymbol{\lambda}}_{\mathsf{T}} = S^{\boldsymbol{\lambda}}_{\mathsf{T}\succeq}/S^{\boldsymbol{\lambda}}_{\mathsf{T}\succ}$ is a graded $\mathscr{H}_n^{\Lambda}$-module. By choosing any total order on $\mathrm{Std}[\boldsymbol{\lambda}]$ that extends the partial order $\succeq_B$, it is easy to see that $S^{\boldsymbol{\lambda}}$ has a filtration with subquotients being precisely the modules $S^{\boldsymbol{\lambda}}_{\mathsf{T}}$, for $\mathsf{T} \in \mathrm{Std}[\boldsymbol{\lambda}]$.

For $\boldsymbol{\lambda} \in \mathcal{P}_n^{\Lambda}$ let $\mathsf{T}^{\boldsymbol{\lambda}} = \{\, t \in \mathrm{Std}(\boldsymbol{\lambda}) \mid t \sim_B t^{\boldsymbol{\lambda}} \,\}$. In view of (3.7.2), if $s, t \in \mathrm{Std}(\boldsymbol{\lambda})$ and $\langle B_s, B_t \rangle \neq 0$ then $s \sim_B t^{\boldsymbol{\lambda}} \sim_B t$ so that $s, t \in T^{\boldsymbol{\lambda}}$. Therefore, $\dim D^{\boldsymbol{\lambda}} \leq |\mathsf{T}^{\boldsymbol{\lambda}}|$. Of course, if $\boldsymbol{\lambda} \notin \mathcal{K}_n^{\Lambda}$ then this bound is not sharp because $D^{\boldsymbol{\lambda}} = 0$ whereas $|\mathsf{T}^{\boldsymbol{\lambda}}| \geq 1$.

Conjecture 4.4.1. *Suppose that F is a field of characteristic zero and that $\boldsymbol{\lambda} \in \mathcal{P}_n^{\Lambda}$. Then $S^{\boldsymbol{\lambda}}_{\mathsf{T}}$ is an irreducible $\mathscr{H}_n^{\Lambda}$-module, for all* $\mathsf{T} \in \mathrm{Std}[\boldsymbol{\lambda}]$.

As discussed in [57, §3.3], and is implicit in (4.1.4), by fixing a composition series for $S^{\boldsymbol{\lambda}}$ and using a Gaussian elimination argument, it is possible to construct a basis $\{C_{\mathsf{t}}\}$ of $S^{\boldsymbol{\mu}}$ such that (1) each module in the composition series has a basis contained in $\{C_{\mathsf{t}}\}$, and (2), if $\mathsf{t} \in \operatorname{Std}(\boldsymbol{\lambda})$ then $C_{\mathsf{t}} = \psi_{\mathsf{t}}$ plus a linear combination of "higher terms" with respect to some total order on $\operatorname{Std}(\boldsymbol{\lambda})$. This defines a partition of $\operatorname{Std}(\boldsymbol{\lambda}) = T_1 \sqcup \cdots \sqcup T_z$ (disjoint union), where the tableaux in the set T_k are in bijection with a basis of the kth composition factor. Therefore, there exists an equivalence relation on $\operatorname{Std}(\boldsymbol{\lambda})$, together with an associated composition series, such that the analogue of Conjecture 4.4.1 holds for this equivalence relation. Our conjecture attempts to make this equivalence relation on $\operatorname{Std}(\boldsymbol{\lambda})$ explicit and canonical.

If $\mathcal{T} \subseteq \operatorname{Std}(\boldsymbol{\lambda})$ define its character to be $\operatorname{ch}_q \mathcal{T} = \sum_{\mathsf{t}\in\mathcal{T}} q^{\deg \mathsf{t}} \cdot \mathbf{i}^{\mathsf{t}} \in \mathcal{A}[I^n]$. The point of this definition is that $\operatorname{ch}_q \mathcal{T}$ is a purely combinatorial invariant of $\mathcal{T}$. As two examples, $\operatorname{Ch}_q S^{\boldsymbol{\lambda}} = \operatorname{ch}_q \operatorname{Std}(\boldsymbol{\lambda})$ and $\operatorname{Ch}_q S^{\boldsymbol{\lambda}}_{\mathsf{T}} = \operatorname{ch}_q \mathsf{T}$.

Proposition 4.4.2. *Suppose that Conjecture 4.4.1 holds when $F = \mathbb{C}$.*

a) Suppose that $\boldsymbol{\mu} \in \mathcal{K}^\Lambda_n$. Then $D^{\boldsymbol{\mu}}_{\mathbb{C}} \cong S^{\boldsymbol{\mu}}_{\mathsf{T}^{\boldsymbol{\mu}}}$ and $\operatorname{Ch}_q D^{\boldsymbol{\mu}}_{\mathbb{C}} = \operatorname{ch}_q \mathsf{T}^{\boldsymbol{\mu}}$.

b) If $\boldsymbol{\lambda} \in \mathcal{P}^\Lambda_n$ and $\mathsf{T} \in \operatorname{Std}[\boldsymbol{\lambda}]$ then there is a unique pair $(\boldsymbol{\nu}_{\mathsf{T}}, d_{\mathsf{T}})$ in $\mathcal{K}^\Lambda_n \times \mathbb{N}$ such that $\operatorname{ch}_q \mathsf{T} = q^{d_{\mathsf{T}}} \operatorname{Ch}_q D^{\boldsymbol{\nu}_{\mathsf{T}}}_{\mathbb{C}} = q^{d_{\mathsf{T}}} \operatorname{ch}_q \mathsf{T}^{\boldsymbol{\nu}_{\mathsf{T}}}$. Moreover,

$$d_{\boldsymbol{\lambda}\boldsymbol{\mu}}(q) = \sum_{\substack{\mathsf{T}\in\operatorname{Std}[\boldsymbol{\lambda}]\\ \boldsymbol{\nu}_{\mathsf{T}}=\boldsymbol{\mu}}} q^{d_{\mathsf{T}}}.$$

Proof. By Corollary 3.2.7, $D^{\boldsymbol{\mu}}_{\mathbb{C}} \neq 0$ since $\boldsymbol{\mu} \in \mathcal{K}^\Lambda_n$. The irreducible module $D^{\boldsymbol{\mu}}_{\mathbb{C}}$ is generated by $B_{\mathsf{t}^{\boldsymbol{\mu}}} + \operatorname{rad} S^{\boldsymbol{\mu}}_{\mathbb{C}} = \psi_{\mathsf{t}^{\boldsymbol{\mu}}} + \operatorname{rad} S^{\boldsymbol{\mu}}_{\mathbb{C}}$, so $D^{\boldsymbol{\mu}}_{\mathbb{C}} \cong S^{\boldsymbol{\mu}}_{T^{\boldsymbol{\mu}}}$ since both modules are irreducible by Conjecture 4.4.1. Hence, (a) follows.

For part (b), $S^{\boldsymbol{\lambda}}_{\mathsf{T}} \cong D^{\boldsymbol{\nu}}_{\mathbb{C}}\langle d\rangle$, for some $\boldsymbol{\nu} \in \mathcal{K}^\Lambda_n$ and $d \in \mathbb{Z}$, because $S^{\boldsymbol{\lambda}}_{\mathsf{T}}$ is irreducible by Conjecture 4.4.1. Therefore, $\operatorname{Ch}_q S^{\boldsymbol{\lambda}}_{\mathsf{T}} = q^d \operatorname{Ch}_q D^{\boldsymbol{\nu}}_{\mathbb{C}}$. The uniqueness of $(\boldsymbol{\nu}_{\mathsf{T}}, d_{\mathsf{T}}) = (\boldsymbol{\nu}, d) \in \mathcal{K}^\Lambda_n \times \mathbb{Z}$ now follows from Theorem 3.7.1. Moreover, $d \geq 0$ by Corollary 3.5.27. As every composition factor of $S^{\boldsymbol{\lambda}}_{\mathbb{C}}$ is isomorphic to $S^{\boldsymbol{\lambda}}_{\mathsf{T}}$, for some $\mathsf{T} \in \operatorname{Std}[\boldsymbol{\lambda}]$, the formula for $d_{\boldsymbol{\lambda}\boldsymbol{\mu}}(q)$ is now immediate. □

Proposition 4.4.2 shows that Conjecture 4.4.1 encodes closed formulas for the characters and graded dimensions of the irreducible $\mathscr{H}^\Lambda_n$-modules and for the graded decomposition numbers of $\mathscr{H}^\Lambda_n$. For this result to be useful we need to first verify Conjecture 4.4.1 and then to explicitly determine the equivalence relation $\sim_B$. Our last result is a step in this direction.

Lemma 4.4.3. *Suppose that $\mathsf{s}, \mathsf{t} \in \operatorname{Std}(\boldsymbol{\lambda})$ and that $\mathsf{t} = \mathsf{s}(r, r+1)$ such that $\mathbf{i}^{\mathsf{s}}_{r+1} \neq \mathbf{i}^{\mathsf{s}}_r \pm 1$, where $1 \leq r < n$ and $\boldsymbol{\lambda} \in \mathcal{P}^\Lambda_n$. Then $\mathsf{s} \sim_B \mathsf{t}$.*

Proof. By assumption, either $\mathsf{s} \rhd \mathsf{t}$ or $\mathsf{t} \rhd \mathsf{s}$. Without loss of generality we assume that $\mathsf{s} \rhd \mathsf{t}$. It follows from (4.3.5), and Theorem 3.6.2, that

$$B_{\mathsf{s}}\psi_r = \psi_{\mathsf{t}} + \sum_{\mathsf{u}} a_{\mathsf{u}}\psi_{\mathsf{u}} = B_{\mathsf{t}} + \sum_{\mathsf{u}} b_{\mathsf{u}} B_{\mathsf{u}},$$

where $a_{\mathsf{u}}, b_{\mathsf{u}} \in F$ are non-zero only if $\ell(d(\mathsf{u})) < \ell(d(\mathsf{s}))$. Therefore, $\mathsf{s} \succeq_B \mathsf{t}$. If $\mathbf{i}^{\mathsf{s}}_{r+1} \neq \mathbf{i}^{\mathsf{t}}_r$ then $e(\mathbf{i}^{\mathsf{s}})\psi_r^2 = e(\mathbf{i}^{\mathsf{s}})$ by (2.2.3), so $\mathsf{s} \sim_B \mathsf{t}$. Now consider the more interesting case when $\mathbf{i}^{\mathsf{s}}_{r+1} = \mathbf{i}^{\mathsf{s}}_r$ or, equivalently, $\mathbf{i}^{\mathsf{s}}_r = \mathbf{i}^{\mathsf{t}}_r$. Using (2.2.2),

$$B_{\mathsf{t}} y_{r+1} = \Big(B_{\mathsf{s}}\psi_r - \sum_{\mathsf{u}} b_{\mathsf{u}} B_{\mathsf{u}}\Big) y_{r+1} = B_{\mathsf{s}}(y_r\psi_r + 1) - \sum_{\mathsf{u}} b_{\mathsf{u}} B_{\mathsf{u}} y_{r+1}.$$

In view of Proposition 3.2.10(c), B_{s} appears on the right-hand side with coefficient 1. Hence, $\mathsf{t} \succeq_B \mathsf{s}$ implying that $\mathsf{s} \sim_B \mathsf{t}$ as claimed. □

Finally, we remark that it is easy to check that Conjecture 4.4.1 is true in the trivial cases considered in Example 3.7.7 and Example 3.7.8. With considerably more effort, using [26, Lemma 9.7] and results of [55, Appendix], it is possible to verify the conjecture when $\Lambda \in P^+$ is a weight of level 2 and $e > n$. In all of these cases, the conjecture can be checked because $B_{\mathsf{st}} = \psi_{\mathsf{st}}$, for all $(\mathsf{s}, \mathsf{t}) \in \mathrm{Std}^2(\mathcal{P}_n^\Lambda)$.

The B-basis, and hence Conjecture 4.4.1 and all of the results in this section (except that in positive characteristic we can only say that $d_{\mathsf{T}} \in \mathbb{Z}$ in Proposition 4.4.2, rather than $d_{\mathsf{T}} \in \mathbb{N}$), make sense over any field. We restrict our conjecture to fields of characteristic zero because it would be foolhardy to venture into the realms of positive characteristic without strong evidence. This said, whether or not our conjecture for the B-basis is true, we are convinced that, in *all characteristics*, there exists a "canonical" graded cellular basis $\{C_{\mathsf{st}}\}$ of $\mathscr{R}_n^\Lambda$ such that the analogous version of Conjecture 4.4.1 holds for the $\sim_C$ equivalence classes.

To put it another way, the results of [57, §3.3] show that the KLR-tableau combinatorics is rich enough to give closed combinatorial formulas for both the graded decomposition numbers and the graded dimensions of the irreducible representations of $\mathscr{H}_n^\Lambda$. We believe that over any field the graded Specht modules have a distinguished homogeneous basis that "canonically" determines these combinatorial formulas.

References

[1] H. H. Andersen, J. C. Jantzen, and W. Soergel, *Representations of quantum groups at a pth root of unity and of semisimple groups in characteristic p: independence of p*, Astérisque, 1994, 321.

[2] S. Ariki, *On the semi-simplicity of the Hecke algebra of* $(\mathbb{Z}/r\mathbb{Z}) \wr \mathfrak{S}_n$, J. Algebra, **169** (1994), 216–225.

[3] ——, *On the decomposition numbers of the Hecke algebra of* $G(m,1,n)$, J. Math. Kyoto Univ., **36** (1996), 789–808.

[4] ——, *On the classification of simple modules for cyclotomic Hecke algebras of type* $G(m,1,n)$ *and Kleshchev multipartitions*, Osaka J. Math., **38** (2001), 827–837.

[5] ——, *Representations of quantum algebras and combinatorics of Young tableaux*, University Lecture Series, **26**, American Mathematical Society, Providence, RI, 2002. Translated from the 2000 Japanese edition and revised by the author.

[6] ——, *Proof of the modular branching rule for cyclotomic Hecke algebras*, J. Algebra, **306** (2006), 290–300.

[7] ——, *Graded q-Schur algebras*, 2009, preprint. arXiv:0903.3453.

[8] S. Ariki, N. Jacon, and C. Lecouvey, *The modular branching rule for affine Hecke algebras of type A*, Adv. Math., **228** (2011), 481–526.

[9] ——, *Factorization of the canonical bases for higher-level Fock spaces*, Proc. Edinb. Math. Soc. (2), **55** (2012), 23–51.

[10] S. Ariki and K. Koike, *A Hecke algebra of* $(\mathbf{Z}/r\mathbf{Z}) \wr \mathfrak{S}_n$ *and construction of its irreducible representations*, Adv. Math., **106** (1994), 216–243.

[11] S. Ariki, V. Kreiman, and S. Tsuchioka, *On the tensor product of two basic representations of* $U_v(\widehat{\mathfrak{sl}}_e)$, Adv. Math., **218** (2008), 28–86.

[12] S. Ariki and A. Mathas, *The number of simple modules of the Hecke algebras of type* $G(r,1,n)$, Math. Z., **233** (2000), 601–623.

[13] S. Ariki, A. Mathas, and H. Rui, *Cyclotomic Nazarov-Wenzl algebras*, Nagoya Math. J., **182** (2006), 47–134. (Special issue in honour of George Lusztig), arXiv:math/0506467.

[14] E. Backelin, *Koszul duality for parabolic and singular category* $\mathcal{O}$, Represent. Theory, **3** (1999), 139–152 (electronic).

[15] A. Beilinson, V. Ginzburg, and W. Soergel, *Koszul duality patterns in representation theory*, J. Amer. Math. Soc., **9** (1996), 473–527.

[16] M. Broué and G. Malle, *Zyklotomische Heckealgebren*, Astérisque, **212** (1993), 119–189. Représentations unipotentes génériques et blocs des groupes réductifs finis.

[17] M. Broué, G. Malle, and J. Michel, *Towards spetses. I*, Transform. Groups, **4** (1999), 157–218. Dedicated to the memory of Claude Chevalley.

[18] J. Brundan, *Modular branching rules and the Mullineux map for Hecke algebras of type A*, Proc. London Math. Soc. (3), **77** (1998), 551–581.

[19] ——, *Centers of degenerate cyclotomic Hecke algebras and parabolic category* $\mathcal{O}$, Represent. Theory, **12** (2008), 236–259.

[20] J. Brundan and A. Kleshchev, *Schur-Weyl duality for higher levels*, Selecta Math. (N.S.), **14** (2008), 1–57.
[21] ——, *Blocks of cyclotomic Hecke algebras and Khovanov-Lauda algebras*, Invent. Math., **178** (2009), 451–484.
[22] ——, *Graded decomposition numbers for cyclotomic Hecke algebras*, Adv. Math., **222** (2009), 1883–1942.
[23] ——, *The degenerate analogue of Ariki's categorification theorem*, Math. Z., **266** (2010), 877–919. arXiv:0901.0057.
[24] J. Brundan, A. Kleshchev, and P. J. McNamara, *Homological properties of finite-type Khovanov–Lauda–Rouquier algebras*, Duke Math. J., **163** (2014), 1353–1404. arXiv:1210.6900.
[25] J. Brundan, A. Kleshchev, and W. Wang, *Graded Specht modules*, J. Reine Angew. Math., **655** (2011), 61–87. arXiv:0901.0218.
[26] J. Brundan and C. Stroppel, *Highest weight categories arising from Khovanov's diagram algebra III: category $\mathcal{O}$*, Represent. Theory, **15** (2011), 170–243. arXiv:0812.1090.
[27] J. Chuang and H. Miyachi, *Runner removal Morita equivalences*, in Representation theory of algebraic groups and quantum groups, Progr. Math., **284**, Birkhäuser/Springer, New York, 2010, 55–79.
[28] J. Chuang and R. Rouquier, *Derived equivalences for symmetric groups and $\mathfrak{sl}_2$-categorification*, Ann. of Math. (2), **167** (2008), 245–298.
[29] R. Dipper and G. James, *Representations of Hecke algebras of general linear groups*, Proc. London Math. Soc. (3), **52** (1986), 20–52.
[30] ——, *The q-Schur algebra*, Proc. London Math. Soc. (3), **59** (1989), 23–50.
[31] R. Dipper, G. James, and A. Mathas, *Cyclotomic q-Schur algebras*, Math. Z., **229** (1998), 385–416.
[32] R. Dipper and A. Mathas, *Morita equivalences of Ariki-Koike algebras*, Math. Z., **240** (2002), 579–610.
[33] J. Du and H. Rui, *Specht modules for Ariki-Koike algebras*, Comm. Algebra, **29** (2001), 4701–4719.
[34] B. Elias and G. Williamson, *Soergel Calculus*, 2013, preprint. arXiv:1309.0865.
[35] A. Evseev, *On graded decomposition numbers for cyclotomic Hecke algebras in quantum characteristic 2*, 2013. arXiv:1309.6917.
[36] M. Fayers, *Weights of multipartitions and representations of Ariki-Koike algebras*, Adv. Math., **206** (2006), 112–144.
[37] ——, *Core blocks of Ariki-Koike algebras*, J. Algebraic Combin., **26** (2007), 47–81.
[38] ——, *James's Conjecture holds for weight four blocks of Iwahori-Hecke algebras*, J. Algebra, **317** (2007), 593–633.
[39] ——, *Decomposition numbers for weight three blocks of symmetric groups and Iwahori-Hecke algebras*, Trans. Amer. Math. Soc., **360** (2008), 1341–1376 (electronic).
[40] ——, *An LLT-type algorithm for computing higher-level canonical bases*, J. Pure Appl. Algebra, **214** (2010), 2186–2198.
[41] M. Fayers, *Dyck tilings and the homogeneous Garnir relations for graded*

Specht modules, 2013, preprint. arXiv:1309.6467.
[42] A. R. Francis and J. J. Graham, *Centres of Hecke algebras: the Dipper-James conjecture*, J. Algebra, **306** (2006), 244–267.
[43] M. Geck, *Brauer trees of Hecke algebras*, Comm. Algebra, **20** (1992), 2937–2973.
[44] ——, *Representations of Hecke algebras at roots of unity*, Astérisque, 1998, Exp. No. 836, 3, 33–55. Séminaire Bourbaki. Vol. 1997/98.
[45] F. M. Goodman and J. Graber, *Cellularity and the Jones basic construction*, Adv. in Appl. Math., **46** (2011), 312–362.
[46] F. M. Goodman and H. Wenzl, *Crystal bases of quantum affine algebras and affine Kazhdan-Lusztig polynomials*, Internat. Math. Res. Notices, **5** (1999), 251–275.
[47] R. Gordon and E. L. Green, *Graded Artin algebras*, J. Algebra, **76** (1982), 111–137.
[48] J. J. Graham and G. I. Lehrer, *Cellular algebras*, Invent. Math., **123** (1996), 1–34.
[49] I. Grojnowski, *Affine $\widehat{sl}_p$ controls the modular representation theory of the symmetric group and related algebras*, 1999. arXiv:math/9907129.
[50] I. Grojnowski and G. Lusztig, *A comparison of bases of quantized enveloping algebras*, in Linear algebraic groups and their representations (Los Angeles, CA, 1992), Contemp. Math., **153**, Amer. Math. Soc., Providence, RI, 1993, 11–19.
[51] I. Grojnowski and M. Vazirani, *Strong multiplicity one theorems for affine Hecke algebras of type A*, Transform. Groups, **6** (2001), 143–155.
[52] T. Hayashi, *q-Analogues of Clifford and Weyl algebras—spinor and oscillator representations of quantum enveloping algebras*, Comm. Math. Phys., **127** (1990), 129–144.
[53] A. E. Hoffnung and A. D. Lauda, *Nilpotency in type A cyclotomic quotients*, J. Algebraic Combin., **32** (2010), 533–555.
[54] J. Hu and A. Mathas, *Graded cellular bases for the cyclotomic Khovanov-Lauda-Rouquier algebras of type A*, Adv. Math., **225** (2010), 598–642. arXiv:0907.2985.
[55] ——, *Cyclotomic quiver Schur algebras I: linear quivers*, 2011, preprint. arXiv:1110.1699.
[56] ——, *Graded induction for Specht modules*, Int. Math. Res. Not. IMRN, **2012** (2012), 1230–1263. arXiv:1008.1462.
[57] ——, *Seminormal forms and cyclotomic quiver Hecke algebras of type A*, 2013, preprint. arXiv:1304.0906.
[58] N. Iwahori and H. Matsumoto, *On some Bruhat decomposition and the structure of the Hecke rings of $\mathfrak{p}$-adic Chevalley groups*, Inst. Hautes Études Sci. Publ. Math., **25** (1965), 5–48.
[59] G. James, *The representation theory of the symmetric groups*, Lecture Notes in Mathematics, **682**, Springer, Berlin, 1978.
[60] ——, *The decomposition matrices of* $\mathrm{GL}_n(q)$ *for* $n \leq 10$, Proc. London Math. Soc. (3), **60** (1990), 225–265.

[61] G. JAMES, S. LYLE, AND A. MATHAS, *Rouquier blocks*, Math. Z., **252** (2006), 511–531.

[62] G. JAMES AND A. MATHAS, *The Jantzen sum formula for cyclotomic q-Schur algebras*, Trans. Amer. Math. Soc., **352** (2000), 5381–5404.

[63] ——, *Equating decomposition numbers for different primes*, J. Algebra, **258** (2002), 599–614. arXiv:math/0111140.

[64] G. JAMES AND G. E. MURPHY, *The determinant of the Gram matrix for a Specht module*, J. Algebra, **59** (1979), 222–235.

[65] M. JIMBO, K. C. MISRA, T. MIWA, AND M. OKADO, *Combinatorics of representations of $U_q(\widehat{\mathfrak{sl}}(n))$*, Comm. Math. Phys., **136** (1991), 543–566.

[66] V. G. KAC, *Infinite-dimensional Lie algebras*, Cambridge University Press, Cambridge, third ed., 1990.

[67] S.-J. KANG AND M. KASHIWARA, *Categorification of highest weight modules via Khovanov-Lauda-Rouquier algebras*, Invent. Math., **190** (2012), 699–742.

[68] S.-J. KANG AND E. PARK, *Irreducible modules over Khovanov-Lauda-Rouquier algebras of type A_n and semistandard tableaux*, J. Algebra, **339** (2011), 223–251.

[69] M. KASHIWARA, *On crystal bases of the Q-analogue of universal enveloping algebras*, Duke Math. J., **63** (1991), 465–516.

[70] M. KASHIWARA, *Global crystal bases of quantum groups*, Duke Math. J., **69** (1993), 455–485.

[71] ——, *Biadjointness in cyclotomic Khovanov-Lauda-Rouquier algebras*, Publ. Res. Inst. Math. Sci., **48** (2012), 501–524.

[72] M. KASHIWARA AND Y. SAITO, *Geometric construction of crystal bases*, Duke Math. J., **89** (1997), 9–36.

[73] D. KAZHDAN AND G. LUSZTIG, *Representations of Coxeter groups and Hecke algebras*, Invent. Math., **53** (1979), 165–184.

[74] M. KHOVANOV AND A. D. LAUDA, *A diagrammatic approach to categorification of quantum groups. I*, Represent. Theory, **13** (2009), 309–347.

[75] ——, *A diagrammatic approach to categorification of quantum groups II*, Trans. Amer. Math. Soc., **363** (2011), 2685–2700.

[76] A. KLESHCHEV, *Branching rules for modular representations of symmetric groups. II*, J. Reine Angew. Math., **459** (1995), 163–212.

[77] ——, *Branching rules for modular representations of symmetric groups. III. Some corollaries and a problem of Mullineux*, J. London Math. Soc. (2), **54** (1996), 25–38.

[78] ——, *Branching rules for modular representations of symmetric groups. IV*, J. Algebra, **201** (1998), 547–572.

[79] ——, *Linear and projective representations of symmetric groups*, Cambridge Tracts in Mathematics, **163**, Cambridge University Press, Cambridge, 2005.

[80] ——, *Representation theory of symmetric groups and related Hecke algebras*, Bull. Amer. Math. Soc. (N.S.), **47** (2010), 419–481.

[81] A. KLESHCHEV, A. MATHAS, AND A. RAM, *Universal graded Specht modules for cyclotomic Hecke algebras*, Proc. Lond. Math. Soc. (3), **105** (2012), 1245–

1289. arXiv:1102.3519.
[82] A. Kleshchev and D. Nash, *An interpretation of the Lascoux-Leclerc-Thibon algorithm and graded representation theory*, Comm. Algebra, **38** (2010), 4489–4500. arXiv:0910.5940.
[83] A. Kleshchev and A. Ram, *Homogeneous representations of Khovanov-Lauda algebras*, J. Eur. Math. Soc. (JEMS), **12** (2010), 1293–1306.
[84] ——, *Representations of Khovanov-Lauda-Rouquier algebras and combinatorics of Lyndon words*, Math. Ann., **349** (2011), 943–975.
[85] A. S. Kleshchev, J. W. Loubert, and V. Miemietz, *Affine cellularity of Khovanov-Lauda-Rouquier algebras in type A*, J. Lond. Math. Soc. (2), **88** (2013), 338–358. arXiv:1210.6542.
[86] S. König and C. Xi, *On the structure of cellular algebras*, in Algebras and modules, II (Geiranger, 1996), CMS Conf. Proc., **24**, Amer. Math. Soc., Providence, RI, 1998, 365–386.
[87] ——, *Affine cellular algebras*, Adv. Math., **229** (2012), 139–182.
[88] B. Kostant and S. Kumar, *The nil Hecke ring and cohomology of G/P for a Kac-Moody group G*, Adv. in Math., **62** (1986), 187–237.
[89] A. Lascoux, B. Leclerc, and J.-Y. Thibon, *Hecke algebras at roots of unity and crystal bases of quantum affine algebras*, Comm. Math. Phys., **181** (1996), 205–263.
[90] A. Lascoux and M.-P. Schützenberger, *Polynômes de Schubert*, C. R. Acad. Sci. Paris Sér. I Math., **294** (1982), 447–450.
[91] A. D. Lauda and M. Vazirani, *Crystals from categorified quantum groups*, Adv. Math., **228** (2011), 803–861.
[92] G. Li, *Integral Basis Theorem of cyclotomic Khovanov-Lauda-Rouquier Algebras of type A*, PhD thesis, University of Sydney, 2012.
[93] G. Lusztig, *Some problems in the representation theory of finite Chevalley groups*, in The Santa Cruz Conference on Finite Groups (Univ. California, Santa Cruz, Calif., 1979), Proc. Sympos. Pure Math., **37**, Providence, R.I., 1980, Amer. Math. Soc., 313–317.
[94] G. Lusztig, *Canonical bases arising from quantized enveloping algebras*, J. Amer. Math. Soc., **3** (1990), 447–498.
[95] G. Lusztig, *Introduction to quantum groups*, Progress in Mathematics, **110**, Birkhäuser Boston Inc., Boston, MA, 1993.
[96] S. Lyle and A. Mathas, *Blocks of cyclotomic Hecke algebras*, Adv. Math., **216** (2007), 854–878. arXiv:math/0607451.
[97] ——, *Cyclotomic Carter-Payne homomorphisms*, Represent. Theory, **18** (2014), 117–154. arXiv:1310.7474.
[98] R. Maksimau, *Canonical basis, KLR-algebras and parity sheaves*, 2013, preprint. arXiv:1301.6261.
[99] ——, *Quiver Schur algebras and Koszul duality*, 2013, preprint. arXiv:1307.6013.
[100] G. Malle and A. Mathas, *Symmetric cyclotomic Hecke algebras*, J. Algebra, **205** (1998), 275–293.
[101] L. Manivel, *Symmetric functions, Schubert polynomials and degeneracy loci*, SMF/AMS Texts and Monographs, **6**, American Mathematical Society,

Providence, RI, 2001.
[102] A. MATHAS, *Decomposition matrices of Hecke algebras of type* **A**, 1997.
[103] ——, *Simple modules of Ariki-Koike algebras*, in Group representations: cohomology, group actions and topology (Seattle, WA, 1996), Proc. Sympos. Pure Math., **63**, Amer. Math. Soc., Providence, RI, 1998, 383–396.
[104] ——, *Iwahori-Hecke algebras and Schur algebras of the symmetric group*, University Lecture Series, **15**, American Mathematical Society, Providence, RI, 1999.
[105] ——, *Tilting modules for cyclotomic Schur algebras*, J. Reine Angew. Math., **562** (2003), 137–169.
[106] ——, *Matrix units and generic degrees for the Ariki-Koike algebras*, J. Algebra, **281** (2004), 695–730. arXiv:math/0108164.
[107] ——, *The representation theory of the Ariki-Koike and cyclotomic q-Schur algebras*, in Representation theory of algebraic groups and quantum groups, Adv. Stud. Pure Math., **40**, Math. Soc. Japan, Tokyo, 2004, 261–320.
[108] ——, *Seminormal forms and Gram determinants for cellular algebras*, J. Reine Angew. Math., **619** (2008), 141–173. With an appendix by Marcos Soriano, arXiv:math/0604108.
[109] ——, *A Specht filtration of an induced Specht module*, J. Algebra, **322** (2009), 893–902.
[110] H. MATSUMOTO, *Générateurs et relations des groupes de Weyl généralisés*, C. R. Acad. Sci. Paris, **258** (1964), 3419–3422.
[111] K. MISRA AND T. MIWA, *Crystal base for the basic representation of* $U_q(\mathfrak{sl}(n))$, Comm. Math. Phys., **134** (1990), 79–88.
[112] G. E. MURPHY, *The idempotents of the symmetric group and Nakayama's conjecture*, J. Algebra, **81** (1983), 258–265.
[113] ——, *The representations of Hecke algebras of type* A_n, J. Algebra, **173** (1995), 97–121.
[114] C. NĂSTĂSESCU AND F. VAN OYSTAEYEN, *Methods of graded rings*, Lecture Notes in Mathematics, **1836**, Springer-Verlag, Berlin, 2004.
[115] M. NAZAROV, *Young's orthogonal form for Brauer's centralizer algebra*, J. Algebra, **182** (1996), 664–693.
[116] O. V. OGIEVETSKY AND L. POULAIN D'ANDECY, *Jucys-Murphy elements and representations of cyclotomic Hecke algebras*, 2012, preprint. arXiv:1206.0612.
[117] A. OKOUNKOV AND A. VERSHIK, *A new approach to representation theory of symmetric groups*, Selecta Math. (N.S.), **2** (1996), 581–605.
[118] A. RAM, *Seminormal representations of Weyl groups and Iwahori-Hecke algebras*, Proc. London Math. Soc. (3), **75** (1997), 99–133.
[119] M. J. RICHARDS, *Some decomposition numbers for Hecke algebras of general linear groups*, Math. Proc. Cambridge Philos. Soc., **119** (1996), 383–402.
[120] R. ROUQUIER, *Derived equivalences and finite dimensional algebras*, in International Congress of Mathematicians. Vol. II, Eur. Math. Soc., Zürich, 2006, 191–221.
[121] ——, *2-Kac-Moody algebras*, 2008, preprint. arXiv:0812.5023.
[122] ——, *Quiver Hecke algebras and 2-Lie algebras*, Algebra Colloq., **19** (2012),

359–410.

[123] R. Rouquier, P. Shan, M. Varagnolo, and E. Vasserot, *Categorifications and cyclotomic rational double affine Hecke algebras*, 2013, preprint. arXiv:1305.4456.

[124] S. Ryom-Hansen, *The Schaper formula and the Lascoux, Leclerc and Thibon algorithm*, Lett. Math. Phys., **64** (2003), 213–219.

[125] ——, *Grading the translation functors in type A*, J. Algebra, **274** (2004), 138–163.

[126] P. Shan, *Graded decomposition matrices of v-Schur algebras via Jantzen filtration*, Represent. Theory, **16** (2012), 212–269.

[127] P. Shan, M. Varagnolo, and E. Vasserot, *Koszul duality of affine Kac-Moody algebras and cyclotomic rational DAHA*, 2011, preprint. arXiv:1107.0146.

[128] C. Stroppel and B. Webster, *Quiver Schur algebras and q-Fock space*, 2011, preprint. arXiv:1110.1115.

[129] The Sage-Combinat community, *Sage-Combinat: enhancing Sage as a toolbox for computer exploration in algebraic combinatorics*, 2008. combinat.sagemath.org.

[130] W. Turner, *RoCK blocks*, Mem. Amer. Math. Soc., **202** (2009), viii+102.

[131] D. Uglov, *Canonical bases of higher-level q-deformed Fock spaces and Kazhdan-Lusztig polynomials*, in Physical combinatorics (Kyoto, 1999), Progr. Math., **191**, Boston, MA, 2000, Birkhäuser Boston, 249–299.

[132] M. Varagnolo and E. Vasserot, *Canonical bases and KLR-algebras*, J. Reine Angew. Math., **659** (2011), 67–100.

[133] M. Vazirani, *Parameterizing Hecke algebra modules: Bernstein-Zelevinsky multisegments, Kleshchev multipartitions, and crystal graphs*, Transform. Groups, **7** (2002), 267–303.

[134] B. Webster, *Knot invariants and higher representation theory*, 2013, preprint. arXiv:1309.3796.

[135] G. Williamson, *Schubert calculus and torsion*, 2013, preprint. arXiv:1309.5055.

[136] ——, *On an analogue of the James conjecture*, Represent. Theory, **18** (2014), 15–27. arXiv:1212.0794.

[137] A. Young, *On Quantitative Substitutional Analysis I*, Proc. London Math. Soc., **33** (1900), 97–145.

[138] X. Yvonne, *A conjecture for q-decomposition matrices of cyclotomic v-Schur algebras*, J. Algebra, **304** (2006), 419–456.